ENERGY EFFICIENT BUILDING
A Design Guide

ENERGY EFFICIENT BUILDING

A Design Guide

Edited by

Susan Roaf
BA Hons, AA Dipl, PhD, RIBA

AND

Mary Hancock
BA, BSc (Arch), ARCUK

OXFORD
BLACKWELL SCIENTIFIC PUBLICATIONS
LONDON EDINBURGH BOSTON
MELBOURNE PARIS BERLIN VIENNA

Blackwell Scientific Publications
Editorial Offices:
Osney Mead, Oxford OX2 0EL
25 John Street, London WC1N 2BL
23 Ainslie Place, Edinburgh EH3 6AJ
3 Cambridge Center, Cambridge, Massachusetts 02142, USA
54 University Street, Carlton Victoria 3053, Australia

Other Editorial Offices:
Librairie Arnette SA
2, rue Casimir-Delavigne
75006 Paris
France

Blackwell Wissenschafts-Verlag
Meinekestrasse 4
D-1000 Berlin 15
Germany

Blackwell MZV
Feldgasse 13
A-1238 Wien
Austria

First published 1992

Set by Best-set Typesetter Ltd., Hong Kong
Printed and bound in Great Britain by
Hartnolls, Bodmin, Cornwall

DISTRIBUTORS

Marston Book Services Ltd
PO Box 87
Oxford OX2 0DT
(*Orders*: Tel: 0865 791155
Fax: 0865 791927
Telex: 837515)

USA
Blackwell Scientific Publications, Inc.
3 Cambridge Center
Cambridge, MA 02142
(*Orders*: Tel: 800 759-6102
617 225-0401)

Canada
Oxford University Press
70 Wynford Drive
Don Mills
Ontario M3C 1J9
(*Orders*: Tel: 416 441-2941)

Australia
Blackwell Scientific Publications
(Australia) Pty Ltd
54 University Street
Carlton, Victoria 3053
(*Orders*: Tel: 03 347-0300)

British Library
Cataloguing in Publication Data
Energy efficient building: a design manual.
I. Roaf, Susan II. Hancock, Mary
721

ISBN 0-632-03245-6

Library of Congress
Cataloging in Publication Data
Energy efficient building: a design guide/edited by Susan Roaf and Mary Hancock.
p. cm.
Includes bibliographical references and index.
ISBN 0-632-03245-6
1. Buildings—Energy conservation.
2. Architecture and energy conservation.
I. Roaf, Susan. II. Hancock, Mary.
TJ163.5.B84E5435 1992
720′.472—dc20 91-46439
CIP

The illustrations in Chapter 12 by Geoffrey Pitts are reproduced with kind permission of Timber Research and Development Association

This book is dedicated to:

Michael Humphreys, in appreciation for his meticulous and original work on comfort which has opened new doors for us in the search for responsible thermal standards for indoor environments.

and to:

Designers of buildings, every single one of whom, from student to senior partner, has an important role to play in the reduction of energy consumption in buildings around the world.

Contents

HOUSING

NON-DOMESTIC BUILDINGS

CONCLUSIONS

Authors' biographies

Nicholas Baker (Bsc, MA, PhD) is currently Director of the Martin Centre for Architectural and Urban Studies, Department of Architecture, Cambridge University. A physicist by training, he has worked in the field of Architecture and Building in teaching, consultancy and research. His interest in energy conservation stems from a broader concern for environmental issues. He has published widely on energy conservation in buildings and has done seminal work on energy simulation using laboratory studies and computer programs – the latest of which he is developing is the Lighting and Thermal Method program. He sits on various National and International committees including the RIBA's Environment and Energy Committee.

Robert Bell (BSc, CEng, MCIBSE) is a Chartered Engineer who qualified in electrical engineering, specializing in Power Utilization and Illuminating Engineering. His main interests lie in the elements which make up good lighting. He believes that the quality of the design process is the most important of these and that many installations fail because they were designed to match the building rather than the needs and activities of the people who occupy it. He is the Company Chief Lighting Engineer with Thorn Lighting, at its International Headquarters at Borehamwood, and is responsible for all its lighting engineering. He has worked in research, development, engineering, sales and marketing and has written many scientific papers, articles and books. In 1981 he was presented with the Leon Gaster Award for his work on lighting design parameters.

Loren Butt (BSc(Eng), CEng, MIMechE) studied engineering at University College London and from 1959 to 1965 was an air-conditioning systems commissioning engineer with Carrier Engineering Co Ltd, London. In 1965 he changed jobs to become a specialist ventilation design engineer with George Wimpy Co Ltd. London and from 1965 to 1968 returned to air-conditioning system design as an Associate of D. W. Thompson and Company in Vancouver, Canada. From 1968 to 1970 he was an executive engineer with G. N. Haden & Sons Ltd London and between 1970 and 1987 was first Associate and then a Director of Foster Associates, responsible for engineering concepts, design and technical co-ordination on all projects. Since 1987 he has been in private practice as an independent consultant trading as Loren Butt Consultancies, in which capacity he has completed his involvement with the design and building of Stanstead Airport.

Peter F. Chapman (BA, MA, PhD, FRSA) trained as a research physicist at Cambridge and moved to the Open University in 1970 to establish a career as a teacher. He started and directed the Energy Research Group between 1973 and 1979 and published widely on energy policy and technology, including writing *Fuels Paradise* for Penguin Books. Since 1975 he has been involved with energy use in dwellings, field trials, developing BREDEM and promoting various energy labelling initiatives. He is currently Professor of Energy Systems at the Open University, Director of Energy Advisory Services Ltd which produces and markets all the commercially available versions of BREDEM, and Technical Consultant to the National Energy Foundation responsible for the design of the software and training scheme for the National Home Energy Rating Scheme.

John Connaughton (BSc, ARICS) is a Chartered Quantity Surveyor with extensive experience in the construction industry including the management and supervision of consultancy projects. For several years he has undertaken and managed consultancy work in the field of energy use in buildings. Much of this work has been concerned with the costs, economics and marketability of energy efficiency and low energy design, and with the energy embodied in building construction. He is currently a Royal Institution of Chartered Surveyors representative on the Construction Industry Council Energy Group and a member of various RICS committees on research and energy efficiency. He is an Associate of Davis Langdon Consultancy.

Michael Corcoran (MSc(Arch), CEng, CIBSE) is a Building Services Engineer with a Masters in Architecture. He joined the Building Design Partnership in 1972 and from 1978 has headed the practice's Energy Group. In 1986 the Energy Group became BDP Energy and Environment of which he is Director. The work of BDP Energy and Environment includes applied research, design support and energy consultancy. His projects have included the Low Energy Hospital Study for the Department of Health, and the concept design for the recently completed low energy hospital in the Isle of Wight, and many studies for the Department of Energy ranging from the application of heat pumps to the potential for passive solar design in non-domestic buildings. He is currently working on atria, assessment of building performance and a feasibility study on a proposed Design Advice Scheme.

Stephen Curwell (BSc, MSc, RIBA) graduated in architecture from Queen's University of Belfast in 1973 and subsequently spent sometime in private practice. From 1979 to 1986 he lectured on Surveying, Building and Environmental Science courses at Salford University. From 1986 he has taught at Leeds Polytechnic where he is now a Principal Lecturer and Architectural Professional Group Leader. He is co-author and editor of *Hazardous Building Materials: a Guide to the Selection of Alternatives*, *CFCs in Buildings*, and of *Buildings and Health: the Rosehaugh Guide* and has published numerous papers in the field of health and the environment. He is a member of the RIBA's Environment and Energy Committee.

John Ellis (CEng, MIMechE, MCIBSE, MInstE) is a professional engineer and Chairman of the mechanical and electrical services engineering profession of Building Design Partnership (BDP). BDP is one of Europe's largest practices of architects and engineers. His design interests centre on integrated low energy buildings and high technology facilities for pharmaceutical and microelectronics companies.

Mary Hancock (BA(Hons), BSc(Arch), ARCUK) read architecture at Bristol University, studying environmental sciences under Peter Burberry. After graduating as an architect she specialized in housing and housing for the disabled with Middlesborough Borough Council. She has worked in private practice for a number of years and has done some consultancy specializing in thermal design. For the last three years she has taught environmental science at Oxford Polytechnic where she is teaching in both the graduate and undergraduate schools.

Philip Haves (MA, PhD, CEng, MCIBSE, MASHRAE, MIEE) read physics at Oxford University and carried out research work in radio astronomy before moving to Texas to teach and research into passive cooling. On his return to the UK he added a daylighting and lighting control facility to the building thermal simulation program SERI-RES and collaborated on the development of a link between the architectural CAD program SCRIBE MODELLER and SERI-RES. After a move to Oxford University he has worked on self-tuning controls for HVAC systems and on methods of assessing building control systems. His present work involves the development of automated commissioning techniques for building control systems. He is a member of the ASHRAE technical committees concerned with energy calculations and controls as well as being a member of CIBSE and the IEE. Until recently he was Chairman of the BEPAC task group on plant and controls. He works also as a consultant in the field of building and HVAC system simulation and teaches part-time at Oxford Polytechnic School of Architecture and the Architectural Association Graduate School.

George Henderson (BSc, MSc, CEng, MIEE) was trained in physics, receiving a first class honours degree from Aberdeen University where he worked subsequently on Instrumentation for his MSc. From 1969 to 1974 he worked as an electronic design Engineer with Marconi Instruments Ltd and in 1974 he joined the Instruments Section of the BRE of which he became Head in 1979. In 1983 he transferred to work on energy use in buildings. As Head of the Energy Economics and Statistics Section he has been involved in many aspects of this subject including the development and testing of energy models, building energy use databases, cost effectiveness of energy efficiency measures, energy labelling and greenhouse gas emissions. This work has routinely involved the provision of advice to both the Department of the Environment and the Department of Energy.

Michael Humphreys (BA, BSc, Dip Ed) graduated in Physics from Hatfield College, Durham in 1958 and subsequently took a Diploma in Education. He was a school master in Worcestershire before joining the staff of the Building

Research Establishment where he became Head of the Human Factors Section. His research at the BRE dealt largely with Comfort Studies and his field work in educational buildings led to seminal studies on the relationship between indoor comfort and outdoor climate. His involvement in the work of the churches led to his training for the Christian Ministry at Regent's Park College, Oxford, where he graduated in Theology in 1981. He was Minister of the United Free Church (Baptist) in Tring from 1975 to 1989, and is currently General Secretary to the Hertfordshire Baptist Association. He is married to Mary and they have four children and three grandchildren.

John Littler (BA, MA(Oxon), PhD, MCIBSE, MCIOB, MASHRAE) read Chemistry at Wadham College Oxford and after completing his PhD on Photochemistry at Nottingham University travelled on a post-doctoral fellowship to Case Western Reserve, Cleveland, USA and Kingston, Canada. He was Assistant Professor of Chemistry in Marshall University, USA and then SERC Visiting Fellow in the Chemistry department and Senior Research Fellow in the School of Architecture at Cambridge University. In 1979 he was made London Master Builders' Reader in Building at the PCL where since 1987 as Professor of Building at the Polytechnic of Central London he has directed the Research Development Group working on superinsulation, various modes of passive solar building design, air movement and infiltration, atria, the development of simulation tools, IT, advanced window design, moisture movement and condensation and domestic heat recovery systems. He has published widely on these subjects and sits on committees of the International Energy Agency and the Research Committee of the PCL.

David Lush (BSc(Eng), CEng, MIEE, FCIBSE) is an electrical engineer who has been closely involved in the application and design of control and automation systems for building services for the last 30 years. He has a wide experience in the co-ordination and interfacing between the electrical and mechanical design disciplines. He has specialized in all facets of energy conservation and utilization in services including 'smart' and 'intelligent' buildings. He is particularly interested in the environmental criteria necessary to create a comfortable and healthy working environment. He has worked on many engineering projects and has published widely on many aspects of energy and services in building. He joined Ove Arup Partnership in 1970 and since 1971 has worked in Arup Research and Development of which he is now a Technical Director. He has acted as a specialist adviser to the House of Commons Environment Committee's Report on Indoor Pollution and is currently Vice President of CIBSE.

Joe Lynes (BSc, MA, CEng, MIEE, FCIBSE) trained as an electric lighting designer, and as a discharge lamp development engineer, before heading the Pilkington Daylight Advisory Service and subsequently the Interior Lighting Applications section of Jules Thorn Lighting Laboratories. He has taught at Schools of Architecture at Manchester University, Plymouth Polytechnic and the Technical University of Nova Scotia, Canada. He is now Senior Lecturer at Hull School of Architecture, and is associated with the ETSU Passive Solar Programme. He is co-author of the HyperLight computer-based tutorial package on lighting design.

Ian McCubbin (BSc, PhD) trained as a chemist at Southampton University in 1980. He went on to do research for his Doctorate at the Royal Institution in London, studying light-initiated chemical and physical processes. He joined the Energy Technology Support Unit at Harwell Laboratory in 1986, working on renewable energy programmes on behalf of the Department of Energy. His work involves planning, procurement, and technical management of solar energy research projects, and in particular he has been responsible since 1988 for investigating the potential benefits of passive solar design in non-domestic buildings.

Anthony Mould (Dipl Arch, RIBA) trained in architecture at the North London Polytechnic and has worked widely in the energy industry in this country and in Germany and Iraq, where he established the Building Research Centre in Baghdad. He has wide interests and expertise in energy in design specializing in heat flow and thermal insulation, ventilation, heat stress in building materials, and building hazards. His interests include external insulation which he first pioneered in 1973 and domestic mechanical ventilation. He has worked for the Building Research Establishment, The United Nations, the Midlands Housing Consortium and The Electricity Research Council and in private practice. He has also worked with the Pilkingtons Fibreglass Ltd, the National Energy Foundation, ETSU, BRECSU and is currently self-employed as an energy consultant.

Tadj Oreszczyn (PhD, MInstE, CEng, MCIBSE) is currently a lecturer in Environmental Design and Engineering at the Bartlett School of Architecture, Building Environmental Design and Planning, University College London. His main teaching commitment is in the area of energy and health in buildings on the Masters Course in Environmental Design and Engineering. Before joining the Bartlett he was a Senior Energy Consultant for the Energy Conscious Design Partnership. His education included a PhD in solar energy while at the Energy Research Group of the Open University, and an honours degree in physics from Brunel University. He is a chartered Engineer and a Member of the Institute of Energy, and the Chartered Institute of Building Services Engineers. He has carried out research and published papers on condensation in buildings and energy use in greenhouses and conservatories.

Geoffrey Pitts (BArch, MPhil) graduated from the University of Melbourne, Australia with Honours before going on to receive his MPhil at Edinburgh University. In 1978 he joined TRADA to research timber frame housing with particular emphasis on energy efficiency. He was project architect for a scheme which won the 1986 North West Energy Group Award. In 1984 he joined the

South London Consortium Energy Group working on the assessment of energy efficiency in Local Authority building. In 1985 he rejoined TRADA as Deputy Chief Architect and was involved in the Lifestyle 2000 house design at the Milton Keynes Energy World Exhibition. He has published widely on the subject and recently produced a book on *Energy efficiency in housing – a timber frame approach*.

Susan Roaf (BA(Hons), AADipl, PhD, RIBA) trained in architecture at Manchester University and the Architectural Association, London before going on to do a PhD at the School of Architecture at Oxford Polytechnic. She spent three years researching the traditional buildings of Iran and is currently publishing books on the windcatchers of the Middle East, the architecture of the nomads of Luristan in south-western Iran, and the desert houses of the Central Iranian Plateau. She lived with her husband Dr Michael Roaf, a Near Eastern Archaeologist, on excavations in rural Iraq and in Baghdad for seven years not only as an archaeologist but also recording the vernacular architecture of the country which will eventually be published. She is co-author of *The Ice-houses of Britain*, is a Senior Lecturer and Professional Training Adviser at the School of Architecture, Oxford Polytechnic, does some private practice on the conservation of ice-houses, and sits on the RIBA's Environment and Energy Committee.

Graeme Robertson (BArch, ANZIA, MNZIA, FRSA) is a Senior Lecturer in the Department of Architecture, University of Auckland, New Zealand. After a decade as an architect in private practice, he has, in recent years, been researching and teaching in the general subject area of energy efficient buildings. He is a member of the International Union of Architects (UIA) Project Group 'The Implications of the Greenhouse Effect for Architecture and the Built Environment' and in his current research he is concentrating on 'Commercial Building Responses to Climate Change'. He has lectured widely, published and presented papers in Europe, North America and Japan, as well as in his home territory of Australia and New Zealand. He is past Chairman of the NZ Institute of Architects (Auckland) and a past Chairman of Solar Action NZ. He has been closely associated with the formulation of the recently adopted NZIA Environmental Policy and its National Co-ordinator.

Paul Ruyssevelt (BA, BArch, RIBA, PhD) trained as an architect and worked in private practice before spending six years researching into low energy and passive solar buildings at the Polytechnic of Central London, and then three years as Director of the Energy Monitoring Company involved in applied energy research in buildings. He joined the consulting engineers Halcrow Gilbert Associates in 1988 and now manages the Building Environment and Energy Services business. His current work includes management of major government R & D programmes and provision of energy design advice to architects and clients.

John Willoughby (BSc, MPhil, CEng, MCIBSE) is an energy and environmental design consultant based near Cheltenham. His work covers a wide range of energy design and energy management activities. He is currently part-time energy manager for Cheltenham and Gloucester College of Higher Education and Cheltenham Borough Council. His recent contracts have included work on the ETSU Passive Solar R & D programme; a strategic energy study for a large District General Hospital in Wales, work for the National Energy Foundation on its National Home Energy Rating Scheme and the engineering design for a new visitor centre for the Royal Horticultural Society. He has been involved in the design of a wide range of low energy housing schemes.

Introduction

Susan Roaf
BA(Hons), AADipl, PhD, RIBA
AND
Mary Hancock
BA(Hons), BSc(Arch), ARCUK

Energy efficiency is no longer an optional extra in design – it has become a basic requirement for the designing professions.

This need to design buildings that consume less energy arises from a variety of external pressures: developments in legislation on building performance; increasing demands of the professional institutions in the field of energy performance; the importance, when working with developing technology, of limiting liability risk; and the wishes of clients who increasingly seek 'low energy' buildings for economic or moral reasons. Coupled with this is the personal commitment of many designers who believe it imperative to build green buildings because they appreciate the importance to all of us of stabilizing emissions of greenhouse and ozone depleting gases, and minimizing our pollution of this planet.

The significant commercial benefits of energy efficiency are also being increasingly recognized by both designers and clients alike. Energy efficiency means developers can build cheaper and simpler buildings, landlords can increase rents due to the added benefits of lower running and maintenance costs, tenants pay lower running and maintenance costs, and building users are more productive and healthier in the better indoor environments which can accompany energy efficiency in buildings. Clients are also influenced by the fact that many of the best of modern buildings have energy efficiency high on their list of design priorities, and the ability to produce good energy buildings is becoming a hallmark of first class designers.

In particular, basic standards of energy design in housing have improved substantially over the last decade and many developers, from the largest to the smallest, are designing houses to high standards of energy efficiency in response to public demand for low running and maintenance costs. The success of the housing energy labelling schemes bear witness to this change in attitude towards energy standards in new houses. The ability to analyse and predict building and running costs through computer simulation during design provides a very effective tool for improving the performance of a building, and one that by the end of the century will be used in virtually all design offices.

On a national scale, governments are also under great pressures to reduce energy consumption in buildings. For instance, the European Parliament has set out a number of strategies to be adopted in the Community to reduce energy consumption within this decade. It has outlined three areas of future action in its proposals written in February 1991. These are technological, financial and consumer behaviour measures, and include the introduction of new laws on boiler performance, voluntary agreements on the improvement of energy efficiency of domestic appliances, minimum performance of cars, cogeneration of powers, financial and taxation measures and the education and training of consumers in efficient energy use. We can anticipate in the next ten years radical programmes of equipment labelling, energy rating schemes for buildings, higher fuel costs, and stricter legislation on the energy performance of buildings, and perhaps taxes on energy consumption such as a carbon tax.

So where does this leave the designer? We can only answer from our own point of view. This leaves us wanting and needing to know more, which is the reason for this book. At the School of Architecture at Oxford Polytechnic we listed those subjects we wanted to know more about, identified leaders in each of the various fields, and invited them to lecture here. From this beginning the 1990/1991 Oxford Energy Lectures developed and areas of information crystallized from lecture topics into sections and chapters of this book. One challenge in choosing a limited range of topics from the vast subject of low energy architecture is getting the balance of information right. We have included chapters in this book that range in scope from issues of basic competence, to advanced design, to enable designers to get an insight into the whole gamut of the subject, and at the same time provide sufficient back–up references for individuals to be able to follow up areas of special interest.

Nearly half of all energy consumed in Britain is used in buildings. It has been calculated that this energy expenditure could be almost halved if the existing building stock was adequately insulated. This sounds easy, but the task ahead of us is enormous if we are to meet the global targets for reductions in energy consumption this century.

In the period between 1973 and 1985, through two oil crises, the ratio of final energy consumption to GDP improved by more than 20% in Europe, showing that major reductions in energy consumption are possible. In this case these reductions were largely due to high energy prices and a major restructuring of industry. In 1986 energy efficiency objectives for the period of 1986 to 1995 were set a further 20% target for improvement in the energy intensity of final demand, a figure based on evaluations of the potentials that exist for energy savings.

We cannot possibly meet these targets, for current estimates show that, at best, if improvements in energy consumption continue at their present levels, then a maximum of only 14% improvement could be achieved by 1995, 6% short of the target figure of 20%.

Yet between 1986 to 1989 many countries in the European Community actually ceased to fund energy efficiency programmes and there was a significant slow-down in the improvement of energy intensity figures. Since 1989 many member states have recognized the urgency of taking action, and increased their funding and activity in this field as a result of serious concerns over climatic changes and the threat to energy supplies of the conflicts in the Middle East. It is extraordinary that governments are allowed to gamble so short-sightedly with the future of our planet (Fig. i).

Energy efficiency is a primary mechanism to limit the environmental damage caused to our planet by energy use; environmental damage that now threatens the

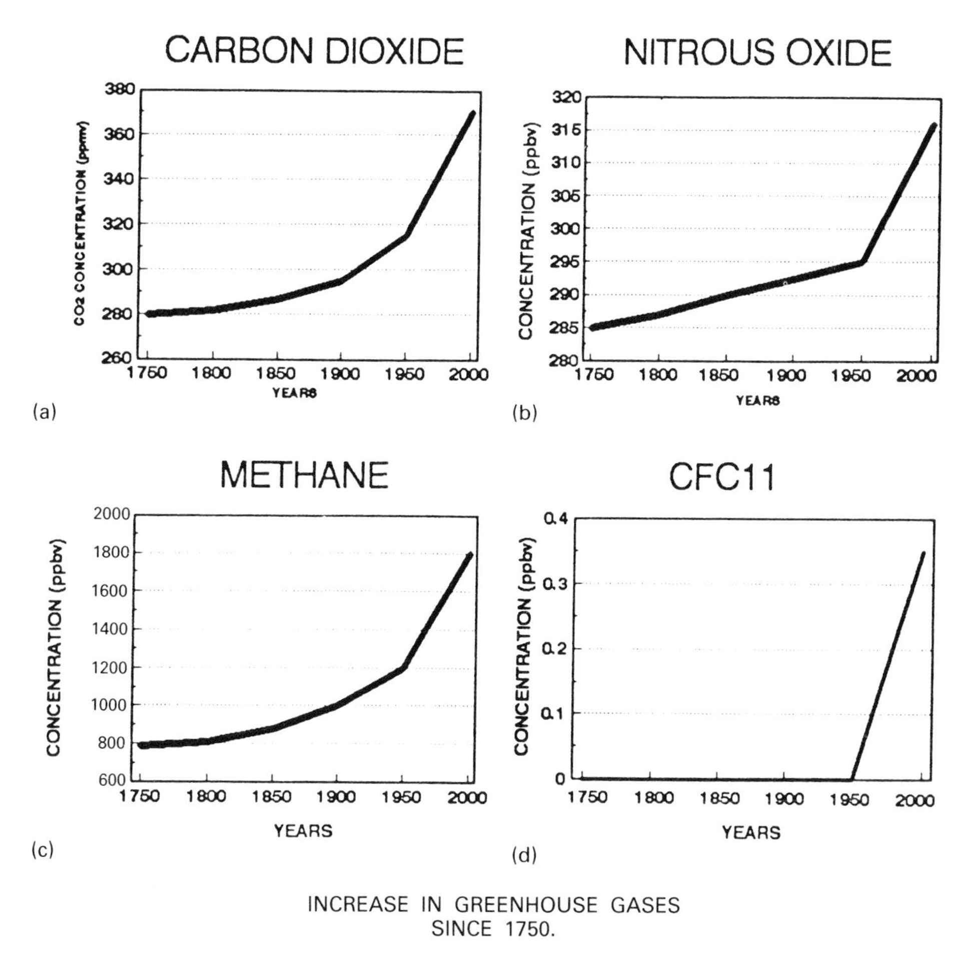

Fig. i These graphs give an indication of the severity of the damage being caused to the Earth's atmosphere by emissions of gases that pollute, destroy the ozone layer, and increase global warming.

survival of many thousands, and perhaps millions, of people on Earth through global climate change.

> 'During the last few centuries, human society has entered into a new and momentous relationship with the global environment. For the first time in history, we have become a geological agent comparable to some of the other forces influencing our planet. We have altered the face of the earth by clearing forests, building cities and converting wild land into agriculture. During the past century industrial production has gone up by more than a 100-fold and the use of energy has risen by a factor of about 80. We have changed the composition of the Earth's hydrosphere and atmosphere through the use of fossil fuels, the expansion of agriculture and the production and release of industrial compounds. Almost without recognising it, we have embarked on an enormous,

unplanned, planetary experiment that poses unprecedented challenges to our wisdom, our foresight and our scientific capability.

We are just beginning to understand this marvellously complex planet that we live on and the ways in which we are changing it, and we need to devote our best efforts to the task. There is not time to waste.'

Dr D. Allan Bromley, Director of the USA Office of Science and Technology Policy, 1989

For such reasons there is no doubt that energy efficiency has moved from being largely the preserve of specialists and cranks, to being a required skill of each and every design professional.

We hope that this book will help you in developing this skill.

Susan Roaf & Mary Hancock
Oxford
September 1991

Acknowledgements

This book is based on 'The Oxford Energy Lectures 1990–1991'. This series was initially made possible through the generosity of Ian Pollard who first sponsored it, and Christopher Cross, Head of the School of Architecture at Oxford Polytechnic who gave us his valuable support. We are grateful to both of these, and to William Bordass whose suggestions influenced many of the decisions on topics and speakers for the series.

We also owe a debt of gratitude to all the members of staff at the School who helped us organize the lectures: Vivian Ebbs, Elaine Barnett, Humphrey Truswell, Ann Henderson, Rhiannon Astill, Iradj Parvaneh, Gordon Nelson, the staff of EMU, and particularly to Carolyn Gulliver who provided the refreshments every week.

We would also like to thank Van Nostrand Reinhold Publishers, New York, for permission to use chapter 17; Cambridge University Press for permission to use the graphs in this introduction; Harper & Brothers Publishers, New York, for permission to reproduce Fig. 1.1, and The Frauenhoffer Institute of Stuttgart, for permission to reproduce Fig. 11.7.

We are also grateful to Julia Burden and Sue Moore at Blackwell Scientific Publications Ltd, for their help and encouragement with this book.

Finally our thanks goes to all the contributors to this book, to whose expertise and generosity it is a tribute.

The Indoor Environment: Design Standards

Chapter 1

Thermal comfort in the context of energy conservation

Rev. Michael A. Humphreys
BA, BSc, Dip Ed

There are two types of study of preferred indoor temperature or comfort. Climate chamber studies show that people doing the same activity, in the same clothing all like the same temperature irrespective of their age, sex, culture, race, season, colour of the room or climate to which they are accustomed. The second method is to conduct field studies of people performing normal activities, in their normal environments and ask them their opinion of their thermal environment. The environment measured is not altered by the researcher. These field studies show that the temperature which people find comfortable can be estimated from the average room temperature during the survey, which is also related to the monthly mean outdoor temperature of the country and season where the survey was conducted. Field studies report that people are comfortable in temperatures that range from at least 15 to 30°C. It is proposed that there is no need for uniformity of indoor temperatures world wide and by adopting indoor design temperatures suitable to prevailing climate and seasons significant saving in energy consumption can be made.

1.1 Introduction

With the present worry about the use of energy, not only because of cost or scarcity, but also because of its implications for the emission of carbon dioxide and global warming, a grasp of the dynamics of thermal comfort has become of increased importance. It seems useful in this chapter to give a retrospective comment on my research in this subject, conducted at the Building Research Establishment between 1966 and 1978. My aim is to pick out what seems either interesting or important for our broad understanding of thermal comfort in relation to the built environment. This chapter is based on a lecture delivered at Oxford Polytechnic, and retains something of the spoken tone of that occasion. Much of the detail, and some of the substantiation of what I say, is necessarily omitted, and all I can do is refer the reader to earlier and more detailed results and arguments (Humphreys, 1976, 1978, 1981).

Consider this quotation:

'(Laura) was cold. She was too cold to wake up. The covers seemed very thin. She snuggled closer to Mary and burrowed her head under the thin covers. In her sleep she was tight and shivering, till finally she grew cosy and warm. The next she knew, she heard Pa singing, . . . Laura opened one eye and peeked out from under the covers. Snow fell softly on her face, a great lot of snow.

Fig. 1.1 'Pa shovelling snow off Laura'.

"Ow!" she said.

"Lie still, Laura!" said Pa. "All you girls lie still. I'll shovel you out in a minute. Soon as I get this fire started and the snow off Ma."'

(Ingalls Wilder, 1939)

The house had been unfinished, and there was an unexpected spring blizzard. Clearly, the 'bedroom' temperature was well below freezing, for the snow did not melt when it fell on Laura's bed (Fig. 1.1). Once the snow had fallen she was cosy and warm, snuggled up against Mary. What is relevant for Laura's comfort is the temperature within the cocoon of her bed. The snow provided her with a low-energy solution for her thermal comfort requirement. Let's just say that this incident casts doubt on the need for a bedroom temperature of not less than 15°C!

A second quotation, this time fictional:

'The Commissaris shivered, and Silva immediately showed concern.

"You haven't caught a chill, have you? It's this damned air-conditioning. It's a comfort, of course, but a danger at the same time. This isn't the best season and the heat hits you like a hot towel when you step outside, but here, in the offices, it's too cool. I'll turn the machine down a little."

"No, no," the Commissaris said quickly, "I feel quite well, better in fact than I have for a long time. But I did, probably, shiver because of the change of temperature."'

(Van de Wetering, 1976)

This is from a novel set in a hot and fairly humid island. Outside, your skin is moist with evaporating sweat. When you enter the air-conditioned space, the sweat evaporates, giving a powerful transient cooling. Was it sensible to set the indoor temperature so low? Shouldn't it have born some relation to the temperature outside? It calls into question the need to use fuel for refrigeration in such circumstances. Could not the (fictional) designer have designed a building to maintain a steady, slightly warmer temperature, more cheaply and more satisfactorily?

Or consider the following exchange, which took place between Thomas Bedford, the father of thermal comfort surveys, and, I believe, Charles Webb, well known for his work on thermal comfort in equatorial and tropical climates (Webb, 1959). Dr Bedford had been very uncomfortable when visiting India. This was largely because he did not adapt his dress appropriately. Charles Webb explained that, if one removed the jacket and tie, and arranged one's open-neck shirt outside the trousers, all could be well. Bedford replied 'But there are limits, Webb'. Here a viable strategy for comfort is rejected on cultural grounds. Such a rejection is not unusual, for although the physical effect of dress is chiefly thermal insulation, it has powerful social and cultural functions too. So the architect and engineer need to know not only the climate but the culture, or even the subculture of the occupants, before the temperature can be specified.

1.2 The body's temperature control

The human thermal regulation system is astonishing. The body reacts to control its internal temperature, and especially of the brain, whose temperature is regulated to within a few hundredths of a degree on a diurnal cycle of waking and sleeping.

A small increase in brain temperature induces sweating, although the pattern and quantity of sweat production is influenced also by conditions of temperature and pressure nearer the body surface. There are pressure-points on the surface which control which surfaces will sweat, so that, for example, a person lying on one side will sweat chiefly on the upper surfaces of the body, well demonstrated by Kuno's pioneering work on human perspiration (Kuno, 1956). Clothing suitable for a hot climate must allow free circulation of air to the skin. Evaporation of sweat or other moisture from the skin is an intensely powerful cooling force. When you start to dry your hands under a hot-air dryer, the initial feeling is cold. When the hands have dried, the airstream is too hot to bear. So in a really hot climate there is an optimum windspeed. Less windspeed than this does not evaporate all the sweat, while more than this windspeed evaporates all the sweat and then heats the skin. It was shown as early as the 18th century (in a delightfully reported experiment) that it was possible to stay in a dry oven with a temperature above boiling point for lengthy periods, so long as sweat was freely evaporating from the skin (quoted by Leithead and Lind, 1964). (A note of caution: this is not true of an oven saturated with moisture. In this case temperatures of above about 36°C are lethal, there being no possibility of evaporative cooling.)

A small decrease of brain-temperature increases muscle tension and shivering. The process begins, for a naked person at rest, at a room temperature of about 28°C. Shivering increases the metabolic heat production, and so tends to stabilize the body temperature. A shivering person may double or even treble his heat production.

Along with the sweating and shivering goes the dilation and constriction of blood vessels near the body surface. Thus, if the body core is cold, the blood flow to the surface is reduced, while if the core is hot, the blood flow to the surface is increased. This gives extra regulation, increasing the band of room temperature at which there is no obvious sweating or shivering, but just a variation of the temperature of the skin.

During exercise the body-core temperature increases, inducing sweating, which dissipates the heat generated by the exercise. During sleep, the set temperature of the brain 'thermostat' is lower, and so when one falls asleep the brain temperature is above its new set-point, and sweating takes place until the new equilibrium is reached.

Broadly speaking, conditions which tend to move the body away from equilibrium are unpleasant, while those tending to restore the equilibrium are pleasant. So people will, when free to do so, seek conditions which tend to restore their body's thermal equilibrium.

Within overall equilibrium the body becomes used to moderately different distributions of surface temperature, so quite different patterns of clothing can be comfortable, provided that the overall thermal balance is correct. The change from one clothing pattern to another produces very noticeable sensations of warmth or coolness, as, for example, when one changes from long to short trousers. But the different sensation is not noticed for long, because temperature perception at the skin is much more sensitive to change than it is to steady conditions.

This is a very simplified description of a marvellous and subtle system of control. In very young babies the system is not fully operational, and in the elderly it is less effective. So it is the babies and the aged who are most prone to die from extremes of heat or cold.

1.3 Behavioural regulation

People are not passive in relation to their thermal environment. We have already noticed that people will tend to seek out comfortable conditions. Outdoors they seek shade or sunshine, wind or shelter, as may be most comfortable – though they may trade-off comfort for some other benefit (Humphreys, 1979). They also regulate their posture, activity and clothing to obtain comfort. An Antarctic survey team spent the winter in upturned tractor crates, as their supply-ship carrying their proper huts had stuck in the ice (Palmai, 1962 and Goldsmith, 1960). A survey showed that they were comfortable at very low temperatures, their clothing being adequate to the conditions. At the other pole, eskimos reached a different equilibrium within their traditional igloos. The inside temperature was kept at about 30°C by oil lamps and body heat, and the clothing was very light. They opted for a 'tropical' indoor environment (Wulsin, 1949).

Building design itself can be regarded as part of this regulatory procedure. Over the centuries styles have been developed which perform well in tropical climates, hot dry climates, temperate climates and cold climates. These buildings can be thought of as a long-term human adaptation to the environment. They may be adjustable by means of blinds, shutters, ventilation, and active heating or cooling systems, and they may use their mass to reduce temperature variation, so as to maintain the desired temperature or at least ameliorate the rigours of the outdoor climate. We clearly have a wide-ranging regulatory mechanism in operation, with physiological, behavioural, cultural and technical aspects all meshed together. Discomfort arises when the total effect of all these regulationary actions, for one reason or another, fails to yield easy thermal equilibrium.

It follows that in general, those things which reduce the number of strategies which a person can adopt in the quest for comfort, are likely to cause discomfort, while less restrictive circumstances assist comfort. Soldiers on parade have virtually no mechanisms available except the basic physiological ones. The posture, activity level, clothing and the environment are all outside their control. So it is hardly surprising that fainting on parade is common, or that military history has many instances of soldiers dying from heat stress, not because of the rigours of the climate, but because of the rigours of military discipline. Contrast this with the situation of someone in their own home. They have many strategies available, for activity, posture and clothing, and often the room temperature and the amount of radiant heat are matters of choice. Home is likely to be comfortable.

Discomfort also arises when the environment is much warmer or cooler than had been expected. This can happen because of a sudden change in weather, and one has chosen clothing which is unsuitable, and there is no opportunity to change it. It can also happen if a building is maintained at an unexpected temperature such as when a hotel bedroom is much warmer than home, and they provide a megatog duvet which can result in a sweaty night; or when you visit friends, and they keep their house cooler than you do at home.

Conversely, comfort is greatly helped if the thermal environment is predictable. The church where I ministered for fourteen years occupied a large Victorian-style chapel with a high, uninsulated roof. There was little spare money for heating, so I suggested that we should try to maintain a winter heating temperature of 15°C during services. People soon learned that it was necessary to dress fairly warmly and there were few complaints. In fact they complained that a nearby church with whom we later occasionally shared services kept their chapel too warm. It was about 21°C.

The point I am trying to make is very simple. Within a wide range of circumstances, thermal discomfort is not caused by the room temperature itself, but by a mismatch between actual temperature and the desired temperature: the desired temperature is variable, depending on a complex web of the factors already mentioned.

1.4 Research methods and results

Two principal research methods have been used in thermal comfort research during this century; the climate chamber and the field-study. The former is a controlled experimental procedure, while the latter is observation accompanied by measurement. The difference of approach, although not absolute, becomes clear when the methods are considered in more detail.

1.4.1 Climate-chambers

A climate-chamber is a specially constructed room whose thermal environment can be controlled. Ideally it should be possible to control independently the air temperature, the temperature of the room surfaces as they radiate to the occupants, the humidity of the air, and its velocity. A sophisticated chamber can also provide temperature gradients from floor to ceiling, and different temperatures on different surfaces. It is possible to put volunteers in the chamber, in standard clothing, and performing standard activities. Either they are asked to adjust the room temperature until it is just right, or they can be asked to assess how warm or cool they feel. Good work on climate–chambers has been done by Ole Fanger in Denmark, the late Ralph Nevins in the USA, and Ian Griffiths and Don McIntyre in this country.

A short summary of climate-chamber research results is as follows. People with the same activity and clothing all like the same temperature (plus or minus a degree or so in standard deviation), irrespective of sex, age, culture, race, season, the colour of the room, or the climate to which they are accustomed (Fanger, 1970 and McIntyre, 1980).

From results obtained in climate-chambers, a physical-physiological equation can be constructed to show which thermal environment (expressed in terms of air temperature, thermal radiation, humidity, and air movement) will be most comfortable. One must specify the activity and the clothing insulation.

1.4.2 Field-studies

The alternative technique is to meet people in their normal environments, while performing their normal activities, and ask their opinion of the thermal environment. The environment is measured, but not altered, by the researcher. In both types of work rather similar rating scales have been used to assess people's

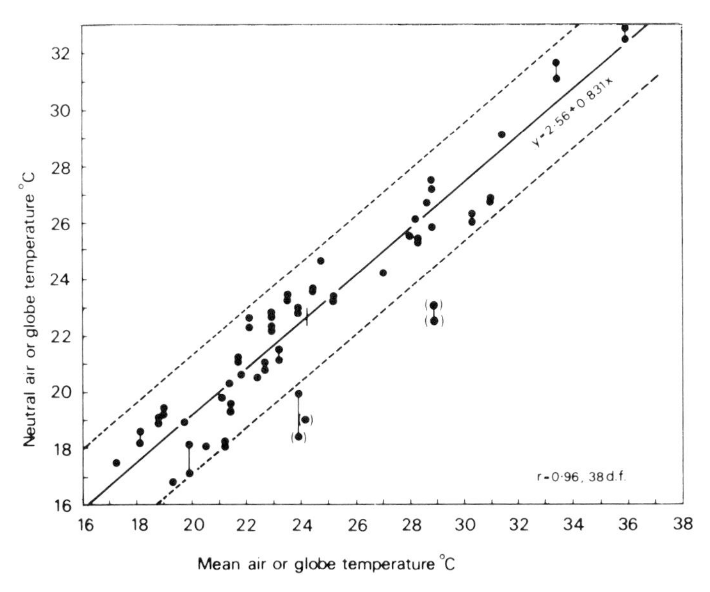

Fig. 1.2 Scatter diagram of mean temperature and neutral temperature (the paired observations arise from two different methods of analysis, see Humphreys (1976)). (Reproduced by permission of the Controller of HMSO.)

responses to the thermal environment. Typically people are asked to say how they feel by responding to a seven-point scale ranging from much too cool to much too warm, with a central point indicating neutrality or comfort. The number of categories, the manner of presentation, and the labelling of the scales have varied considerably. The classic field-study was conducted by Thomas Bedford during the 1930s, and his original full research report is still well worth reading (Bedford, 1936). In the late 1970s I correlated the results of all the thermal comfort field-studies that I could find. Today, several more studies could be added (e.g. Fishman & Pimbert, 1981; Hunt & Gidman, 1982 and Oseland & Raw, 1990), but as far as I can see without detailed analysis, they all tend to confirm and extend the overall picture which had emerged from the comparison of the studies available to me in the 1970s.

In brief, the results can be summarized as follows. The temperature which people find comfortable, irrespective of age, sex or race, can be estimated (plus or minus a degree or so in standard deviation) from the average room-temperature during the survey (Fig. 1.2). The temperature found to be comfortable is also related to the monthly mean outdoor temperature of the country and season where the survey was conducted (plus or minus a little over a degree in standard deviation) (Fig. 1.3). In most circumstances a simple measure of the thermal environment is sufficiently good, such as the air temperature, or, if a source of thermal radiation is present, the temperature of a globe thermometer. But in hot climates it is common to use the cooling effect of natural or induced air move-

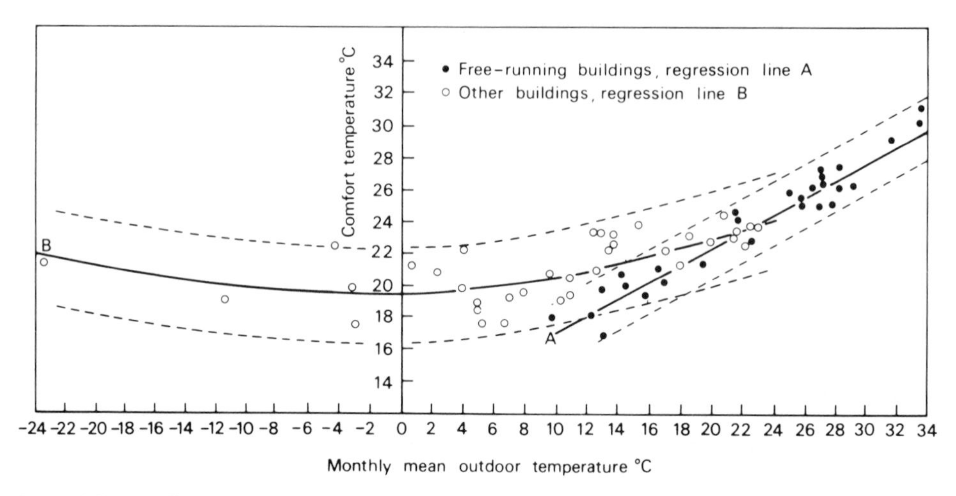

Fig. 1.3 Scatter diagram for neutral temperature, see Humphreys (1978). (Reproduced by permission of the Controller of HMSO.)

ment, and this must be allowed for. In these climates humidity becomes important if it limits the evaporation of moisture from the skin (Fig. 1.4).

1.4.3 Conflict of results

There is nothing essentially contradictory in the above statements of the results of the two methods. The climate-chamber discovers what happens in controlled conditions; the field-study what happens in uncontrolled conditions.

However, a puzzling conflict does appear when we notice that in the field-studies the range of temperatures which people report as comfortable is much wider than would have been expected from the results of climate-chamber research. It is not possible to explain this range, from about 15 to about 30°C, by means of the difference between the clothing common in cool and warm climates. People who are used to high temperatures report them to be acceptable, suggesting a substantial acclimatization which alters the temperature that they like or accept. Physiological adaptations to heat and to cold (one may simultaneously be acclimatized to both) are well established phenomena. However, when people are flown from hot climates to Denmark, they prefer, according to Ole Fanger's results (Fanger, 1970), the same temperature as the Danes. As far as I am aware, this conflict has not yet been satisfactorily resolved.

In the present state of knowledge it would be unwise to dogmatize. I believe that the climate-chamber studies yield a possible combination for comfort, but not the only one, and perhaps not the best one when the climate is hot and variable, and people are moving between outdoors and indoors. It seems to me that in hot countries light sweating is normal, acceptable and not uncomfortable, provided

Set of measurements	Restrictions upon their use
1 Air temperature	Difference between air temperature and mean radiant temperature is less than 2°C. Air movement less than 0.2 metres/sec. Sweat is freely evaporated from the skin.
2 Globe temperature	Air movement less than 0.2 metres/sec. Sweat is freely evaporated from the skin.
3 Globe temperature* Air velocity	Sweat is freely evaporated from the skin.
4 Globe temperature* Air velocity Wet-bulb temperature	None.

* If the difference between the air temperature and the mean radiant temperature is small, the air temperature may be used instead of the temperature of the globe thermometer.

Fig. 1.4 Measurement of the thermal environment. The table shows the simplest sets of measurements necessary to specify the warmth of the environment. They apply to indoor conditions where there is no great radiant asymmetry or air temperature gradient.

that the clothing is suitably loose and permeable to moisture. A room temperature which is cool enough to make sweating unnecessary would also be acceptable if one were used to it. However, this temperature would seem cold initially.

1.5 Design temperatures for different climates and seasons

If the above analysis is correct, it means that it is unnecessary to specify unvarying indoor temperatures for buildings for use throughout the year. Rather it ought to be possible to relate the indoor temperature to the average monthly outdoor temperature. Such a policy would not reduce comfort, provided it was done gradually enough (it is the unexpected changes of temperature which are prime sources of discomfort). This would in fact reduce the discomfort entailed in moving between indoors and outdoors. It could also reduce the amount of fuel used for heating or cooling. Although in theory one could lower winter indoor temperatures below those shown in the figure, this would require a change in attitudes to modern clothing, and would probably cause strong complaint.

It also follows that there is no need for uniformity of indoor temperatures worldwide – each region of the world could adopt temperatures suitable to the prevailing climate and season. This would reduce fuel used for heating and cooling, without causing discomfort to local people. There would perhaps be some sacrifice to the comfort of international travellers. Figure 1.3 gives guidance on the temperatures appropriate to different outdoor temperatures. Diurnal swings of indoor temperature should preferably be kept within fairly close limits – plus or minus two degrees might cause some discomfort, while plus or minus one degree would not produce noticeably more discomfort than an unvarying temperature.

1.6 Concluding comments

Here I am going beyond my expertise, and stray into the design of buildings.

It seems to me that, with judicious design, making use of solar heat gain in winter, and night-time ventilation in summer, the need to use fuel for heating and cooling of buildings can be virtually eliminated in many parts of the world.

When heating or cooling are needed, thought can always be given to the possibility of localized heating and cooling. The native Americans used to point out that European settlers built a big fire and sat a long way from it, while they built a small fire and sat close to it. The level of comfort would be the same, but the fuel consumption different. (It is even more efficient to place the heating or cooling inside the clothing, but this is not often convenient.)

There are few places in the world where the mean monthly outdoor temperature (as opposed to the mean daily maximum temperature) in the hottest month is unacceptably hot. On this basis, a massive building in almost any climate, provided the internal heat gains were small, should not need cooling.

Now that it seems at least probable that the carbon dioxide we are generating is producing a warming of the earth (global warming), it seems to me that the reduction of the burning of fossil fuel is no longer just economically desirable, but perhaps a matter of survival for substantial numbers of the world's population. (The arguments are set out in a semi-popular form in Gribbin, 1990). It seems to me that we should leave the world in reasonable condition for the next generation, if it is within our power to do so.

Acknowledgements

Much of the work described in this chapter formed part of the research programme of the Department of the Environment Building Research Establishment, and is published by permission of the Chief Executive. Figures 1.2 and 1.3 are reproduced by permission of the Controller of HMSO: Crown Copyright 1991.

References

Bedford, T. (1936) *The Warmth Factor in Comfort at Work*. MRC Industrial Health Board Report No. 76, HMSO, London.

Fanger, P. O. (1970) *Thermal Comfort*, Danish Technical Press, Copenhagen.

Fishman, D. S. & Pimbert, S. L. (1982) The thermal environment in offices. *Energy and Buildings*, **5**, pp. 109–116.

Goldsmith, R. (1960) Use of clothing records to demonstrate acclimatisation to cold in man. *Applied Physiology*, **15**, pp. 776–780.

Gribbin, J. (1990) *Hothouse Earth*, Bantam Press, London.

Humphreys, M. A. (1976) Field studies of thermal comfort compared and applied. *Building Services Engineer*, **44**, pp. 5–27.

Humphreys, M. A. (1978) Outdoor temperatures and comfort indoors. *Building Research and Practice*, **6**(2), pp. 92–105.

Humphreys, M. A. (1979) The variation of comfortable temperatures. *Energy Research*, **3**, pp. 13–18.

Humphreys, M. A. (1981) The dependence of comfortable temperature upon indoor and outdoor climates. In: *Bioengineering, Thermal Physiology and Comfort*, (Ed. K. Cena & J. A. Clark) *Studies in Environmental Science*, **10**, Elsevier, Amsterdam.

Hunt, D. R. G. & Gidman, M. I. (1982) A National field survey of house temperatures. *Building and Environment*, **17**, pp. 107–124.

Ingalls Wilder, Laura. (1967) *By the Shores of Silver Lake*, Puffin, p. 190. (First published in the USA, 1939).

Kuno, Y. (1956) *Human Perspiration*, Charles C Thomas.

Leithead, C. S. & Lind, A. R. (1964) In *Heat Stress and Heat Disorders*, Cassell.

McIntyre, D. A. (1980) *Indoor Climate*, Applied Science Publishers, London.

Oseland, N. & Raw, G. (1990) *Thermal Comfort in starter homes in the UK*, Proceedings of the Environmental Design Research Association's 22nd Annual Conference on Healthy Environments. (BRE note PD 186/90)

Palmai, G. (1962) Thermal comfort and acclimatisation to cold in a sub-antarctic environment. *The Medical Journal of Australia*, Jan, pp. 9–12.

Van de Wetering, Janwillem. (1976) *Tumbleweed*, Heinemann, London.

Webb, C. G. (1959) An analysis of some observations of thermal comfort in an equatorial environment. *British Journal of Industrial Medicine*, **16**, pp. 297–310.

Wulsin, F. R. (1949) Adaptations to climate among non-european peoples, In: *Physiology of Heat Regulation and the Science of Clothing*, (Ed. by L. H. Newburgh), W.B. Saunders, Philadelphia.

Chapter 2

Control of health and comfort in the built environment

David Lush

BSc(Eng), C Eng, MIEE, FCIBSE
Technical Director, Ove Arup Partnership

Sick building syndrome (SBS) is increasingly experienced in British buildings and this chapter examines the many variables that are thought to contribute to its occurrence. Design engineers and architects often omit vital specifications for comfort requirements, indoor air quality and health and safety factors. While there is no proof as to which of the many criteria are responsible for SBS there are steps that designers can take to minimize the potential for SBS in buildings through good design, commissioning and maintenance procedures.

2.1 Introduction

In any discussion on the control of environmental comfort and health, including sick building syndrome (SBS), there is often an implication that it is the fault of the designer, or, to be more specific, the building services engineer. In fact the engineers of various disciplines and their architectural colleagues in the design team are all involved. However, the occurrence of SBS and other environmental and health problems may or may not be their responsibility. What needs to be said at the outset is that there are many questions and very few specific answers.

The traditional environmental parameters are still vital for comfort and while a common topic for discussion is SBS, it is important to remember that it is only part of the wider issue of a comfortable and healthy working environment. In modern society a large proportion of the population spend up to 90% of their lives indoors and creating a suitable internal environment, whether for work, play or domestic situations, is a crucial element of building and building services design.

This chapter looks first at the traditional parameters and then at the criteria which are, or may be, important for a comfortable and healthy environment in future buildings and refurbishment situations.

2.2 The traditional parameters

In the first instance, consider the normal working environment where overall comfort is specified only by temperature, relative humidity – if the space is air conditioned, lighting level (illuminance) and noise criteria. If these parameters are catered for in the detailed design they will only be achieved if the systems are correctly installed, fully commissioned and properly maintained. It has to be said that the last two elements are all too frequently in default and the question then becomes 'when does genuine discomfort turn into an SBS statistic?' As buildings become more complex, these problems are obviously exacerbated.

What can be done is to examine good practice in areas suspected of contributing to SBS, to ensure that only the best is provided – at design, during installation and commissioning, and by proper maintenance after completion. The four traditional specified criteria should be examined first, together with the problems associated with the air in the working environment which can be considered from familiar and alternative view points.

2.2.1 Temperatures and humidities

Working environments may be comfortable or uncomfortable, but what is acceptable to some is often rejected by others. This is a statement of the obvious, but is extremely relevant to any environment provided by building services systems. Extremes of temperature or humidity, in environmentally controlled buildings can be uncomfortable and annoying, but have not been associated with medically defined illness. The thermal illnesses of hyperthermia and hypothermia do not occur within the normally accepted range of temperatures for the working environment and if temperature is in any way related to SBS it can only be assumed that it is in combination with one or more of the many possible causes identified and discussed later.

Office environments, which are the prime concern here, would normally be designed by service engineers with comfort dry bulb temperatures selected from the range of 20°C to 24°C with limits of ±1°C or ±2°C. With heating only systems, one figure would be chosen, but with air conditioning systems two figures could be selected, the higher one for summer (cooling) conditions. These figures are taken to apply generally in the United Kingdom (UK) but in the USA and Far East higher figures would often be used and in the Middle East, where external temperatures reach 40°C, an internal temperature of 27°C to 30°C would be considered comfortable.

Comfort conditions for relative humidity (rh) are generally considered to be between 40% and 60% rh. In heated buildings there is normally no humidity control, but the UK climate is such that internal conditions tend to fall within these limits except in prolonged cold spells or hot and 'sticky' summer conditions.

In air conditioned buildings there should be humidity control and the common specification would be 50% ± 10% rh. Limits of ±5% rh should be used only when there is a particular reason or process e.g. paper or card sorting machines.

Below 40%, unpleasant electrostatic effects may be experienced in modern offices due to the nylon and other plastics present in fittings and furniture. There is also evidence of dry throats and noses (Gilperin, 1973; Green 1984). Above about 65% rh evaporation from the skin is inhibited and a feeling of stickiness or sultriness may be experienced.

In offices served only by heating systems the humidity is uncontrolled and temperature may be markedly affected by solar gain. The systems however, have not basically altered for at least half a century and there is no reason why they should suddenly cause SBS. What can be said is that co-ordinating the design of the building fabric and services to minimize the effect of unwanted solar gain is extremely important to reduce temperature swings.

In air conditioned offices both temperature and humidity should be specified and controlled. Again, over the period when the incidence of SBS has apparently been increasing, the temperature control of air conditioned buildings has been moving from global or zone control to individual space control. This has provided much tighter temperature control, with fewer variations or extremes, and the facilities for local occupant control. It is therefore strange that variations in temperature are quoted as a common complaint, implying that they are a source of SBS. Completed SBS studies should perhaps be re-evaluated to determine the level of temperature control on the building services systems and new studies should examine this very closely.

Humidity control for comfort is over such a wide range that there should be no complaints unless the system is inoperative or, possibly, where comfort cooling has been passed off as air conditioning.

Having provided arguments as to why there should be no problems with temperature and humidity control it is still necessary to examine why complaints about these parameters do recur. There are valid reasons, none of which invalidate the previous arguments:

(1) Older buildings were rarely provided with local space control for temperature.
(2) Modern buildings do have a much higher content of local temperature control but the level is governed by the relative quality of individual buildings and the cost constraints, cf the Ford and the Rolls Royce which are both cars. Building services designers should always choose to have individual control wherever possible.
(3) Before occupancy took place, the controls (and the services plant) may not have been fully checked, commissioned and set up to ensure that the system met the design intent.
(4) Controls are hidden/inaccessible.
(5) Maintenance has not been carried out in accordance with the operating and maintenance manuals provided at the end of the construction phase of the project.
(6) There may be a design fault.
(7) People have increasing expectations from buildings, and their threshold of discomfort may be lower for other reasons.

2.2.2 Lighting

The lighting level (illuminance) for a particular task is a well-defined parameter (CIBSE, 1984) and if properly selected is not of itself likely to create any problems for the occupant. If the designer ensures that glare is kept below a specified level and takes into account the initial loss of light output from the selected lamp over the first 200 hours of use and the reduction in output which can occur between routine maintenance visits, the basic lighting system is straightforward.

The basic lighting calculations use the surface reflectances of walls, ceiling and floor and the illuminance is therefore a function of the colour and finish of the surfaces. The illuminance increases with higher reflectance. Lighting has to be designed for most office situations reasonably early in the overall process, in order to meet contractual and installation obligations. The values of reflectances have to be agreed with the architect at this early stage, but his final decision on finishes frequently take place much nearer the completion of the project and the reflectances may differ markedly from those used in the lighting calculations, normally to the detriment of the illuminance. Because the eye is very adaptive such changes in the illuminance may be acceptable for the task, but may no longer meet the specification. This may occasionally be a major problem and simply emphasizes the need for true multi-disciplinary designs.

2.2.3 Acoustics

Noise levels in the workplace are one of the normally specified environmental parameters. This review is general rather than a dissertation on the complexities of noise and acoustics, but Section A1 of the Chartered Institution of Building Services Engineers *Guide Book A* identifies the criteria.

Given such figures and the external acoustic environment, the building services engineer needs to perform two tasks. Firstly, to work with the architect to ensure that fabric is designed to attenuate the external noise so that the internal design figures are not exceeded. Secondly, the building services design and associated equipment selection have to be carried out to ensure that noise from the services also meet the specified conditions.

Whether noise created by building services simply annoys the occupants, or is a possible cause of SBS, is irrelevant – the former should be avoided by good design and the latter then becomes a non-event. Equipment needs careful selection to meet declared levels of permitted noise and good design also has to ensure that when connected into piped and ducted distribution systems the noise transmission, whether direct or indirect, does not cause problems in the occupied spaces.

Indirect transmission can be created by resonant effects from primary plant vibration, or from noise generated by water in pipework or, more commonly, by air in ductwork and at discharge orifices. It is very easy to write down what should happen, but achieving the correct solutions is not always simple given the constraints imposed by architecture, structure, services, distribution and cost – even in properly co-ordinated projects. Specialist acoustic advice is required in the design.

Noise from lifts and other vertical transport systems should not be objectionable and proper siting and acoustic insulation of the system plantrooms is part of the design process. Noise from communications and information technology (IT) systems also has to be considered.

2.2.4 Air problems

The problems associated with the air in the workplace, both real and alleged, are one of the most common areas cited in health and comfort studies, and designers can certainly ensure that many of the apparent problems are alleviated or eliminated. It is not exclusively, or even necessarily, a design solution which will improve the situation; occupant training as to the proper use of systems and the right level of maintenance are just as important. The best designs, properly installed and commissioned will exhibit, in time, all forms of air faults unless properly operated and maintained. Lack of fresh air, stuffiness and too much air movement can occur in heated and naturally ventilated buildings as well as in air conditioned (or mechanically ventilated) buildings. The reasons may be different in the two cases.

In the heated and naturally ventilated situation the necessary minimum fresh air rate is provided by infiltration and manual control of the windows and this is still the traditional method used in the majority of UK buildings. The use of windows for this purpose permits a wide degree of local control for all three factors mentioned above and is frequently also used for temperature control – the window is sometimes described as the 'British thermostat'. While this appears a satisfactory solution to many occupants it is possible to postulate several pertinent statements:

(1) What is satisfactory ventilation for the person opening the window may be a gale to an adjacent occupant.
(2) The energy required to heat excessive fresh air input is wasteful.
(3) Any control system used in the system design is almost certain to be incapable of controlling the internal environment – however good the design.

As SBS is reported in such buildings it must be assumed that the air problem parameters are not necessarily contributing to the outbreaks.

In air conditioned buildings the design will normally be based on the assumption that the windows will remain closed and the designers need to cater for the correct quantities of fresh air, proper distribution and circulation of air and the avoidance of draughts. The minimum quantities of fresh air for particular types of building and occupancy are well defined (CIBSE, 1986, Guide Books A and B) and can be selected and/or modified from experience. Multiplying factors are available for various levels of pollution caused by smokers, but the effects of other pollutants are not yet classified sufficiently for similar factors to be introduced to cover them.

The effect of filters on fan sizing is important as a dirty filter will reduce the volume of air as compared to operation in the clean condition. Perhaps more important than this, in the context of SBS, is the maintenance of such filters. If left in a dirty condition some of the trapped material will be carried into the airstream and distributed in the space.

Variable air volume (VAV) systems in particular, because of their operating characteristics, have to be carefully designed and controlled to ensure that they provide sufficient fresh air when operating on low loads and that individual terminals at minimum output do not create stuffiness or erratic air velocities. Fan-assisted VAV terminals can overcome this problem. Where SBS is reported in air-conditioned (or mechanically ventilated) buildings it is clear that fresh air quantities, recirculation systems and controls should be carefully checked.

Stuffiness may be due to incorrect air distribution, lack of fresh air, or a combination of the two. The air distribution case should be eliminated by good design which should also accommodate a degree of flexibility for changes in occupancy and usage patterns. There are however limits to this adaptability and final occupancy or gains can sometimes be as much as double the figures in the design brief. Such changes, coupled with occupants injudiciously altering the discharge characteristics of terminal units and commissioning and maintenance defects, can result in stuffiness. Any stuffiness arising from such causes will be exacerbated if there is a lack of fresh air.

There is some evidence that a general increase in fresh air quantities reduces the incidence of SBS complaints. Apart from the obvious dilution of the many pollutants which are known to be present, which is beneficial, there is no obvious reason for the 'cure'.

Excessive air movement, draughts perhaps, are an interesting phenomenon. What is considered quite acceptable with open windows in a heated and naturally ventilated building is considered to be a problem in an air conditioned building. What reasonable explanations are there for this? Possibly a general draught from an open window is expected, whereas those in air conditioned situations are often said to be localized i.e. round the legs or neck. Secondly, occupants moving into an air conditioned office from premises which are heated and naturally ventilated have high expectations about their new environment, which are frequently not met. Thirdly, designers have a fairly limited range of choice in terms of air velocities for comfort and are normally careful about their selection, which needs to be related to the final air distribution and discharge terminals. If the designer follows the rules and complaints still occur the most likely reasons are:

(1) The occupants were not expecting air movement.
(2) The distribution of air is incorrect due to incomplete commissioning.
(3) The building was occupied before the plant was fully commissioned, particularly in terms of air balancing and correct air discharge quantities.
(4) The control system is not adequate.
(5) After commissioning the plant has not been properly maintained.
(6) The complainant is part of the statistical sample falling outside the normal distribution for satisfaction.
(7) Finally, the designer may not have followed all the rules – for a variety of reasons; some of which he cannot control, but should record.

2.3 Widening the environmental specification

There is much debate on the need to extend the criteria for defining a modern working environment which is comfortable and healthy. The range of additional possibilities is vast and it will take time before any of them become fully recognized and accepted nationally and internationally. There are several comfort criteria which are commonly accepted as being important but do not form part of the environmental specification.

2.3.1 Fresh air changes and air movement

Air problems have already been mentioned, but fresh air change rates and air movement are rarely part of the project brief and are commonly selected by the designer. Design guidance is available, quantitatively defined in professional handbooks, and selection is a function of user needs and design experience. As both of these parameters are identified as areas of concern in SBS studies, their more explicit specification should be addressed.

2.3.2 Glare, veiling reflections, daylighting, luminance, etc.

This group of criteria, like fresh air quantities and air movement, are rarely specified in a project brief or specification. They too are part of the designer's evaluation and selection process and suitable guidance is available. Quite apart from SBS there are continuing research and study projects examining a range of postulated potential problem areas relating to light sources, systems and the health of the working population. The findings should be correlated with work on SBS, where the quality of light may be important. Unsuitable lighting arrangements for occupants using display terminals are situations familiar to many of us. There is a publication from the Chartered Institution of Building Services Engineers (CIBSE) entitled *Lighting for Visual Display Terminals*, which addresses this problem.

2.3.3 Additional elements

There are yet more parameters which need consideration, some within the control of architect, others the responsibility of the building services engineer, and all of them require a co-ordinated approach by both disciplines. Colour finishes in a building and their reflectances, i.e. the amount of light which strikes the surface and is then reflected, are extremely important in terms of occupant satisfaction, and this applies as much to the furniture and equipment as it does to the walls, ceilings and floors. The same point could be made about the colour tint chosen for many window solutions, with the added factor that it can detrimentally affect the occupants' perception of the outside environment. This group of criteria may sound somewhat esoteric in terms of comfort and health, but they do require objective review and further research.

The recognizably engineering factors which may enter more directly into future environmental standards are:

- Mean radiant temperature – the effect of large hot or cold surfaces.
- Infra sound – the possible effect of low frequency vibration.
- Mains flicker and ultra-violet radiation from luminaires – current studies and improvements in technology suggest that these are not, and need not be, problems.
- Ionization – the effect of negative or positive ions in the air; an area of continuing debate and argument.
- Information technology (IT) acoustics – mentioned earlier. The effects of the various levels and frequencies of noise emanating from IT equipment may have a marked effect on occupancy comfort. From the traditional telephone ringing, through copiers, word processors/keyboards and on to computers, printers and graphics plotters there is an almost infinite range of acoustic effects which create a background noise level. In this area the building services engineer currently, has virtually no say, or control, over selection or noise levels. However, it is in this field that there is an ever increasing mix of various sounds which may contribute to the incidence of SBS. Rather like air pollutants, very little is known about the long-term exposure to low ambient noise levels comprising a multiplicity of discordant sounds.

The extra criteria listed above are probably the most likely candidates to have a direct bearing on conventional environmental comfort, but there are other factors which may affect our comfort and health at work. They are broadly reviewed below.

2.3.3.1 ***Pollutants***

Modern society recognizes the multiplicity of pollutants in our environment generally, but the following identified groups specifically affect the built environment and comfort.

- Tobacco smoke.
- Fibres and dusts – mineral, paper, ink from printers, etc.
- Volatiles and organic vapours – from adhesives, etc.
- Micro-organisms – bacteria, viruses, spores, dust mites, etc.

All these pollutants are created within the working environment, and very little is known about their concentrations, and the short or long-term exposure effects at differing concentrations. Even less is known about mixtures of these contaminants and the possible 'cocktail' effect. It may be that a very small percentage of the working population are allergic to one or more of these pollutants. As the counted and identified pollutants can exceed three figures, any such allergies could conceivably affect a relatively large percentage of the occupants. On a longer term view, if any pollutants are identified as causing such problems, their source materials could be banned (COSHH, 1988), or fresh air change rates could be increased to lower the concentrations to safe levels – but only to a limited extent. A recent study and tests on a UK building with a history of SBS has indicated that a reduction in the dust mite population has also reduced the level of SBS complaints in the treated area.

2.3.3.2 *Physiological v. psychological*

The debate on comfort and health covers both the obvious and the less frequently addressed criteria which affect our physical and physiological reactions to our environment. It also extends to consideration of the psychological factors which are outside the professional competence of the architect and engineer. The designers may in future be guided by medically orientated research into design solutions which could alleviate the effect of factors, such as:

- Lack of job fulfilment.
- Repetitive/boring work.
- Lack of privacy.
- Lack of individual identity.
- Perceived lack of control over 'personal' environment.
- Poor management.

2.3.3.3 *Medical risks*

Until recent times the design of building services (HVAC systems in particular) was a function of an environmental specification which had to be met to achieve comfort. Apart from the, as yet, unresolved problems associated with SBS the designer now has to deal with several medically classified criteria. Engineering designers are not professionally qualified to assess medical risks but they can design to alleviate or eliminate such risks once they have been identified. Thus, design of buildings and building services can certainly contain the risks associated with Legionnaires' disease (and its associated family of illnesses) and radon, and they must always be considered in the design process. Less clearly defined, but possibly hazardous, are the effects from electromagnetic radiation generally and high voltage overhead cables in particular and design teams may need to use consultant medical input as part of the design processes of the future.

At this point it is relevant to address SBS in more detail.

2.4 What is SBS?

Everybody, or nearly everybody, has heard of sick building syndrome but unless you have been involved with an actual study of a particular building you may still be unsure of what it is supposed to be. Firstly, it is a misnomer, because the buildings are not sick, although occupants of such buildings exhibit symptoms associated with being ill.

The World Health Organization identifies a range of symptoms for SBS which cause genuine distress to some building occupants, but cannot be clinically diagnosed and therefore cannot be medically treated. They typically include:

- Stuffy nose.
- Dry throat.
- Chest tightness.
- Lethargy.
- Loss of concentration.
- Blocked, runny or itchy nose.
- Dry skin.
- Watering or itchy eyes.
- Headache.

Affected individuals may suffer from one or more of these symptoms and the syndrome is characterized by the additional feature that the symptoms are said to disappear soon after the affected people leave the building.

2.4.1 Is it important?

The whole issue is, understandably, emotive since even in the best of buildings up to 10% of the occupants may suffer from some problems. Unless the dissatisfaction level is around 20% of the occupants, the problems are not likely to be considered as SBS, and if the percentage dissatisfied rises above 20% in one part of the building, where the environment is said to be common throughout, the problem is probably much more mundane.

Sick building syndrome was reported as long ago as thirty years, but its significance, or apparent tendency to increase, has only been apparent over the last few years. Recent research (Wilson & Hedge, 1987; Wilson et al. 1987) has indicated that it occurs more often in air-conditioned buildings, but as it does occur in heated and naturally ventilated buildings, the cause cannot simply be ascribed to air-conditioning. Overall comfort and health in buildings cannot be satisfied by the HVAC industry in isolation.

2.4.2 What is known about SBS?

Let's be honest. While SBS may be classified as an illness, we don't have any absolute proof that it is caused by any one factor or combination of factors. A large number of field studies are being carried out (Kroeling, 1988), often initiated because of the high level of occupant dissatisfaction or SBS symptoms in the working environment. Other research work is being done internationally to try to identify actual causes. There is a wide range of reasons and factors which need further examination. Further, it is possible to identify areas where the lack of current best practice may be a contributory cause to complaints categorised as SBS.

2.4.3 What contributes to SBS and other aspects of comfort and health?

There is a body of evidence which suggests that levels of comfort, health or SBS may be caused by a combination of factors, but the mix and the weighting against each is indeterminate at present. It is appropriate to summarize, in broad terms, those systems and effects which are under suspicion and it should be no surprise that the list includes areas already mentioned:

- Internal environmental specification.
- Internal ambience – colour/reflectances.
- External awareness – size, colour and type of glazing.
- Pollutants created internally.
- Level of boredom with the work tasks.
- Quality of management.
- Maintenance – or lack of it.
- Perceived lack of local environmental control.
- Acoustic problems.
- Lack of fresh air, stuffiness, too much air movement.
- Unacceptable temperatures and, possibly, humidities.
- Lighting problems.

2.4.4 What areas of design improve comfort and health and minimize the risk of SBS?

A pedantic view could be that design is not a problem, because there is no absolute proof that any particular factor is the cause of the illness. However, most of the SBS studies categorize the evaluated buildings in a global sense as good or bad and it has to be accepted that design is one of the areas which needs attention. It cannot be simply a matter of saying that 'good' design will alleviate or eliminate SBS since there is always a degree of subjectivity in any design approach which may result in an admired building – inside and out – whose occupants still suffer.

2.5 What else matters?

In discussing what constitutes a comfortable and healthy working environment, and SBS, there are several areas which are all too often ignored but should be considered as they may actually be very relevant to the situation. They can be listed:

(1) While SBS is not new, what has brought it into prominence? Its importance has apparently arisen from the noticeable increase in complaints from occupants in office environments.
(2) In the past such complaints would have put occupants at risk – so they were not made.
(3) In older heated and naturally ventilated buildings, conditions were variable and expectations were fairly low, so the situation was accepted.
(4) Studies of the environmental conditions, where the occupants were questioned, were rare.
(5) Staff associations are now consulted about working conditions, or make complaints anyway.
(6) Insufficient information is provided to occupants moving from heated to air conditioned buildings about the different sensory perceptions and the local controls available.
(7) In air conditioned buildings, expectations are higher and the complaint threshold is lower.
(8) The commissioning of the environmental services has not been properly carried out or completed.
(9) The design criteria were either not correctly specified, met, or maintained.

2.6 What can be done?

When the criteria for environmental comfort and health in buildings are considered, together with the symptoms of SBS, the wide range of parameters which need to be researched and/or included in best practice design becomes clear. It is very important to remember that good design, properly supervized construction and comprehensive commissioning will all be brought to nought if the building and its systems are not correctly maintained.

(1) Don't treat maintenance as an unnecessary and costly expense. The operating efficiency of your staff is dependent on a good working environment.
(2) Institute a maintenance programme which includes:
    (a) Regular checking of all temperature (and humidity) control settings and their performance.
    (b) Routine servicing of heating, ventilating and air conditioning (HVAC) plant – clean filters are particularly important.
    (c) Regulation of fresh air quantities, air movement and air distribution – wherever there is mechanical ventilation or air conditioning.
    (d) Proper water treatment in all relevant situations – while Legionnaires' disease is not related to SBS it is a health hazard which proper water treatment can virtually eliminate.
    (e) Ensuring that luminaires are all replaced in bulk at the proper intervals and that faulty luminaires are not left for long periods before remedial action is taken.

(3) Don't occupy working spaces until you are satisfied that the commissioning is complete and the environmental performance meets the design intent.
(4) Inform your staff about the comfort criteria of their working environment and what control they have over their local conditions.
(5) Treat hygiene as a fundamental parameter for comfort and health. Regular cleaning of normally inaccessible spaces (HVAC terminals, ducts, ceiling and floor voids) is essential as dusts, by themselves or as breeding grounds for micro-organisms, can give rise to complaints not apparently associated with them.
(6) Consider management and administrative systems to improve job satisfaction.

2.7 Summary and conclusions

The examination of parameters which need to be specified to extend the specification of the internal environment for comfort and health and in order to reduce or prevent SBS, leads inevitably to a series of conclusions:

(1) Insufficient parameters are presently specified for a modern comfortable and healthy environment, or in respect of SBS.
(2) The range of possible criteria which could affect the incidence of SBS is enormous and covers many elements, only some of which directly relate to building services.
(3) There is currently no absolute proof that any or all of the possible criteria do or do not contribute to SBS, or to what extent.
(4) In order to identify parameters contributing to, and their level of contribution to, SBS a vast amount of further study, measurement and evaluation is required.
(5) When contributory parameters have been positively identified there will need to be national and international agreement on how such parameters are defined and quantified before they can be generally specified. This will be easier for those parameters which are directly related to building services e.g. air movement, than to those indirectly related to the environment e.g. colour tints in glazing or 'cocktails' of volatiles.
(6) In the interim, clients, designers and building users who wish to produce comfortable and healthy buildings will need to:
    (a) Be more explicit in their environmental requirements in the project brief and design specifications.
    (b) Design on a totally integrated basis.
    (c) Closely monitor the installation to ensure that environmental co-ordination is achieved.
    (d) Ensure that operating and maintenance procedures are adequate to retain the design intent and operational performance over a sustained period.

It would be satisfying to be able to define the perfect working environment, or conditions where SBS and its variants could not occur, but this is not possible, certainly at this time. Hopefully, this review of the current situation will assist in understanding the problems, and the summary given above should go a long way to minimising any current or potential problems in this complex field.

References

Chartered Institution of Building Services Engineers (1984) *CIBSE Code for Interior Lighting*, CIBSE, Balham, London.

Chartered Institution of Building Services Engineers (1986) *CIBSE Guide Book A*, Section A1–8/9. CIBSE, Balham, London.

Chartered Institution of Building Services Engineers (1986) *CIBSE Guide Book B*, Section B2–4 *et seq*. CIBSE, Balham, London.

Gilperin, A. (1973) Humidification and Upper Respiratory Infection Incidence. *Heating, Piping and Air Conditioning*, March.

Green, G. H. (1984) Health Implications of the Level of Indoor Air Humidity. Swedish Council for Building Research, *Indoor Air* – **1**.

Health and Safety Executive (1988) *The Control of Substances Hazardous to Health Regulations (COSHH)*, Statutory Instrument 1988, No 1657, HMSO, London.

Kroeling, P. (1988) Health and Well-Being Disorders in Air Conditioned Buildings: Comparative Investigations of the 'Building Illness' Syndrome, *Energy and Buildings*, **11**(1–3), March 22, pp. 277–282.

Wilson, S. & Hedge, A. (1987, May) *A Study of Building Sickness*, Building Use Studies Ltd, London.

Wilson, S., O'Sullivan, P., Jones, P. & Hedge, A. (1987, December) Sick Building Syndrome and Environmental Conditions, Building Use Studies, London.

Chapter 3

Daylight and energy

Joe A. Lynes
BSc, MA, CEng, MIEE, FCIBSE
Senior Lecturer, Hull School of Architecture, Humberside Polytechnic

Window design calls for a logical sequence of decisions to be used sensibly as required during design. This sequence of decisions is outlined, showing clearly the responsibilities of each of the specialists concerned, enabling each to contribute without trying to do another's job.

3.1 Introduction

Windows impinge, directly or indirectly, on all our senses. Their design may well affect each of the items listed in Fig. 3.1. This has been the nub of the problem for architects and for engineers. There have been plenty of specialists in heating and in lighting; but how is one to reconcile their conflicting advice? Textbooks have seemed bent on treating different facets in isolation. One chapter has dealt with heat loss, another with solar gain. A separate book has covered daylight and sunlight, again in different chapters. And who is the expert on *view*? Or are all windows equally suitable for visual communication?

3.2 View

Obviously some windows provide a better outlook than others (Lynes, 1974). The bow window in Fig. 3.2 breaks most of the rules of environmental control. The extra glass exacerbates winter heat loss. The convex geometry invites solar gain. Sealing problems leave occupants exposed to draught, dirt and noise from outside. Natural lighting suffers through reduction in the solid angle of visible sky. Every one of the classical criteria cries out against bow windows; so if we are to accept the outlook as a justification for a window we are forced to reconsider our environmental priorities.

The view through a window depends on its shape, its size and its position in relation to the observer. Sometimes one can identify a critical viewpoint inside a room, for instance the patient's head in a hospital bed. Sometimes one must settle for the most disadvantaged occupant.

Draw a straight line from the eye to an attractive object outside. If the line passes through a window the object is visible; if not, it is hidden. The best vehicle for this analysis is an interior perspective of the window superimposed on a perspective of the view beyond. Failing this, one can use a plan and cross-section through a rectangular window.

The law of gravity tends to stratify an outdoor view into three layers, see Fig.

Admission of sunlight
Admission of skylight (daylight factor)
Heat loss
Solar gain
Evenness of natural lighting indoors
View
Sky glare
Privacy
Natural ventilation
Composition of facade

Fig. 3.1 Some effects of window design.

Fig. 3.2 A bow window at Queens' College, Cambridge.

3.3. The topmost layer is the sky. The second layer mainly comprises upright objects such as trees or buildings. The bottom layer is the foreground, more or less horizontal; this layer contains visual cues about the distance of elements in the second layer, helping to give scale to the outdoor scene.

Fig. 3.3 The view through alternative windows from a given point indoors: (a) revealing the whole skyline; (b) some sky visible; (c) no foreground visible.

Dividing lines between the three strata are marked by sharp changes in luminance, outline and texture. This is where visual information is most concentrated.

Marked drops in satisfaction occur when the skyline is cut by the window head and especially when it is eliminated from view. This helps to explain the importance of the 'no-sky line'. This 'line' (strictly it is a surface in three-dimensional space) divides a room into two parts – one part is exposed to a view of the sky, the other part receives no direct skylight at all. Behind the no-sky line occupants feel deprived of both daylight and view.

Here one encounters a conflict between the competing demands of view, comfort and illumination. Above the roofline any addition to the area of visible sky enriches the view very little, but it does increase visual distraction; in illuminating engineering terms, the source area expands while the background luminance remains constant. This can only mean glare. The other side of the coin was pointed out by the Roman architect Vitruvius some two thousand years ago: the amount of daylight reaching a point indoors depends mainly on how much sky can be seen from that point. To design a window rationally one must weigh the conflicting imperatives of outlook, visual comfort and the admission of daylight.

3.3 Daylight factors

In the UK daylight design has been based upon overcast sky conditions for three reasons:

(1) It is prudent: if the daylight level indoors is adequate on an overcast day, it is likely to be sufficient on a sunny day too.
(2) It avoids the complication of taking orientation into account: on an overcast day the sky luminance pattern is independent of the compass-point.
(3) Although the illuminance, inside and outside, varies continually, the illuminance at a given point indoors under overcast conditions is a fairly constant fraction of the prevailing outdoor illuminance.

This constant ratio of the indoor illuminance to outdoor illuminance under an unobstructed overcast sky is known as the 'daylight factor'. It is expressed as a percentage. Traditionally daylight factors have been specified as minima. This minimum was hard to predetermine, witness the junkyard of abandoned protractors, charts and pepperpots. The present trend (Crisp & Littlefair, 1984), supported by the *CIBSE Window Design Guide* (CIBSE, 1987), is to average the daylight factor across the whole room.

The average daylight factor *df* is given by the equation:

$$df = \frac{T\,W\,\theta\,M}{A(1 - R^2)} \quad \% \tag{3.1}$$

where T = diffuse transmittance of glazing material (clear single glazing =0.85; double glazing =0.75)

W = net area of window

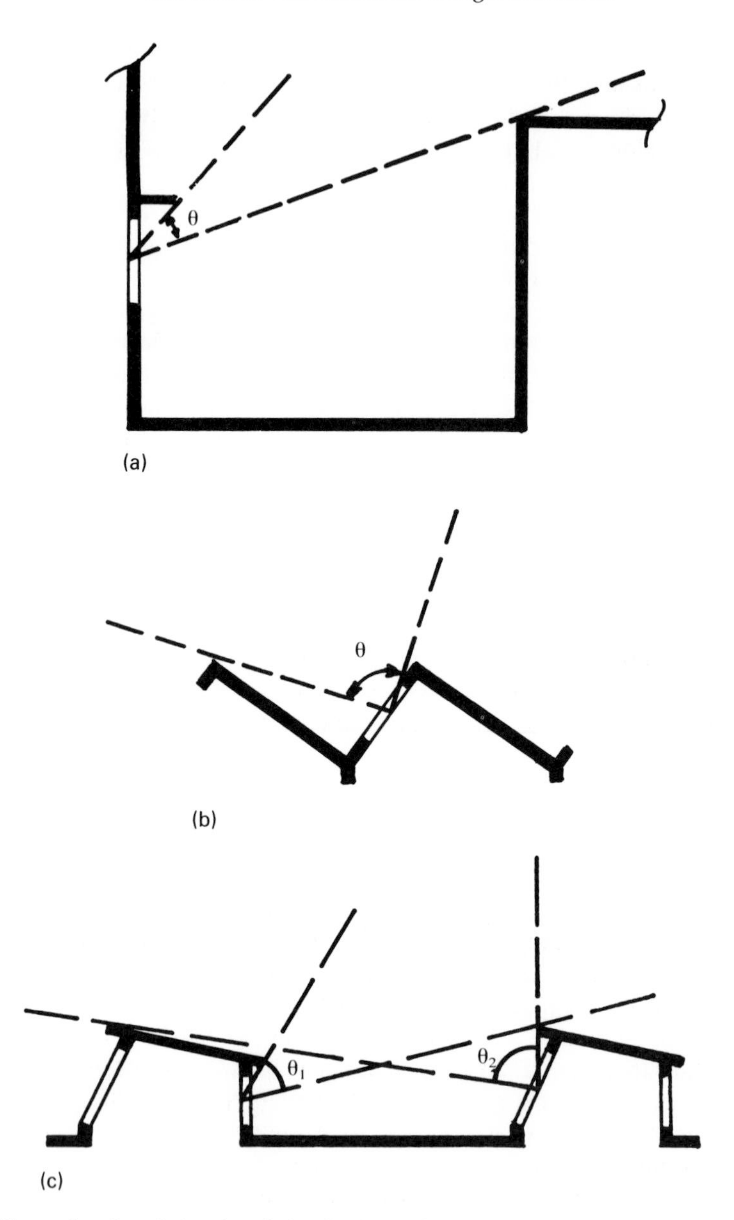

Fig. 3.4 The angle subtended at the window by the visible sky.

θ = angle subtended in the vertical plane by sky visible from centroid of window (degrees). See Fig. 3.4.
A = total area of interior wall surfaces: floor and ceiling and walls, including windows.
R = area-weighted average reflectance of indoor surfaces (fairly light =0.6; fairly dark =0.4)

M = maintenance factor to allow for dirt and deterioration.

Experience indicates that if the average daylight factor exceeds about 5% the interior will look cheerfully daylit. If the average daylight factor drops below 2%, natural lighting alone will seldom be satisfactory and electric lighting is likely to be in constant use.

3.4 A framework for window design

Equation 3.1 can be decomposed into elements which correspond to stages in the window design process (Lynes, 1979).

3.4.1 Stage 1: Site layout

The angle θ in Equation 3.1 and Fig. 3.4 is the angle subtended at the window by the visible sky. This angle is determined by the spacing and massing of the neighbouring buildings. The block layout is effectively determined very early in the design process, usually before fenestration can be considered at all. Access, orientation, site utilization and microclimate are implicitly optimized at this stage. Because these decisions can act as a tight constraint on window design, the determination of the angle θ is singled out as Stage 1 of the window design process.

3.4.2 Stage 2: Area of window

The term W in Equation 3.1 stands for the net area of the windows. If windows are along only one wall the determination of W is preceded by two quick tests on the likely evenness of the natural lighting. The initial check is on the no-sky line, invoking Vitruvius' contention that the amount of daylight reaching a point indoors depends mainly on how much sky can be seen from that point. Beyond the no-sky line there is no direct skylight and hence little chance of adequate natural lighting. If the no-sky surface seriously impinges on the working plane then the room is too deep to be well daylit. Whatever the size of the window, the back of the room will seem relatively deprived of natural light and occupants will mostly switch their electric lighting on whatever the weather.

If the room survives the first check it may still prove too deep for good natural lighting, so a second overriding check must now be applied. The maximum room depth for satisfactory natural lighting will be exceeded when

$$l/w + l/h > 2/(1 - R_B) \tag{3.2}$$

where l = depth of side-lit room (see Fig. 3.5)
w = width
h = height of window head above floor level
R_B = area-weighted average reflectance in back half of room.

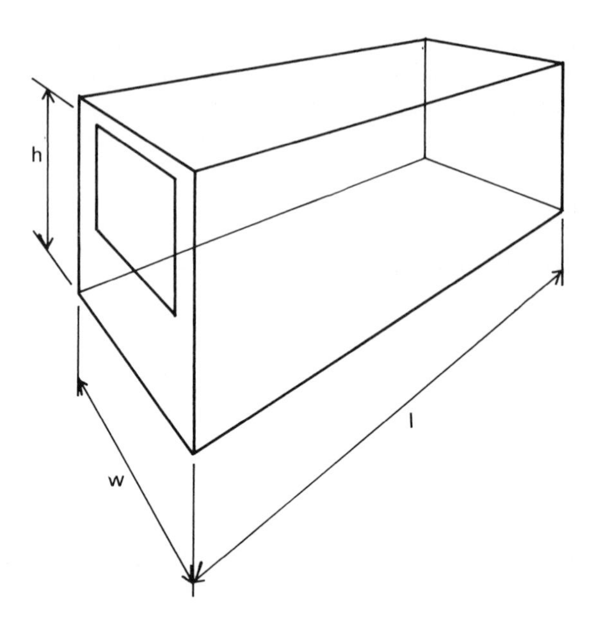

Fig. 3.5 The limiting depth of a side-lit room.

If both the uniformity criteria are satisfied, i.e. if the room depth is not excessive, then a suitable average daylight factor should be chosen.

Otherwise the daylight contribution must be considered cosmetic, rather than functional, and the window should be treated as a source of view rather than as a source of light.

The window area W is estimated by inverting Equation 3.1:

$$W = \frac{df\ A\ (1 - R^2)}{T\ \theta\ M} \tag{3.3}$$

where df is the required average daylight factor.

At this point the designer has another conflict to resolve. The average daylight factor is proportional to the window area W. But so is the winter heat loss through the windows, and so (other things being equal) is the daily mean solar cooling load through the windows.

The conflict must be acknowledged and reconciled at this stage, perhaps by juggling the variables in Equation 3.3, perhaps by adjusting the thermal properties of the window: the U-value or the solar gain factor. It is important to settle the window area W at this juncture, before the designer enters the Stage 3 scenario, which is concerned with window shape and position.

3.4.3 Stage 3: Window shape and position

By Stage 3 either the window area is fixed and the average daylight factor known, or else the room has been identified as too deep for satisfactory natural lighting.

In either case the shape and position of the window remain to be fixed, so a further dilemma must be faced. This arises from the competing claims, outlined in Section 3.2 above, of view, uniformity and freedom from sky glare. Its resolution is obviously facilitated by the decisions taken in Stage 2, namely the freezing of the window area or a ruling that the room is too deep for satisfactory natural lighting.

Uniformity is promoted by ensuring that the back of the room sees as much sky as possible. Visual comfort is promoted by ensuring that as little sky as possible is visible. The view is best exploited by making sure that its most attractive elements can be seen from indoors. Conflict is resolved, in Stage 3 as in Stage 2, by balancing priorities.

In Stage 3 the possible merits of multilateral fenestration may also be reviewed. Windows in more than one wall should improve the uniformity of natural lighting by increasing the solid angle of sky visible from the worst-lit parts of the interior. They will also mitigate the luminance contrast between the sky and the window surrounds. In naturally ventilated buildings they will promote cross-ventilation.

3.5 The credibility gap

Obviously this is not the way that windows normally get designed. Indeed if an architect were to follow the sequence slavishly the building elevations would be chaotic, if not unbuildable. There can be no substitute for informed professional judgement.

The key decision is to identify relevant crunch-points. What is the most important or the most critical opening on a given facade? Design this window properly, and perpetuate variants of the solution across the elevation. Then the building will work, and the fine-tuning of its services can follow.

3.6 Conclusion

Window design calls for a logical sequence of decisions, not to follow slavishly, but to fall back on when stuck. The sequence outlined here clarifies the responsibility of each of the specialists concerned, enabling each to contribute without trying to do another's job. This is the key to good daylighting. Might it also have a message for architecture as a whole?

References

CIBSE (1987). *Applications Manual: Window design*, Chartered Institution of Building Services Engineers, London.

Crisp, V. H. C. & Littlefair, P. J. (1984). Average daylight factor prediction. In *Proceedings of National Lighting Conference*, Cambridge, Chartered Institution of Building Services Engineers, London.

Lynes, J. A. (1974). The window as a communication channel. *Light and Lighting*, **67**, Nov./Dec., pp. 284–8.

Lynes, J. A. (1979). A sequence for daylight design. *Lighting Research and Technology*, **11**(2), pp. 102–6.

The Indoor Environment: Controls and Health

Chapter 4

Environmental control in energy efficient buildings

Philip Haves

MA, PhD, CEng, MCIBSE, MASHRAE, MIEE

Research Associate, Department of Engineering Science, University of Oxford

Daylighting and natural ventilation have the potential to reduce significantly the energy consumption of commercial buildings.Climate-accepting designs are also preferred by occupants, who respond to being given an active role in the environmental control of the building by being more tolerant of variations in conditions inside the building. However, the design of environmental control systems for climate-accepting buildings is more difficult than for conventional, climate-rejecting buildings. This chapter discusses the need for the designer to integrate the behaviour of the envelope, the occupants and the mechanical systems. Techniques for postponing the transition from natural ventilation to mechanical ventilation and then to air-conditioning are indicated. Different HVAC systems are described briefly and their energy efficiency discussed. The effects of different control strategies are considered, as is the use of building energy management systems to facilitate the detection of inefficient operation. The role of computer-based tools in the prediction of performance is considered and the assessment of environmental control performance is discussed in terms of the trade-off between discomfort, energy costs and maintenance costs. The advantages of an integrated approach to CAD involving design, commissioning and operation are discussed.

4.1 Introduction

A well established priority in design is to meet the needs of the building owner or occupier by producing a design that will provide a comfortable and productive environment with energy and maintenance costs that are as low as practically possible throughout the life of the building. There are two, complementary, ways in which the environmental control performance of buildings can be improved:

(1) Improved design.
  (a) Better envelope – reduced heating and cooling loads for the whole building.
  (b) Better internal planning – minimize simultaneous heating and cooling loads in different zones.
  (c) Better heating, ventilating and air-conditioning (HVAC) systems – more efficient equipment more carefully sized.
  (d) Better control systems – operate the building and plant as efficiently as possible.
(2) Improved operation.
  (a) Better commissioning – ensure equipment is installed and set up correctly.

(b) Better maintenance – routine inspection and servicing.
(c) Better energy management – continuously checking that the performance is as expected.

It is widely accepted that the performance of the existing building stock is limited as much by poor operation as by poor design. However, improved design and improved operation are not independent and there is an increasing concern with 'designing for operability'. The key points are to:

(1) Keep the design as simple, and hence as comprehensible, as possible.
(2) Plan to communicate the design intent in a way that will be understood by those who will be responsible for the operation of the building, both immediately and in the future.

The starting point is a clear conceptual design that results in a straightforward operating strategy. In particular, the operating strategy must be matched to the competence of the people who will be available to operate the building.

4.2 Energy efficient environment control

Amongst the issues for the designer of an energy efficient building are:

- The effect of the building form on the heating and cooling loads.
- The role of the envelope in providing daylight, solar heat gain and natural ventilation.
- The role of the occupants in the operating strategy for the environmental control of the building.

A basic question is whether the building will be predominantly 'climate-accepting', i.e. utilize daylighting, natural ventilation and passive solar heating, or be predominantly 'climate-rejecting' and rely on artificial lighting, heating, ventilation and air-conditioning. In particular, whether the building will use:

- Natural ventilation.
- Mechanical ventilation.
- Comfort cooling.
- Full air-conditioning.

'Comfort cooling' involves the use of chiller plant to limit the temperature in the occupied spaces without any direct control of the humidity. 'Full air-conditioning' indicates that humidity is also controlled, and possibly that the concentrations of particulate matter and chemical pollutants are also controlled or reduced in some manner. In temperate climates, full air-conditioning, as opposed to comfort cooling, is not generally required to maintain thermal comfort, although reducing high levels of humidity allows thermal comfort to be maintained at slightly higher temperatures. Full air-conditioning may be required in such applications as libraries and art galleries where there is a specific requirement to control humidity. The term 'air-conditioning' is often used in situations where what is being provided is actually comfort cooling.

The site will determine whether fixed fenestration is required to keep out noise, dirt or intruders. The use to which the building will be put, together with the site ('high rent' or 'low rent'), will determine the occupant density and the internal loads. If we assume a commercial or institutional building with no special environmental control requirements, the building can be climate-accepting and naturally ventilated in temperate climates if the site permits operable windows and a shallow plan and if the loads are not too high. Even if the loads are high enough to require comfort cooling during the summer, the building can be naturally ventilated for the rest of the year; an example of this type of 'mixed mode' building is described in chapter 17. If the site requires a sealed building then it must use mechanical ventilation, comfort cooling or full air-conditioning depending on the magnitude of the loads and the stringency of the environmental control requirements.

As discussed in chapter 16, the heating load is not a major issue in UK commercial buildings unless the internal gains are quite low. The cooling load is usually much more significant, and, as discussed in chapters 3 and 17, daylighting has the potential to reduce the cooling load as well as the electricity consumed by the lights. Daylighting requires a shallow plan, which as just noted, is also required for natural ventilation. There are indications (e.g. Robertson, 1990) that occupants may be satisfied with a wider range of environmental conditions in a climate-accepting building than in a climate-rejecting one. All this suggests that the climate-responsive, naturally ventilated, shallow plan concept should be investigated first, and only rejected if the constraints imposed by the site or the loads are too great.

If dirt and noise levels do not necessitate a sealed building, the requirement for mechanical ventilation or comfort cooling depends upon the cooling load. Several strategies can be employed to postpone the transition from natural ventilation to mechanical ventilation or from mechanical ventilation to comfort cooling:

(1) Careful control of solar gain using a combination of blinds and external shading devices.
(2) Use of ceiling fans or other means of increasing the air speed over the occupants, allowing thermal comfort at higher temperatures.
(3) Use of thermal mass to reduce peak temperatures, either by simple smoothing or by nocturnal ventilation.
(4) Planning the use of the interior spaces so as to:
    (a) Facilitate separate venting of major sources of localised heat.
    (b) Concentrate spaces with higher loads (e.g. conference rooms, lecture theatres) in one part of the building, allowing the rest of the building to be naturally ventilated.

Where possible, the first two strategies should be implemented so that the occupants can operate the fans and blinds, since it appears that the ability to intervene in the control of the environment is a major factor in the acceptability of an increased range of environmental conditions in climate-responsive buildings. Tong (1989) refers to studies that indicate a correlation between sick building syndrome and lack of user control.

If supplying outside air by means of mechanical ventilation is insufficient to meet the cooling load there are three alternatives:

(1) Supply air that is cooler than the outside air.
(2) Reduce both the air temperature and the radiant temperature in the space by means of extended cooled surfaces, usually water-cooled panels on the ceiling (CIBSE, 1991a).
(3) Cool the air in the space by means of water-cooled 'chill beams' on the ceiling (CIBSE, 1991b).

The first alternative is the approach employed in conventional air-conditioning systems. The third approach is new and produces air-flow patterns that are compatible with displacement ventilation. The second and third approaches are only suitable for comfort cooling since any dehumidification would involve condensation in the occupied space. All three require a refrigeration plant or, an option that is attracting increasing interest for comfort cooling, indirect evaporative cooling.

It is possible to list different types of building in order of increasing loads that can be met (albeit with decreasing efficiency):

(1) A naturally ventilated, daylit building – shallow plan, clean quiet site with unobstructed solar access.
(2) As above, but with increased thermal mass and more careful shading to reduce the peak cooling load.
(3) As above, but with ceiling fans to allow comfort at higher temperatures.
(4) A noisy, dirty site with a mechanically ventilated building – shallow plan to allow daylighting.
(5) As above, but with cool air blown through hollow cores in the floor slab during the night in order to reduce daytime cooling loads.
(6) Higher occupant density – indirect evaporative cooling used to cool the air supplied to the zones.
(7) Higher occupant density, shallow plan, comfort cooling in particular zones (e.g. lecture theatres, conference rooms).
(8) High occupant density, expensive site surrounded by tall buildings – fully air-conditioned, deep plan, artificially lit.

It is clear that it is easier to design a climate-accepting, low energy building on a low-rent greenfield site than in the City of London.

4.3 Environmental control requirements

Part of the task of designing a building is deciding which properties of the indoor environment need to be controlled, how they are going to be controlled, and the values at which they will be maintained under different circumstances. The first problem is to determine what range of environmental conditions will meet the client's need for a productive environment. The variables involved include:

- Temperature.
- Air speed.
- Humidity.
- Indoor air quality.
- Lighting.

Since there is an obvious trade-off between thermal comfort and energy consumption (the heating bills can be reduced to zero by switching off the heating system!), the limits on the environmental conditions that are acceptable, and for how long, should be explicitly agreed with the client. It is commonly asserted that a wider range of environmental conditions is acceptable in climate-accepting buildings. If the design relies on this wider range then this must be accepted by the client, otherwise the building may be operated in a way that frustrates the design intent.

Once the limits on the environmental conditions have been agreed these should be incorporated in detail in the specification for the building services engineers who will design the HVAC system. Current specifications are generally vague in this respect and rarely deal with energy targets or how fast the system should recover from load changes (e.g. a sudden change from zero to full occupancy in a conference room).

4.3.1 Temperature control

Temperature control in conventional spaces is required to maintain comfort during occupancy and to prevent damage to the fabric and contents outside occupancy. The principal agent of temperature control is the building envelope, and the main aim of passive (solar) design is to maximise the effectiveness of the envelope in this respect. Some elements of the envelope, such as its thermal mass or fixed shading devices, are purely static and need no controller to operate them. Other elements, such as operable windows and blinds, are usually controlled by the occupants, whilst the heating or HVAC system usually has automatic controls. Problems arise when the occupants and the automatic controls have different objectives.

The room temperature can be controlled by varying:

- The solar gain.
- The natural ventilation rate.
- The conductive heat flow through the building envelope, e.g. movable window insulation.
- The temperature or flow rate of air introduced into the space by mechanical means.
- The temperature or flow rate of water in radiators.

The first three can be most effectively controlled manually, although automatic control is possible. The last two usually involve automatic controls (e.g. thermostats, thermostatic radiator valves) and a major question for the designer is the extent to which the occupants are to be allowed to reset these controls. One approach is to set up the automatic controls to provide a minimum level of comfort by having a wide dead-band between the heating and cooling set-points (e.g. 18°C and 25°C) and allow the occupants to override these set-points within limits, possibly for a fixed length of time. (See Bordass, 1990 for a discussion of the interaction between manual and automatic controls.)

An issue that often leads to conflict between architects and engineers is the positioning of sensors in occupied spaces. The problem is that if the space temperature is not measured, or cannot be estimated accurately enough from other measurements, then it cannot be controlled satisfactorily by automatic means. The temperature of the air extracted from the occupied space is often used as a proxy for the space temperature, but difficulties arise if the radiant temperature in the space does not track the air temperature, or if the air is extracted through the luminaires. If the output of the luminaires is varied by a lighting control system the amount of heat picked up by the extracted air will vary, leading to a variable difference between the room air temperature and the extract air temperature. The situation is further complicated if the flow rate of the extract air varies, as it does in 'variable air volume' systems, currently the most popular type of HVAC system. In principle these effects can be corrected by a modern control system but careful commissioning is required to obtain satisfactory performance. It is fair to conclude that if the occupants are not involved in the environmental control then proper attention must be given to the sensors on which the automatic controls rely.

The degree of thermal discomfort depends on a number of factors in addition to air temperature, including radiant temperature and air speed. However, conventional practice is to control the heating or air-conditioning using thermostats or temperature sensors that respond principally to air temperature, in spite of the fact that radiant temperature is as important at low air speeds. This is reasonable if the air and radiant temperatures track each other, as they will do if the windows are small and the exterior walls well insulated. Problems occur with large windows, particularly with single glazing in cold weather, since the surface temperature of the glass will be closer to the outside temperature than the inside temperature and will depress the inside radiant temperature. Sensors that respond to the appropriate combination of air and radiant temperature should be used in such cases.

4.3.2 Humidity control

Low humidity, i.e. less than 30% relative humidity, may cause respiratory and other physiological problems such as dry throat and skin. High humidity, i.e. greater than 70% relative humidity, encourages mold growth and narrows the range of temperature over which thermal comfort can be attained. Humidification can be achieved by passing heated air through a pad soaked in water or by injecting steam into the air supplied to the space, and in such systems is controlled by varying the flow rate of water or steam.

In temperate climates such as the UK, the humidity of the outside air is usually low enough for dehumidification to be effected by introducing outside air, by either natural or mechanical ventilation. If the outside air is too humid, air-conditioning systems dehumidify the air by cooling it below its dew point, so that moisture is removed from the air stream by condensation. One problem with controlling humidity is that humidity sensors are often unreliable. In particular the outputs of a number of types of sensor drift as they become contaminated with organic compounds. Because of the expense of installing and maintaining humidity sensors it is usual, in office buildings for example, to measure the average humidity of the air returning from a number of different zones, rather than to measure the humidity in each zone.

4.3.3 Indoor air quality

Sick building syndrome and the complementary problem of building related illness has focused attention on the need to limit the concentration of a wide range of impurities in the air in occupied spaces. These impurities include carbon dioxide, chemical pollutants from the fabric and furnishings, micro-organisms, particles and radon. Sensors are now available to measure the concentration of carbon dioxide and of a broad spectrum of organic compounds and are used in a similar way to humidity sensors, i.e. to regulate the average concentration in a number of zones of a building.

The impurity concentration is regulated by varying the rate at which fresh air is introduced into the zone ('demand controlled ventilation') and this strategy is starting to be used as an alternative to the conventional practice of supplying fresh air at a fixed rate per occupant. Since fresh air may have to be heated in winter and/or cooled in summer, reducing the amount of fresh air whenever possible has the potential to save significant amounts of energy.

4.3.4 Air distribution

The most important aspects of air distribution are the supply of adequate amounts of fresh air to each part of the occupied space and the avoidance of cold draughts. The design of forced air systems is usually left to the services engineer, who has well established design procedures and a wide variety of fittings to draw on and hence has relatively little excuse for bad design. By contrast, the design of natural ventilation systems is typically the responsibility of the architect and presents a number of challenges. Firstly, it is rather more difficult to ensure uniform distribution of air flow; it may not be possible for each occupant to have their own window. The provision of individual offices and, in particular, the need for aural privacy often conflicts with the need to ensure a cross flow ventilation path from one side of the building to the other for each window. If the building has substantial internal head gains, such that significant amounts of outside air are required even in cold weather, the incoming cold air must be diffused and then mixed with the inside air before it reaches the occupants in order to avoid cold draughts.

If it is too hot in the space, thermal comfort may be improved by increasing the speed of the air flowing over the occupants. In a naturally ventilated building, low natural air flow due to low wind speed may be compensated for by the use of ceiling fans. Air speeds up to $\sim 0.8\,\mathrm{ms^{-1}}$ can be used in offices (ASHRAE, 1981) and somewhat higher speeds can be used where rustling of papers would not cause problems. Manually switched, multi-speed ceiling fans allow the occupants to control the air speed as required.

4.3.5 Lighting

Lighting accounts for a major part of the energy use in commercial buildings, both directly and also indirectly, if the building is mechanically ventilated or cooled, through the additional cooling load imposed by the heat given off by the lights. Daylight directly displaces the electricity used by artificial lights and, because sunlight has a better ratio of light to heat than artificial lights, it has the potential to reduce the cooling load. This potential can only be realised if the amount of daylight entering is limited to what is required, in order to limit the corresponding solar heat gain. This almost invariably involves excluding the direct beam radiation whilst admitting most of the diffuse radiation from the rest of the sky, typically by using a combination of fixed shading elements and adjustable blinds.

Lighting controls aim to vary the amount of artificial light in such a way that the combination of daylight and artificial light is just sufficient to provide the desired illuminance. The light output of fluorescent tubes can be made to vary smoothly over a range of about four to one by using special control equipment. This equipment adds appreciably to the cost and reduces slightly the efficiency of the lights at full output but has the additional benefit of eliminating flicker. The output of fluorescent lights can also be varied by having several tubes in each luminaire and switching them on or off one tube at a time, although this can produce annoying steps in the illuminance if the switching is too frequent.

Switching may be manual or automatic; a common observation is that people will switch lights on, though at a wide variety of daylight illuminance levels, but will rarely switch them off, especially at work. A common strategy for manual switching is to switch the lights off at regular intervals, at the end of the morning and afternoon for example, and require the occupants to switch the lights on again manually as required (see chapter 17).

4.4 Energy efficient HVAC systems

4.4.1 Overview of HVAC systems

Whether the building is heated and naturally ventilated or fully air-conditioned, the HVAC system is an integral part of the environmental control system and should be selected at a relatively early stage in the design. In particular, the best choice of HVAC system for a particular building depends not only on external factors such as the climate and the source of energy and its pricing structure but also on the distribution of the loads in both space and time, the availability of space for the services in different parts of the building, and many other factors that depend on the rest of the design. There is clearly a need for integrated design teams on all but the smallest and simplest jobs. That said, it is still possible to categorise HVAC systems according to their energy consumption and their other advantages and disadvantages.

In most air-conditioning systems a set of one or more boilers and one or more chillers generate hot and chilled water that is then distributed to a number of air handling units that condition the air that is supplied to the occupied spaces. A moderately large office building might have three boilers and three chillers, tens of air handling units and hundreds of zones. There are three different types of conventional HVAC system in common use in the UK, each with a number of variations.

(1) Constant air volume (CAV). Cold air is supplied from the air handling unit to each zone where it is re-heated as necessary to maintain the desired temperature in the occupied space. The process of cooling and then re-heating the air is inherently inefficient, although this inefficiency can be ameliorated to some extent by using the heat rejected by the chillers to re-heat the air. Little or no re-heating is required if the loads in each zone vary in such a way that they are always approximately proportional to each other. *Advantages*: CAV systems are simple and relatively easy to control; the air flow to each zone is constant, making it easier to ensure proper distribution of the supplied air. *Disadvantage*: Significant re-heating is required if the loads in the zones vary independently to a significant degree.

(2) Variable air volume (VAV). The temperature in each zone is regulated by varying the rate at which cool air is supplied to that zone (see section 4.6.1). Re-heating is limited to perimeter zones. *Advantages*: Unnecessary cooling and re-heating is largely avoided, reducing the energy consumed by the boilers and chillers; reducing the air flow rate as the load decreases also reduces the energy consumption of the fans. *Disadvantages*: VAV systems are more complicated than CAV systems; it is more difficult to ensure that sufficient fresh air is supplied to all zones and that the fresh air is satisfactorily distributed throughout the occupied spaces at lower flow rates. The distribution problem can be addressed by using local fans to mix the air but this increases the energy consumption.

(3) Fan-coil. Chilled water and hot water are supplied to each zone and the air is

cooled or heated locally as required. *Advantages*: No cooling and re-heating; only the fresh air required to maintain indoor air quality is ducted from a central air handling unit to the zone, reducing both the size of the ducts and the energy consumption of the fan(s) in the air handling units. The reduced duct size can be particularly advantageous in retrofits. *Disadvantages*: The heat exchangers in the zones (the fan-coil units) are generally not equipped to deal with condensation so the temperature of the air supplied to the space must be significantly greater than the dew-point temperature, limiting the cooling capacity just when the loads are greatest, or requiring larger air flow rates and hence higher fan energy consumption. Furthermore, the small fans used in each fan-coil unit have lower efficiencies than the large fans used in air handling units, which tends to negate the reduction in the energy consumption of the fans in the air handling units.

Alternative methods of providing comfort cooling using ceiling panels and 'chill beams' on the ceiling have been noted above in section 4.2. These methods are more widely used in Scandinavia and Germany and are starting to be used in the UK. They have the advantages of compatibility with natural ventilation systems (allow in just enough outside air to meet fresh air requirements) and being well matched to evaporative cooling systems that use cooling towers without chillers.

4.4.2 Energy efficient strategies

There are a number of ways in which the energy efficiency of air-conditioning systems can be improved, with varying degrees of complexity and cost-effectiveness. Some examples follow:

(1) Supply air temperature reset. As the cooling load decreases, the temperature of the (cool) air supplied to the zones can be increased. It is then more likely that the outside air will be cool enough to obviate the need to run the chillers.
(2) Chilled water reset. As the cooling load decreases the temperature of the water supplied by the chiller can be increased, possibly to the point where the cooling towers can supply cold enough water without the use of the chiller. This strategy is particularly appropriate for fan-coil systems where it is not possible to meet cooling loads by bringing in large volumes of outside air.
(3) Heat reclaim from chillers. The heat rejected by the chillers can be used to pre-heat hot water or to heat perimeter zones during cold weather.
(4) Variable speed drives. The speeds of fans, pumps and chillers can be reduced as HVAC loads decrease, reducing the considerable amount of electricity consumed by these components.
(5) Evaporative cooling. Chiller use may be reduced or avoided altogether by using direct or indirect evaporative cooling to pre-cool the air supplied to the zones.

The applicability of these and other strategies depends on the size and diversity of the loads, the need for simultaneous heating and cooling in different parts of the building and various other factors that are influenced by the design of the rest of the building. A collaboration is required between architect and engineer to approach the optimum overall design. In particular, presentation to the engineer of a design in which the fabric and the zoning are metaphorically (or even literally) 'cast in concrete' is unlikely to result in a good HVAC design and an energy efficient building.

4.5 'Intelligent' controls

The controls for domestic heating systems are implemented very simply; the thermostat is a simple temperature-sensitive electrical switch. Technical sophistication has appeared only in the use of microprocessors in the programable clocks used to switch the system on and off. Control systems for air-conditioning systems used to be implemented using compressed air ('pneumatic controls'), but are now usually implemented using a network of simple microcomputers. These systems are often referred to as 'building energy management systems' (BEMS). The name reflects the fact that these systems can collect and store information about the behaviour of the system: zone temperatures, electricity consumptions etc. that can be used by an energy manager to check that the performance is satisfactory and detect and diagnose simple faults. Such systems offer much greater possibilities than their predecessors for control in general and energy saving in particular.

The capabilities of BEMS sometimes lead to their being described as 'intelligent'. This is a pure hyperbole; current systems show none of the classical attributes of intelligence and make little or no use of techniques developed in the field of artificial intelligence. Their very significant advantages arise principally from the ability to transmit, record, display and act on information about the state of the building.

4.5.1 Communication

Many of the new facilities offered by BEMS rely on their capacity to communicate information between different parts of the system in different parts of the building. For example, the need for cooling in each of the various zones served by a single air handling unit can be communicated to the controller that serves the air handler, which can then determine whether the zones need to be supplied with colder air or whether warmer air will suffice.

4.5.2 Optimization

Optimization is another over used term in the field of building controls. An 'optimizer', more properly an 'optimum start controller', determines how late the heating system can be switched on and still heat the building up to the required temperature by the start of occupancy. Air-conditioning systems offer plenty of scope for the optimization of their operation, but most work in this area has not progressed past the research stage. For example, in VAV systems, increasing the flow of cool air to the zones increases the electric power required to run the fans, but may allow the chiller plant to be switched off if, and only if, the outside air temperature is low enough. Almost all current control systems fail to minimize overall energy consumption in this situation. Part of the reason why optimal control is not a feature of current products is that the air-conditioning systems for which it would be most beneficial are typically installed in high rent office buildings where the cost of energy is not generally considered to be a major issue.

4.5.3 Self-tuning and adaptive controls

One of the practical problems with controllers is that they require tuning (i.e. matching to the characteristics of the equipment they operate) in order to give their best performance and this process is so time-consuming that it is rarely done properly. The idea of a controller that tunes itself, and possibly retunes itself if necessary, is, therefore, very appealing. Some companies offer self-tuning capabilities in their microprocessor-based controllers, but the application of self-tuning to HVAC applications involves a number of problems and it is not yet clear whether all manufacturers who offer self-tuning have satisfactorily resolved these problems. The main attraction of self-tuning controls is their potential to reduce the time and effort involved in commissioning. Any improvements in control performance compared to careful manual tuning are unlikely to have a significant effect on the overall performance of the building.

4.5.4 Interaction between manual and automatic controls

Whenever there are a number of ways the environment can be controlled there is the possibility that the building will be operated in an inappropriate way. For example, the blinds may be fully closed rather than correctly adjusted to avoid glare, requiring the lights to be switched on. Alternatively, the windows may be wide open with the heating system full on. Building Energy Management Systems (BEMS) have the ability either to detect or to prevent inappropriate operation given the necessary sensors. A photocell mounted on the outside wall can detect whether there is sufficient daylight to obviate the need for artificial lighting. Micro-switches can detect open windows and the operation of the heating system can be monitored in a variety of different ways. Depending on the type of building, the BEMS can either warn an operator that a particular zone is being operated inefficiently or can inhibit energy-consuming devices if their use is unnecessary or wasteful. The diagnostic approach is to be preferred if the occupants are to be kept involved with, rather than alienated from, the environmental control system.

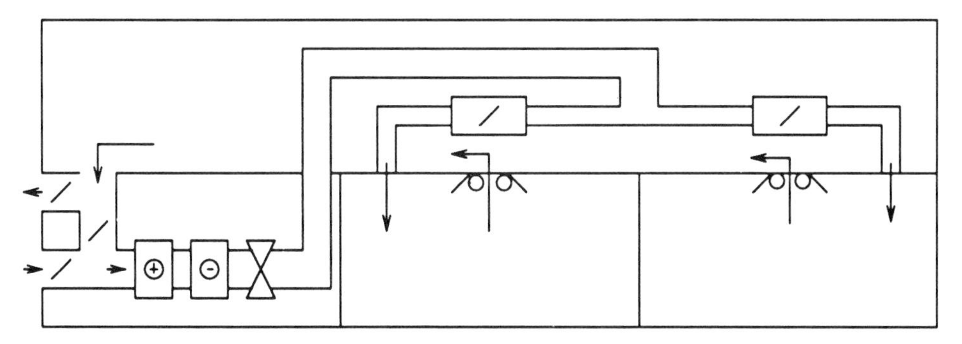

Fig. 4.1 VAV air-conditioning system: section through a typical floor showing the air handling unit, the plenum, the ducts and two zones.

4.6 Example control systems

This section describes an automatic control system typical of those encountered in commercial buildings. The design of such systems is usually done by the controls contractor and occasionally by the mechanical engineer who designs the HVAC system. Rarely if ever does the architect get involved in their design. However, some understanding of control systems is required if they are to be installed in a building where the occupants play an active role in the environmental control, since the proper integration of the manual and automatic controls is essential to the successful operation of the building.

In air-conditioning systems, the basic concept is to 'condition' (i.e. heat or cool, humidify or dehumidify, freshen and filter) the air centrally before supplying it to the zones. In the most common type of air-conditioning system in large buildings, the 'variable air volume' (VAV) system, the zone temperatures are then controlled by varying the flow rate of air (the 'volume') supplied to each zone. Figure 4.1 shows the air handling unit and ductwork in part of a VAV system.

The air that is supplied to the zones is conditioned by the air handling unit. The air is heated with hot water from a central boiler and the rate at which heat is transferred is usually varied by adjusting a valve that varies the flow rate of the water through a heat exchanger. The air is usually cooled by passing chilled water through a similar (though larger) heat exchanger, again controlled by a valve. If the air is cooled below its dew point, condensation occurs and the air is dehumidified. If cooling is required and the outside air is cooler than the air returning from the zones then outside air is used to reduce or obviate the need for cooling using chilled water.

Each air handling unit usually supplies a number of different zones. The loads in the different zones may vary in different ways, for example spaces with east-facing windows require more heating or less cooling in the morning whilst spaces with west-facing windows require less heating or more cooling in the afternoon. The zone temperatures can be controlled by varying either the temperature or the flow rate of the air supplied to each zone. Varying the temperature usually involves cooling the air to the temperature required by the zone with the greatest cooling load and then re-heating the air supplied to each of the other zones in order to avoid over cooling. This cooling followed by re-heating is clearly wasteful of energy, hence the popularity of VAV systems, which reduce the flow rate of the air rather than re-heating it.

4.6.1 Temperature control

Figure 4.2 shows the control scheme used to control the plant shown in Fig. 4.1. There are two kinds of controller used in the control scheme. C1 is a 'supervisory' controller that determines the desired temperature of the air supplied to the zones by the air handler, based on the need for heating or cooling in each of the zones. C2 is a 'local loop' controller that operates the valves and dampers and attempts to produce air at the desired temperature as economically as possible.

The zone temperature is controlled by C3 which varies the air flow rate into the zone by changing the position of the damper in the duct. The set-point for the zone temperature may be fixed or may depend on the occupancy, the time of year or the outside temperature. It is also usual to have different set-points for heating and cooling, separated by a dead-band. For example, heating may be called for if the zone temperature drops below twenty one degrees whilst cooling may not be invoked unless the temperature rises above twenty three. The dead-band can have a significant effect on the energy consumption, as discussed below.

4.6.2 Humidity and indoor air quality control

A humidity sensor is placed either in the return air duct, where it measures the average humidity of the air from the different zones, or in the most critical zone. If the humidity is too low, steam is injected into the supply air as it passes through the air handling unit. If the humidity is too high, the control of the supply air temperature is overridden and the flow of chilled water to the cooling coil is increased until the condensation of water from the air stream supplied to the zones is sufficient to reduce the humidity in the zones to an acceptable level. Indoor air quality is measured using either a carbon dioxide sensor or a sensor that responds to a broad spectrum of organic compounds including formaldehyde and various constituents of cigarette smoke. If the sensor reading rises above the acceptable level, the amount of outside air supplied to the zones is increased by overriding the control of the dampers in the air handling unit.

4.6.3 Lighting controls

The sensor for the automatic lighting controls would ideally be placed on the working plane at the darkest point in the lighting zone. To avoid the sensor being covered or damaged in such a position, it can either be placed on the ceiling or on

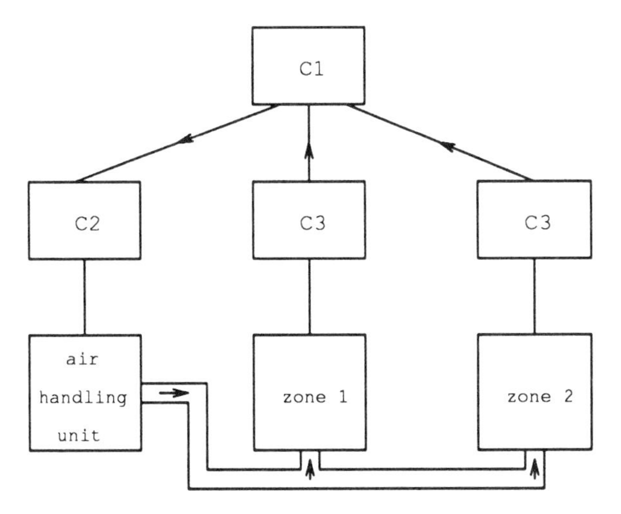

Fig. 4.2 Control scheme for the air conditioning system shown in Fig. 4.1.

the outside wall. The control system is then adjusted so that sufficient illumination, as measured with a portable light meter, is obtained under a range of sky conditions. Setting up lighting control systems so that they avoid annoying changes in luminance is not always a simple task and a number of lighting control systems installed in buildings have been disabled for this reason. Care is needed in selecting a system and a contractor and it would be prudent to observe a similar system in operation in a similar building before making a final selection.

4.7 Commissioning, operation and maintenance

A major cause of poor performance in buildings is inadequate commissioning of the HVAC system. The two major aspects of commissioning are the 'testing, adjusting and balancing' (TAB) of the mechanical equipment and the commissioning of the automatic controls. An important part of the TAB is setting the flow rates of air and water in different parts of the HVAC system to ensure that the design loads can be met in the different zones. Controls commissioning includes checks of the following:

- Calibration of sensors, e.g. temperature sensors.
- Correct operation of valves and dampers.
- Control strategy, in particular the correct sequencing of different items of equipment.
- Tuning the controllers to obtain stable but reasonably responsive control.

Commissioning is often done poorly because delays in construction combined with the desire of the client to occupy the building as soon as possible drastically reduce the time available for commissioning. The commissioning process is further impeded by the involvement of a number of different contractors who tend to take defensive positions rather than co-operating fully in getting the building operating correctly. A further difficulty is that full cooling loads cannot easily be produced until the building is fully occupied and full heating loads only occur in cold weather. Thus it is not usually possible to fully commission the building until after practical completion.

An indication of the importance of commissioning for energy performance comes from a study of the benefits of retrofitting BEMS in existing buildings. Most of the often considerable savings (John & Smith, 1987) were found to be attributable to the recommissioning of the HVAC plant that accompanied the installation of the BEMS.

One consequence of poor commissioning is that the building's environmental control system often does not work as intended, and a common response is then to disable parts of the HVAC system. For example, it is not unusual to find the fresh air dampers in air handling units disconnected from their actuators and fixed in a particular position. This may result in either unnecessary heating of cold outside air or unnecessary cooling of return air, with a significant energy penalty in either case. Ideally, the building operators should be involved in the commissioning process so that they become familiar with the correct operation of the environmental control system, but this rarely happens in practice.

Another desirable development would be for the occupants to become more involved and to take partial responsibility for the operation of the building. This could reasonably be attempted where profit-sharing or some other arrangement gives the occupants some incentive to reduce costs as well as maintain comfort.

As noted in Section 4.5, a BEMS can be used to collect and display information about the operation of the building, including the temperatures of the occupied spaces and the energy consumption of the plant. This information can be used to detect, and even diagnose, problems that manifest themselves as poor comfort or excessive energy consumption. These problems include both faults in the operation of the HVAC system and excessive loads caused by deficiencies in the building envelope or inappropriate action by the occupants. The main limitation of current systems is the requirement for a skilled person to interpret the information collected by the BEMS. In larger buildings it is possible to justify a full time energy manager; in smaller buildings one possibility is to employ a specialist company that monitors a number of buildings remotely using a modem to communicate with the BEMS via a telephone line. Such companies generally negotiate a financial arrangement that gives them an incentive to reduce energy costs.

4.8 Predicting performance

An important question for the designer of a climate-accepting building is how the design will perform on a particular site with particular patterns of use. There are a number of simple tools, mostly computer-based, that are intended to give strategic advice in the early stages of design (see chapter 16). More detailed computer simulation programs can be used to investigate particular design options or sizing questions as the design progresses (see chapter 18), though the skill and effort required to use such programs effectively limits their application to novel designs where the need for such tools has been identified and agreed in advance. There are programs that treat the performance of the envelope and interzone heat flow in considerable detail, others that treat air flow patterns in large spaces such as atria and others that concentrate on HVAC system selection and sizing. The appropriate use of these programs is to compare design alternatives. Their capacity to predict the total energy consumption of a particular building is limited by a number of uncertainties: ventilation rates and ground heat loss rates in shallow plan and smaller buildings, and casual gains and occupant behaviour in all types of buildings.

Predicting the energy consumption of buildings in operation in order to detect operating problems ('energy targeting') generally involves quite simple techniques. Past performance, corrected for outside temperature, can be used to predict current performance, which is then compared to the actual current performance. Alternatively, the performance of an estate of buildings can be monitored and checked for abnormal behaviour on the part of any individual building. These techniques do not take account of changes in use, or differing sensitivity to ambient conditions, but still provide very useful information not easily obtained in any other way.

4.9 Performance assessment – what is good control?

The primary function of the control system is to maintain comfortable conditions in the occupied spaces, and the occupants will generally complain if the environmental control is poor. However, it is well known that changing the inside temperatures has a major effect on the energy consumption, so it is important to know how low, or how high, the temperatures in the occupied spaces can be allowed to go. The idea of a comfort zone with definite edges is a fiction since discomfort changes steadily with temperature. It is necessary to be able to assess discomfort in order to know how to trade off energy costs against discomfort in order to optimise the operation of the building.

There are a number of measures of performance that can be used; in the example presented below the assessment of the overall performance takes account of discomfort, energy costs and maintenance costs. Capital cost would also be considered in cases where designs involving different envelopes or equipment were being compared.

4.9.1 Discomfort

The approach to assessing discomfort used in the example presented below is that of Fanger (1970) who developed a semi-empirical model of human response to specific environmental conditions by recording the response of a large number of people to the thermal environment in a test chamber. Discomfort is taken to be determined by thermal stress, which depends on the air temperature and humidity, the radiant temperature (determined by the surface temperatures) and the air speed, together with the clothing and activity level of the occupant. Fanger also observed the variation in the response of different subjects to specific conditions and his method predicts the fraction of the occupants who can be expected to be dissatisfied with a particular thermal environment. His method assumes identical clothing levels and metabolic rates, and under these circumstances there will always be at least 5% of the occupants who will express dissatisfaction with any thermal environment. Thus the 'dissatisfaction' column in Fig. 4.3 below has an effective zero of 5%.

Fanger's approach takes no account of factors such as the visual environment and psychological state of the occupants. There may also be variations in the extent to which occupants are willing (or are allowed) to change their level of clothing in response to changing environmental conditions during the day and during the year. A more general limitation of Fanger's work is that it is based on observations made in a test chamber rather than a real building. As discussed in chapter 1, there appears to be a greater tolerance to a variety of environmental conditions in real buildings than in test chambers and, as noted above, there appears to be a greater tolerance in climate-accepting buildings than in climate-rejecting ones. A consistent picture emerges if test chambers are seen as the limiting case of a climate-rejecting building.

4.9.2 Energy costs

The energy costs considered in the example below are the costs of operating the oil-fired boiler, the chiller, the cooling tower fan and the supply fan. There is a problem in choosing a reference case to which the total energy consumption should be normalized. In practice an estimate of the energy consumption for a design day or an example year would probably be available. Alternatively, an organization with a large estate of buildings could estimate the energy cost if the building/design in question were similar to others in the estate. Neither of these types of estimates is available in this example so the total energy cost is normalized to that of the first of the two strategies considered.

4.9.3 Maintenance costs

It is generally accepted that frequent switching of mechanical equipment results in higher maintenance costs, shorter lifetimes and higher failure rates. In the absence of any empirical data on the relationship between control activity, wear and maintenance costs, the hourly average number of starts, stops and reversals of every actuator is used as an indicator of probable maintenance costs.

		Energy (MJ)	Energy (%)	Dissatisfaction (%)	Maintenance (h^{-1})
March	OAT reset	886.8	100.0	7.4	11.8
	4°C dead-band	320.2	36.1	11.9	3.0
June	OAT reset	1665.9	100.0	5.6	9.4
	4°C dead-band	1473.9	88.5	6.4	7.9

Fig. 4.3 The results of the tests.

4.9.4 Case study

The results of a simple case study are presented below to illustrate one approach to the evaluation of the control performance of environmental systems in buildings. The study is based on a simulation of an air-conditioned office building with an HVAC system similar to the one shown in section 4.6. Two control strategies were studied, one in which the set-point for the zone temperature depends on the outside temperature ('OAT reset') and one in which there is a four degree Celsius dead-band between the heating and cooling set-points ('4°C dead-band'). The two strategies were chosen to illustrate the trade-off between comfort and energy consumption. Results for two days are presented in Fig. 4.3: a cold sunny March day and a hot sunny June day. A comprehensive study would consider the performance over a typical year. Further details are given in Haves & Dexter (1991).

The significantly greater energy consumption on the June test day is due to the continuous operation of the chiller plant, whereas the cooling load is met by using outside air on the March day. The energy consumption is lower when the large dead-band is employed because the cooling set-point is then higher, resulting in a slightly smaller cooling load and a significantly greater use of free cooling. The dissatisfaction is greater on the March day than the June day because there are larger fluctuations in solar gain, causing greater variations in the zone temperature. The use of the large dead-band strategy also results in lower maintenance costs since the control system is less active.

The question now arises, which of the two control strategies is better? Do the significantly reduced energy and maintenance costs of the 4°C dead-band strategy justify the increased level of dissatisfaction with the thermal environment? The answer is likely to depend on the building owner or tenant and on the nature of the activities pursued in the building. If the question could be reduced to a financial one, the nub would be the economic cost of discomfort. Some work has been done on the effect of the thermal environment on productivity (e.g. Wyon, 1987) but as yet no general approach to the cost of discomfort has been developed.

An approach has been developed that allows linguistic statements from owners, tenants, facilities managers etc. regarding control performance to be related to measurements in real buildings or the results of simulation (Haves & Dexter, *op cit*). This approach allows performance criteria to be obtained implicitly and used to assess different control schemes, something that could not be done if the costs of discomfort and maintenance had to be stated explicitly.

4.10 Integrated design and CAD

The pressure to reduce the effect of buildings on the global environment whilst improving the quality of the indoor environment can be expected to result in requirements to improve the quality of the design, construction and operation of buildings. This will demand improved collaboration between the different design disciplines and will require the architect, if he or she is to retain the role of leader of the design team, to acquire the skills necessary to become more actively involved in the environmental control aspects of design.

One way in which the greater integration of the different aspects of building design may be facilitated in the future is through computer-aided design (CAD). The corner-stone of an integrated CAD system is a database that accumulates all the information relating to the design as it develops. This database would allow different types of analysis, e.g. thermal performance, lighting, costing etc. to be performed without the special data preparation and entry that is required at present. It would also allow the assessment of the design to be continually updated as the design progressed. The benefits of this integration extend beyond design to construction, commissioning and operation. It is particularly important to have a reliable method of preserving an accessible record of the design and subsequent modifications if the building is to be operated efficiently throughout its life.

The application of knowledge-based computer systems (a form of artificial intelligence) to both design and operation is the subject of current research. The effective application of these systems depends on the development of the integrated CAD facilities just discussed.

4.11 Conclusions

Given that the energy costs associated with the lighting and cooling loads dominate the heating energy costs in commercial buildings, two major design issues are whether the building will be naturally ventilated, mechanically ventilated or air-conditioned and whether daylight will make a significant contribution to the lighting requirements. Most occupants prefer to be actively involved in the control of their environment, but care is required in integrating manual and automatic controls to ensure correct operation of the building and hence realization of the design intent. Design of climate-accepting buildings imposes a requirement for additional skills and design activity on the part of the architect but results in lower energy costs and environmental impact and a more positive response by the occupants to the indoor environment.

References

ASHRAE (1981) *Thermal Environmental Conditions for Human Occupancy*, ANSI/ASHRAE Standard 55-1981, American Society of Heating, Refrigerating and Air-Conditioning Engineers, Atlanta, Georgia.

Bordass, W. (1990) The balance between central and local control systems *Proceedings of Environmental Quality 90*, Solihull, Birmingham Polytechnic/RIBA.

CIBSE (1991a) *Building Services*, **13**(5) (May), pp. 17–20.

CIBSE (1991b) *Building Services*, **13**(11) (November), pp. 33–36.

Fanger, P. O. (1970) *Thermal Comfort*, McGraw-Hill, New York.

Haves, P. & Dexter, A. L. (1991) Use of a Building Emulator to Evaluate Control Strategies Implemented in Commercial BEMS, *Proceedings of Building Environmental Performance 1991*, BEPAC, Canterbury.

John, R. W. & Smith, G. P. (1987) Operating experience of building management systems, *Proceedings of ICBEM 1987: the International Congress on Building Energy Management*, Lausanne, Switzerland.

Robertson, G. (1990) The justification of energy efficient multi-storey commercial building design in more temperate climates, *Proceedings of NORTHSUN 1990*, Reading, Pergamon Press, Oxford.

Tong, D. (1989) Intelligent and healthy buildings, *Proceedings of Designing for Environmental Quality 1989*, Solihull, Birmingham Polytechnic/RIBA.

Wyon, D. (1987) The importance of our indoor climate, *Fläkt Review*, **71**.

Chapter 5

Selection of building materials: health and environment

Stephen Curwell
BSc, MSc, RIBA
Principal Lecturer and Professional Group Leader – Architecture, Leeds Polytechnic

Designers are faced with an apparently impossible task in undertaking health and environmental analysis of building materials. The difficulties of identifying clear health standards and problems of inadequate information are discussed along with the use of risk assessment techniques to provide a degree of objectivity in what is currently a fairly subjective analysis. Methods of simplifying the task are identified. The 'cradle to grave' philosophy underlies the suggested methodology of assessment for general environmental hazards.

5.1 Introduction

The need for action in reducing the environmental impact of buildings is now generally accepted. The serious concern over the contribution that CO_2 release from burning fossil fuels is making to global warming, together with the significant international co-operation over the reduction of CFC emissions to protect the stratospheric ozone layer (UNEP, 1987, 1989, 1990), has raised the awareness of building designers. The majority are now well aware of the actions required to mitigate the stress on the atmosphere:

- Use building forms that minimize the need for cooling and the associated use of CFC refrigerants in air conditioning plant.
- Specification of alternatives to insulating foam blown with CFCs.
- Reduce energy consumption of buildings by means of high standards of insulation and the use of high efficiency plant and equipment.

However many will be less sure about deforestation, Legionnaires' disease, asbestos, radon, pesticides and other hazardous chemicals. All have implications for the construction industry.

The RIBA's policy statement on global warming and ozone depletion acknowledges that, with the increasing understanding of the effect of human activity on the viability of the planet, a new dimension has been added to the responsibilities of the design team (RIBA, 1990).

The increasing pressure on designers to address these vitally important global issues may focus attention on energy conservation measures such as the selection of building materials of low process energy consumption. Nevertheless the energy consumed and the associated CO_2 release by manufacture, transport and construction of buildings forms only part of the analysis. It is essential that the health and well-being of individuals is not overlooked; this issue forms the subject of this paper.

In the quest for 'Green' buildings the complexity of the problem facing building designers cannot be overstated. For material selection procedures to effectively review environmental impact, the analysis must trace the material 'from cradle to grave'. In other words it is necessary to consider the potential environmental problems from extraction of the raw materials through manufacture, use, maintenance and ultimately of disposal, in order to fully assess the global, local and personal hazards that may accrue from *all* the materials and components that make up the building. Clearly this is far beyond the resources available to the design team and in most cases beyond the abilities of building designers. In this respect there is a mis-match between the expectations of the public and the skills and information available. How can the mis-match be resolved, and perhaps more significantly, why does it seem impossible at present to apply our developing understanding of the range of hazards and risks to give designers clear answers to these problems? What follows is an attempt to identify an appropriate design methodology for the future.

The following factors are important considerations in the quest for an answer to these problems:

- Greater awareness of what constitutes 'good health' and the interrelationship with 'comfortable' conditions or lack of them in buildings.
- Improved understanding of the hazards in and around buildings and of the risk of harm to the occupants or the local environment.
- Complete information on the detailed content of building products.
- Information on the environmental hazards associated with the extraction of raw material and in the processing of building materials and products.
- Better understanding of where energy is consumed in the manufacture and use of buildings.

5.2 What is good health?

The relationship between poor environmental conditions and ill-health has long been understood. For example Victorian engineering improvements to water and sewage provision were aimed at preventing epidemics of typhoid fever. Regulations enacted in the UK and elsewhere are primarily concerned with the prevention of serious diseases. Clearly, serious or crippling diseases constitute ill-health. However, there is a very wide range of medical conditions that reduce the quality of life of the sufferer, which are far from 'life threatening'. What of these?

The World Health Organization's (WHO) definition of health is extremely wide:

> 'More than the absence of disease or infirmity but including a state of complete physical, mental and social well being'
>
> (WHO, 1961)

An individual suffering a mild irritating complaint such as 'athletes foot' would be classified as ill when otherwise in the best of health. By implication, in many of its other publications, the WHO acknowledges that it is difficult if not impossible to achieve the high standard defined for all of the people all of the time. Doctors Mant and Muir-Gray of Oxford, in their report on their review of health issues for the 1986 building regulations (Mant & Muir-Gray, 1986) used a more robust, two stage definition:

(1) Health issues in which medical research indicates a plausible link between some specific environmental feature and ill-health.
(2) Health issues not associated with disease that may reduce the sense of well being implied by the WHO definition.

Stage (1) above deals with recognized hazards which are the subject of regulation or accepted codes of practice, but by their very nature these measures tend to be retro-active. Stage (2) above is more subjective. Inadequate lighting, temperature or humidity may cause extremely uncomfortable conditions, and if they persist for long periods may result in adverse health symptoms in a proportion of the occupants. This is precisely the problem conditions that pertain in some 'sick buildings'. Minor allergic reactions may also fall into this category, although true allergy would fall into Section 1. The crux issues are:

(1) What constitutes an acceptable percentage of the population suffering an adverse reaction?
(2) What is the risk of the environmental conditions created in any particular building causing this percentage to be exceeded?

In terms of comfort standards the WHO provides guidance on acceptable indoor air quality if (WHO, 1990):

- Less than 50% detect odour.
- Less than 20% experience discomfort.
- Less than 10% suffer mucosal irritation.
- Less than 5% experience annoyance less than 2% of the time.

David Lush's chapter deals with the relationship between comfort and health in greater detail. In order to achieve a better understanding of health risks it is important to have a working knowledge of how science and medical research attempt to identify hazards from materials.

5.3 Health research on materials

Two branches of medical science are involved: toxicology and epidemiology. Toxicology may provide an early indication of possible health effects by means of experimental work on laboratory animals or tissue samples. Interpretation of such experimental work to assess the risk to man remains very difficult.

Factor	Comments
Form and condition of material	Is the material loose and friable – will it be a source of dust? Does it contain volatile elements – will it emit toxic fumes? Is it combustible – again will it emit toxic fumes? Does it contain naturally radioactive elements? What is the concentration of naturally radioactive materials?
Position within the building	Contact with the water supply? Contact with foodstuffs? Internal or external? Exposed or concealed? Is there danger from physical contact?
Means of degradation	Abrasion – normal weathering – normal wear and tear – DIY activities (sanding) Chemical action – corrosion – drying – gas emission – DIY (burning off)
Ventilation	Air change rate – residual properties of dwelling. – normal rates achieved by occupants by opening windows etc.
Lifestyle	Periods of occupation? The time factor governing the period of exposure.
Maintenance cycles	May introduce toxic chemicals or increase dust in home resulting from maintenance.

Fig. 5.1 Factors influencing dose or exposure.

Epidemiology involves observation of a sample population in order to assess some suspected environmental hazard. The work inevitably lags behind material developments and means that the population could be at risk in the intervening period. Figure 5.1 (Curwell & March, 1986) lists the factors that need to be considered in order to assess the dose or exposure to any particular material and indicates the complexity of this aspect of the analysis in real buildings. For reasons of statistical significance any investigation must include an adequate sample of the target population over a reasonable time period. As a result and, because of the greater risk of damage to health in the workplace, the largest proportion of research effort to date has been directed at occupational hazards. Workplace exposure to any hazardous material is relatively 'controlled', i.e. less influenced by other environmental factors, but is usually more severe. For example, workers involved in an unprotected environment using a formaldehyde adhesive to make items of furniture would experience a much greater dose of formaldehyde vapour than an occupant of the space where the furniture was ultimately installed. Therefore careful extrapolation from occupational health information is necessary in order to assess the risk, if any, to the users of buildings. Compared with occupational health information there is a dearth of clear information upon chronic exposure, i.e. low level exposure over a long period of time, to the wide range of vapours and fibres routinely found in indoor air. This is precisely the information designers now find they need in order to assure themselves that their designs satisfy requirements imposed in collateral warranties. The lack of information upon the long term health effects to low level exposure to a range of organic compounds, solvents and certain fibres is a major issue of concern, not necessarily because of any suspected serious hazards, but primarily because there is doubt.

5.4 Sources of information

A large amount of data is produced such as the WHO Environmental Health Criteria, Health and Safety Executive, EH40, *Occupational Exposure Limits*, National Radiological Protection Board, *Statement on Radon in Homes*, Ministry of Agriculture, Fisheries & Food, *Pesticides yearbook*, Health & Safety Commission, Codes of Practice; *Control of Carcinogenic Substances* and *Control of Substances Hazardous to Health*, to name but a few. Most of this data is produced by committees of health experts who review the available research evidence. The latter provide tables of safe occupational exposure limits for a wide range of chemical compounds. For example there has been concern over formaldehyde release from chipboard and furnishings. In Table 1 (not illustrated) of EH40 (Health and Safety Executive, 1991) the maximum occupational exposure limit is 2 ppm. Is this an appropriate safe level for normal exposure? Which materials and components contain formaldehyde and how much gas could be released to the internal air?

Therefore few designers make use of this information. It is not of much help in decision making at the drawing board and without a detailed breakdown of the content of the building materials and components under consideration it is virtually useless in informing design. It is interesting to speculate upon who should be responsible for making information on the content of building products available. Manufacturers claim the need for commercial confidentiality, but it is the author's experience that most manufacturers soon have a very clear idea of the formulation of their competitors new product. It is usually some 'trick' of the competitors process or plant that they may be unable to emulate.

If, as seems likely, the current climate of opinion on health and environmental issues persists then it is inevitable that the detailed information upon the contents of building materials must be made available. The only decision is whether this will be by voluntary code or by means of regulation. An important factor will be the need to establish confidence and trust between manufacturers and designers. Designers must use the information wisely. The simple knee-jerk reaction of avoiding a material that has, for example, a trace of a known carcinogen in its composition is inadequate because the vast majority of materials have some negative health or environmental effects. Conversely manufacturers will need to assure us all that they are honestly addressing these issues. Proper comparative risk assessment of the alternatives seems to be the best option.

5.5 Risk assessment

On the other hand is it really necessary to investigate *every* building material used in a project? The Health and Safety at Work Act (1974) and associated Control of Substances Hazardous to Health requirements (COSHH, 1988) mean that in the UK the majority of hazards to the construction worker are controlled. Therefore serious risk to the occupants from building materials should be avoided by current regulations. The answer to the dilemma of depth of investigation versus limited resources must be risk assessment. Lack of space and expertise permit only some broad observations on this topic, but it is important not to confuse hazard and risk (Bradley, private comm.):

> 'The hazard associated with a material is the ability of the material to do harm to the body.'
>
> 'The risk is how likely the material is to do harm.'

Two examples may serve to illustrate the difference between hazard and risk and the significance for materials selection. Firstly, brown and blue asbestos fibres are very hazardous but if they are kept in a sealed container the risk is very small. Secondly, it may be preferable to use a more hazardous material of lower vapour pressure than one which is less hazardous and of high vapour pressure because of the risk of greater vapour concentrations occurring with the latter. This illustrates the importance of a considered response because a great many innocuous materials may be hazardous if used incorrectly. Effective risk assessment should indicate the likelihood of the hazard being realised and the degree of harm to any individual; however practical experience shows that it is extremely difficult to compare the risks across the main categories of Global: Local: Personal. For example it is impossible to sensibly compare the risks from insulation containing formaldehyde, with potential for respiratory irritation in the occupier and a suspect carcinogen, and a another material blown with CFCs which increase the depletion of the stratospheric ozone layer and therefore the risk of increased skin cancer.

Therefore the question for the future is the form that this analysis might take. The temptation to dismiss the problem as impossibly complex must be resisted on the grounds that experience will be beneficial in the medium to longer term. Every day we undertake, unconsciously, what we would regard as impossibly complex risk assessments if we were to apply our conscious mind to the problem (Shillito, 1990). For example, simply by walking up and down stairs or across a busy road. A key point is that this type of analysis has to be learned.

Building designers are experienced in selecting materials to satisfy function and performance criteria. This may be interpreted as a form of risk assessment. What is the risk that a certain roofing membrane will fail within 20 years? Do the features of the design, such as a large number of penetrations through the roof membrane due to the roof covering being suspended below an exposed roof structure increase the risk of failure? Faced with a new roofing material the

Preparation containing:
Diphenylmethane – 4, 4′ – Di-isocyanate
Harmful by inhalation.
Irritating to eyes, respiratory system and skin.
May cause sensitisation by inhalation.
In case of contact with eyes, rinse immediately with plenty of water and seek medical advice.
HARMFUL
After contact with skin, wash immediately with plenty of water.
In case of insufficient ventilation wear suitable respiratory equipment.
Contains Isocyanate, see information supplied by manufacturer.
Pressurised container; protect from sunlight and do not expose to temperatures exceeding 50°C.
Do not pierce or burn, even after use.
Do not spray on a naked flame or any incandescent material.
Keep away from children and pets.

Fig. 5.2 Hazardous products safety notice.

analysis would be rigorous. A great deal of time would be invested in checking the performance claimed by the manufacturer, also, and very significantly, the designer and his client would expect some form of warranty from the manufacturer of an untried new product. It is to be doubted that manufacturers are in a position to guarantee that their product is free of any hazard to health!

Risk assessment is an essential feature of the COSHH regulations. The Health and Safety at Work regulations require manufacturers to provide basic health and safety information for the protection of the building operative. Hazardous products must carry a safety notice, Fig. 5.2. Designers can obtain this information if they ask for it, but it is essential to ask the right questions of the right people within the manufacturers organization. Sales staff are usually ill-informed about these issues. The problem of addressing the right question is of the utmost importance, however as the H & SW requirements are for the protection of trades people the general environmental hazards are not covered. The information is therefore incomplete from the point of view of 'cradle to grave' analysis. However it will provide an indication both of the seriousness of the hazard from the product in question and therefore of the depth of investigation that might be necessary to assess the risk, if any, to the occupant and possibly to the environment as a whole. The work undertaken for the *Rosehaugh Guide* suggested that a large proportion of manufacturers had not addressed the wider issues. This needs to form part of their 'green' conversion.

The techniques of Environmental Impact Assessment provide a good model. The following questions identified by Paul Tomlinson of Ove Arup and Partners (Tomlinson, 1990) should be considered in evaluating the significance of potential effects:

- Which risk groups will be affected?
- Is the effect reversible or irreversible, repairable or non-repairable?
- Does the effect occur over the long or short term; is it continuous or temporary; and does it increase or decrease with time?

	Potential hazard when in position	Potential hazard when chance of being disturbed	Long-term potential environmental hazard in waste disposal
	A	**B**	**C**
Not reasonably foreseeable	0	0	0
Slight/not yet qualified by research	1	1	1
Moderate	2	2	2
Unacceptable	3	3	3

Fig. 5.3 'ABC' system of risk assessment.

- Is it local, regional, national or global in extent?
- Would the effect be controversial?
- Would a precedent be established?
- Are environmental and health standards and objectives being threatened?
- Are mitigating measures available and how costly are they?

This checklist is primarily aimed at large projects where the consequences of 'getting it wrong' are so much greater. Nevertheless it is a useful aid on smaller projects. Decision weighting, especially when there is incomplete information available is also helpful. Simp!e methods of risk assessment lack precision but have the advantage of adaptability to the information available and therefore of ease of use. For this reason the 'ABC' system adopted in *Hazardous Building Materials* has a lot to recommend it, see Fig. 5.3 (Curwell and March, 1986). The method is subjective and only provides a broad indication of the risk, but this is adequate at the early stages in design when preliminary decisions upon the main construction materials are made.

Finally an individual's perception of the risk is also significant and so it is essential that the designer understands his client's perceptions of these issues from the outset of a project.

5.6 General environmental hazards

Most of the points made above are concerned with aspects of health and environment related to the building in use. Beyond this it is sensible to look at any general environmental hazards in two categories:

(1) 'Upstream' from incorporation into the building, i.e. in terms of extraction and manufacture.
(2) 'Downstream', i.e. in terms of maintenance and ultimately of demolition, recycling and disposal.

As there is general agreement among energy analysts and policy makers that energy conservation will be a major part of a global warming control strategy, energy will therefore form a vital part of the equation, both up and downstream. However, it is important to note that the energy used during the life of buildings constructed to current UK thermal insulation standards exceed by a wide margin that which is embodied in the construction materials. Therefore the first priority must be to use the guidance espoused in the EEO's *Best Practice Programme* (DOE, 1990) and in the other chapters of this book and design energy efficient buildings. Selection of materials on the basis of energy consumed in manufacture is dealt with by John Connaughton in Chapter 7. However from the current perspective it seems inevitable that in the longer term there will be a trend to low embodied energy materials such as timber and timber products, provided the international problems of sensible economic and agricultural strategies for sustainable reforestation, both tropical and temperate, can be established. Growing more trees has the added advantage of fixing CO_2 as biomass, but is there the international political will to effectively address this issue?

There may be some cause for optimism in the international action taken to control the emission of CFCs. Designers will be aware of the need to eliminate the consumption CFCs which form the current generation of refrigerants and blowing agents for insulation foams. Replacement gases with low ozone depletion potential are expected to be available in the UK between 1992–3. These will still, however, be potent greenhouse gases and should therefore be contained and recycled as if they were hazardous materials.

Beyond the issues of CFCs, and some aspects of energy and CO_2 there is relatively little clear data to assist in any comparative analysis. Even if one attempts to use the risk assessment techniques mentioned above one is often involved in basing assumptions on a great deal of relatively subjective analysis. In order to help with decision weighting the best environmental strategy will seek to:

(1) Minimize:
- Effects on ecology.
- The use of raw materials.
- Fossil fuel energy consumption.
- Pollution, both atmospheric, liquid and in terms of land fill.

(2) Maximize:
- The use of recycled materials.
- Renewable or alternative energy sources.
- Proper reprocessing of waste.
- Reinstatement after raw material extraction and/or production ceases.

Together these form a checklist for use in both upstream and downstream environmental analysis.

5.7 Conclusions

The main points may be summarized as:

(1) It is very difficult for designers to select building materials in terms of health and environmental criteria, but the existence of the difficulty should not become an excuse for making no attempt at analysis. In this respect designers need to become better informed about the whole issue of hazard and risk assessment. There is no substitute for experience of risk assessment.
(2) The current Health and Safety legislation aimed at protecting the operative should mean that the vast majority of serious hazards from building materials should cause little risk to the occupants of buildings.
(3) It is not necessary to carry out an investigation of all materials provided the Health and Safety data available from manufacturers under the Health and Safety regulations is used to identify any hazardous content of materials and components, and to indicate whether a more detailed investigation may be necessary. Manufacturers should routinely include this information in technical literature intended for designers.
(4) Risk assessment is necessary to adequately compare hazards to occupants from alternative materials. Due to the incomplete data available simple, adaptable assessment methods are the best option at present, but suffer lack of precision.
(5) In the longer term full information upon the content of building materials and components must be made available, a voluntary code of practice being the most likely system in the current political climate.
(6) Research is required to improve our knowledge and understanding of the health effects from chronic exposure to a wide range of contaminants in indoor air.
(7) In terms of the assessment of more general environmental hazards 'cradle to grave' analysis is still in its infancy. However, use of the techniques of environmental impact assessment do have a part to play in identifying those environmental issues that are of most significance in any project, so that time and effort is not wasted on marginal problems.

References

Bradley, A. Director of Occupational Hygiene, Institute of Occupational Medicine, Edinburgh. Private communication.

Control of Substances Hazardous to Health Regulations (1988), HMSO, London.

Curwell, S. R. & March, C. G. (1986) *Hazardous Building Materials, a Guide to the Selection of Alternatives*, E & FN Spon, London.

Department of Energy (1990) *Best Practice Programme Case Studies*. Energy Efficiency Office, London.

Health and Safety Executive (1991) EH40 *Occupational Exposure Limits*. HMSO, London.

Health and Safety at Work etc. Act (1974), HMSO, London.

Mant, D. C. & Muir-Gray, J. A. (1986) *Building Regulations and Health*. Building Research Establishment.

RIBA (1990) *The Initial Policy Statement on Global Warming and Ozone Depletion*. RIBA, May.

Shillito, D. (1990) The Assessment of Risk. In *Buildings and Health, the Rosehaugh Guide to the Design, Construction, Use and Management of Buildings* (Ed. by S. R. Curwell, *et al.*), RIBA Publications, London.

Tomlinson, P. (1990) Environmental Assessment and Management. In *Buildings and Health, the Rosehaugh Guide to the Design, Construction, Use and Management of Buildings* (Ed. by S. R. Curwell, *et al.*), RIBA Publications, London.

UNEP (1987) *Montreal Protocol on Substances that Deplete the Ozone Layer*, Final Act. (The Montreal Protocol was revised in Helsinki in 1989 and again in London in 1990.)

World Health Organization (1961) Interpretation by Mant, D. C. and Muir-Gray, I. A. – see above.

World Health Organization (1990) *Air Quality Guidelines for Europe*, Geneva, 1987 – from interpretation by Appleby, P. Indoor Air Quality and Ventilation Requirements. In *Buildings and Health, the Rosehaugh Guide to the Design, Construction, Use and Management of Buildings* (Ed. by S. R. Curwell, *et al.*), RIBA Publications, London.

Chapter 6
Efficient and effective lighting

Robert Bell
BSc, CEng, MCIBSE
Chief Lighting Engineer, Thorn Lighting

Most lighting installations today are inefficient, due largely to inadequate briefing at the design stage, too much emphasis on capital cost, lack of thought during design and installation, and poor maintenance procedures. There is great potential for savings through good design of indoor lighting with the added benefit of increased worker satisfaction within the office environment. This chapter outlines objectives and standards, lighting systems, equipment and control patterns of use designed to improve the efficiency and quality of lighting installations.

6.1 Introduction

When we talk about the need for lighting to be efficient and effective some people think immediately of energy conservation or, better still, energy management. But efficient and effective lighting is much more than that. Lighting has a job to do and it must do it well. It's not there for the benefit of the building but for the benefit of the occupants, and good lighting contributes directly to their efficiency and effectiveness. It also contributes to safety.

But let's start with energy. In the UK and other developed countries lighting accounts for less than 4% of the primary energy consumed and about 16% of total electricity consumption. In domestic and industrial premises, lighting isn't a large proportion of the electrical energy consumption, but in commercial premises which include public buildings and offices it accounts for about 50%. With such a large proportion being used for lighting, it would seem, at first sight, a good target for energy reduction. But is it? Would such a cutback really be wise?

6.2 Aims

The principal purpose of lighting is to enable people to see. Unlike temperature and humidity, which must be controlled within fairly close limits for human comfort, lighting conditions can vary over an enormous range, and it's essential to match the lighting standards to the type of work to be carried out. You wouldn't, for example, feel too confident about having an operation if you heard that the lighting levels in the theatre had been reduced to save energy.

Poor lighting contributes to accidents and, more insidiously, reduces the efficiency and effectiveness with which people work. It can decrease morale, increase staff turnover and be a causal factor in industrial action. Lighting in places of work, which is insufficient or unsuitable, directly contravenes the Health and Safety at Work etc. Act. Not only that, but it is bad economics. Even in offices only about 0.5% of the total operating costs are spent on the lighting, with the major share, about 84%, being spent on salaries and associated costs.

It is tempting to conserve energy by reducing lighting standards, but this is naive and uneconomic in real terms. Quite clearly, with most of the operating costs being on staff, any reduction in the efficiency or effectiveness of the work force would be counter-productive. What sane businessman would risk losing a percentage of 84 pence in every pound for a potential saving which is likely to be no more than about one tenth of a penny.

The greatest and most worthwhile improvements are made by correctly designing lighting systems to provide the most efficient and effective solutions and by managing them so as to maintain that effectiveness. No one but a fool will spend good money to reduce their energy consumption unless the savings offer a good return on that investment. But even today, a great many people are short-sighted and are only concerned with the capital cost of the installation, without regard to improvements that can be achieved for a slightly higher investment.

Good lighting works well and uses energy efficiently. The refurbished main gallery of the Royal Museum of Scotland is an attractive interior in which good lighting plays an important part. It was a winner of the National Lighting Award competition for excellence in lighting design. In other words, the lighting design was judged to be the best in its class. Yet this installation also won the EMILAS competition for energy management. So here we have a winning lighting installation with winning efficiency. But neither of these competitions is the real test of efficient and effective lighting.

The museum has a job to do. It must attract visitors and show the exhibits off to the best advantage. The reaction of visitors to the museum has been excellent and is demonstrated by the million people who now visit it each year. That's the real test of efficient and effective lighting.

Efficient and effective lighting means:

(1) Ensuring that the correct lighting standards are provided both in terms of quantity and quality.
(2) Selecting the correct type of lighting system.
(3) Using the most suitable lighting equipment.
(4) Controlling the hours of use.
(5) Maintaining the system in efficient working order.

6.3 Objectives and standards

Let's start with standards. But what are the appropriate lighting standards? The Chartered Insitution of Building Services Engineers (CIBSE) in its current Code for Interior Lighting makes sound and sensible recommendations about lighting standards and good practice, with the emphasis very strongly on the efficient and effective use of energy. It should be compulsory reading for anyone involved in the design or specification of lighting.

Lighting must set out to achieve three things:

(1) People must be able to move about in safety and without risk of accident or injury. This applies at all times and becomes critically important if there is an emergency.
(2) The occupants have things they must see to do. They need to perform certain tasks to an acceptable level. The types of task can vary considerably as can the consequences of errors. Surgeons, for example, may have to see very fine details and an error could be fatal, whereas, watching a television programme is much less demanding.
(3) The lighting must help to create the right appearance and atmosphere and it must create the right sense of comfort. Lighting can, for example, be used to create an intimate atmosphere in a restaurant or a lively one in a fast-food restaurant.

It is easy to think that getting the lighting right is just a matter of achieving the right lighting level, but there is much more to it than that. There are many things that must be considered at the design stage if the lighting is to be correct. The uniformity of illumination, for example, affects our perception of the space and our ability to see and work efficiently and effectively. Discomfort glare determines how uncomfortable we find the lighting and this, in turn, affects our ability to work and to concentrate without distraction or fatigue – and so on down this list. Good lighting starts with the design objectives, and if they are wrong, no amount of good equipment will compensate.

6.4 Lighting systems

The type of lighting system is important. Don't confuse this strategic decision with the details of which equipment to use. That comes later.

6.4.1 General/uniform

The traditional way to light an interior is to use a regular array of lighting to provide uniform illumination over the entire working area. However, a large proportion of the space may be devoted to tasks which are not visually demanding – such as walking about, or filing – and it's wasteful to illuminate these areas to the same standard.

6.4.2 Localized

Localized and local lighting systems reduce this waste by relating the positions of the lighting equipment to the workstations to provide high levels at the workstation and lower levels elsewhere. A localized lighting could be used over a production line with a general lighting system in the roof, to light the space to a lower level for safe movement.

Localized systems are gaining popularity in offices. Uplighter systems, for example, direct light from extremely efficient discharge lamps onto the ceiling, where it is diffused and reflected onto the tasks. A high lighting level is provided for workstations with lower levels in the circulation space. The elimination of a ceiling supply saves costs and the free-standing uplighters can be relocated when the office layout changes.

6.4.3 Local lighting

Unlike localized lighting, local lighting provides illumination only over the small area occupied by the task and its immediate surroundings. Some form of general lighting is also needed to provide sufficient ambient illumination for safe movement. Local lighting must be carefully designed and positioned to minimize shadows, veiling reflections and glare. Units which use conventional tungsten lamps are inefficient and fluorescent fittings are normally a better choice.

6.4.4 VDUs

Visual display units are now commonplace, but can cause many problems. Reflections in the screen from conventional lighting fittings degrade contrast and cause other visual difficulties. A good solution depends upon more than just using the right equipment, and is, in fact, a lighting system problem. One of the solutions to such problems is uplighting. For example, The British Home Office Scientific Research and Development Branch and Thorn Lighting have carried out experiments with different forms of lighting in police control rooms. Specially designed dimmable fluorescent uplighters are now the preferred system for all such emergency services control rooms.

6.4.5 Control

Another system issue is controllability. The occupancy patterns, type of work and availability of daylight will help determine the best control strategy for the space. Perhaps remote control switches could be used to improve switching flexibility in a refurbished office.

6.5 Equipment

When it comes to equipment, the choice of lamp affects the range of suitable light fittings but frequently the performance of the light fitting is ignored. For example, many manufacturers make fittings which look alike and use the same lamps, yet their efficiency can vary by a factor of more than 2:1. So it is very unwise to just select luminaires without comparing their performance.

6.5.1 Luminaires

Luminaires can perform well for their intended purpose (e.g. VDU reflector fittings) and still be exceedingly efficient. And remember, if the fitting is more efficient, fewer will be needed and the installation costs will be lower, so it is often a better economic choice even if at first sight, it seems to cost more. But the real measures of efficiency and effectiveness are how well can operators read the VDUs, is the environment pleasant and does it enable good office productivity?

6.5.2 Fluorescent lamps

The performance of fluorescent lamps has dramatically improved in the last few years. The new Pluslux lamps have the same connections and lengths as the old ones but are only 25 mm in diameter instead of 38 mm. They are also filled with krypton and argon gas rather than just argon. The result is that they provide similar colour quality and light output to conventional cool white, white and warm white lamps but typically use 9% less energy. Before they fail the lamps will have paid for themselves at least three times over in saved energy. Another technological advance made at the same time was to change the coating on these lamps to a mixture of rare earth phosphors. These Polylux lamps not only consume about 9% less power, but they emit 7% more light and the colour quality of the light is excellent and comparable to de luxe lamps.

The two new energy saving lamps (Pluslux and Polylux) can match the colour quality of either white or the better natural fluorescent tubes respectively. But they are as much as 70% more efficient. It's no wonder, therefore, that the Lighting Industry Federation recommends that lamps of these types be used whenever possible because of their clear economic advantages. In fact, the increased light output of the Polylux lamps will also reduce the number of luminaires needed.

6.5.3 Compact fluorescent lamps

About 3% of all electricity is used for domestic lighting but it's a very inefficient use. This could be easily slashed to about 1% if the new generation of energy saving lamps were used. To be completely acceptable in the home the new lamps must have a colour quality indistinguishable from the lamps they replace and they must be a worthwhile investment.

The 16 W 2D lamp is 140 mm square and has roughly the same light output as a 100 W lightbulb but lasts over 5 times longer. The 28 W and 38 W 2D lamps are 205 mm square and have roughly the same light output as 150 W and 200 W

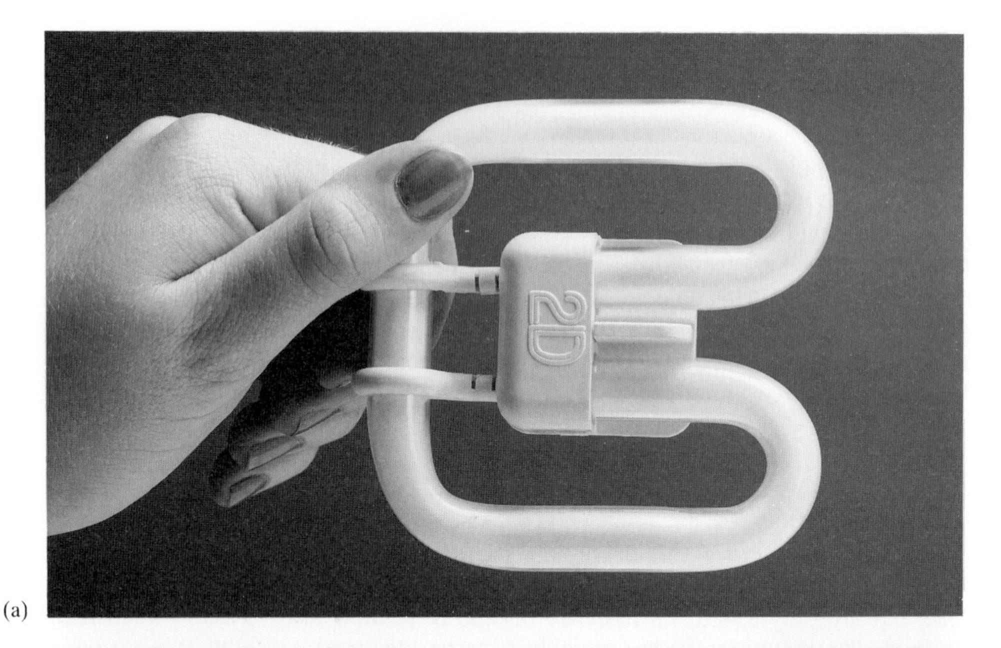

(a)

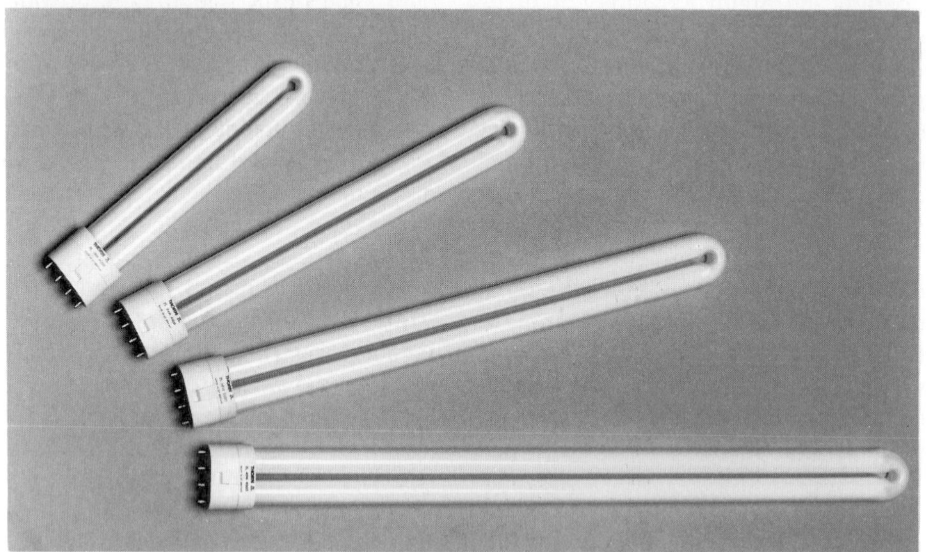

(b)

Fig. 6.1 Examples of (a) 2D and (b) 2L lamp fittings.

lightbulbs, but last over ten times as long (Fig. 6.1a). They all make use of the high quality Polylux phosphors to achieve their high output and excellent colour and they operate at a much safer temperature, and have a lampholder in which the electric contacts cannot be touched. But they do need control gear to operate.

In existing commercial fittings, new control gear could be installed, but special conversion packs make the job quick and simple using existing fixing centres.

However, the new lamps really come into their own with new designs, which can exploit their advantages. Shallow designs are not possible with ordinary lamps and even if they were, the heat build-up would be destructive to most materials. The Halls of Residence at Edinburgh University had 740 conventional fittings with 100 W lightbulbs replaced by brand new fittings with 16 W 2D lamps. In the first year they saved £17 000 which paid for the new equipment in only nine months. Not only that, but lamps had to be replaced five times less often – with obvious reductions in manpower costs. Even allowing for the cost of lamp replacements, the annual savings in subsequent years will be £13 600. Investments such as this which can be recovered in the same financial year cannot be ignored. It's like investing money in a bank at an annual interest rate of over 123% per annum.

The 2L lamp series is similar in colour quality and high performance to the 2D and Polylux series (Fig. 6.1b). But, whereas 2D is designed to produce an area of light, the 2L series is designed to replace fluorescent lamps of twice the length. This means that commercial fittings can be smaller and more efficient. The 40 W 2L lamp, for example, is designed to optimise the popular 600 mm × 600 mm module. The lamp emits more light than two conventional 600 mm lamps or even one 1200 mm lamp. This alone makes the light fitting more efficient. However, the efficiency improvement doesn't stop there. Most conventional lamps are longer than their fitting modules. Not only does this cause design and installation difficulties, but the light from the ends gets wasted. The 2L lamp overcomes all of these problems to produce an efficient result.

6.5.4 Control gear

New electronic starters for fluorescent lamps have the performance and reliability of high quality starterless circuits, but save energy, improve lamp life and permit control gear to be optimized. Completely electronic low loss gear is available today. One electronic ballast can replace five bulky and heavy components in a conventional starter circuit. But the main benefit is high frequency operation. When fluorescent lamps are operated at 20–35 kHz rather than 50 Hz their efficiency is improved and the irritating flicker disappears.

The light which is produced by the mains supply in conventional fluorescent fittings, ripples 100 times a second. This can be irritating to some people and is certainly detectable in the brain, even when we think that we can't see it. High frequency electronic fittings don't cause this problem and are preferred by workers.

It is, of course, possible to design electronic control gear which does more than just simply operate the lamp. It can form part of a controlled visual amenity system, enabling the lamps to switch or dim in response to a number of control decisions. This means that the lighting can be intelligent and dynamic responding to the needs of the occupants as they change and adding an extra dimension to give variety and flexibility. We'll look at this sort of approach in more detail when we come onto controlling the use of lighting.

6.5.5 High pressure sodium

High pressure sodium or SON lamps have gained widespread acceptance in sports halls and for exterior and floodlighting. They have the longest life and highest efficiency of any light source for working interiors, at about 100 lumens of light for each watt of power (that's about 8 times more efficient than a household bulb). But the colour quality is only just acceptable for interiors and it is unsuitable where colour discrimination is important. A few years ago a revolutionary new form of high pressure sodium lamp was introduced – called de luxe high pressure sodium or SON-DL. These lamps are interchangeable with standard SON lamps but sacrifice some life and efficiency to achieve excellent colour rendering. The colour rendering is better in fact than standard white fluorescent tubes.

The colour quality is so good that they are even being used in food stores where good colour is important and has a direct effect on sales. They are well suited to uplighting. An experimental 6 m × 6 m office, for example, is illuminated to current standards by just one 250 W SON de luxe lamp. That's an electrical loading of about 7.5 W per square metre which is less than $\frac{1}{3}$ the value normally required for an office of this size.

6.5.6 Metal halide

Another development of the discharge lamp is the 150 W metal halide arcstream lamp. It has excellent colour rendering, high efficiency and compact size which makes it the ideal choice for commercial uplighting and display lighting.

6.5.7 Low voltage tungsten halogen – lightstream

Low voltage lightstream tungsten halogen spotlights are more efficient and more controllable than their mains counterparts. This means that a 50 W or 35 W lamp can often provide a better display effect than a 150 W conventional spotlight.

However, their main value is the compactness of the fittings and the precision of their performance. This means that displays work better. The lamps not only use less energy, they focus far less heat into the beam, so customers and merchandise stay cooler. There is an added bonus that they last 2–3 times longer than most of the lamps they supercede, they are more robust and can withstand the vibration in a shop window caused by people and traffic.

Electronics has been applied to the transformers that low voltage lamps need, and the result is reduced sized and weight, and improved performance.

6.5.8 Environmental testing and electrical safety

Light fittings have to withstand a variety of physical conditions, involving such things as vibration, moisture, dust, high or low ambient temperature or vandalism. They must also be electrically safe. In a swimming pool, the lighting must be capable of withstanding the harsh environment and must be easy to maintain without draining the pool. But, above all, it must be designed to enable swimmers in difficulty to be easily seen in the water. The latest edition of the IEE wiring regulations states that electrical equipment must satisfy the appropriate British Standard. For lighting fittings in normal use this is BS 4533. Equipment which complies with this standard carries the BSI Safety Mark, which is an independent guarantee of engineering and manufacturing quality and electrical safety. Despite the obvious importance of the Safety Mark scheme, and the legal implications and liabilities in the event of an accident, few specifications call for safety marked equipment. Indeed it is rare to see safety mentioned at all.

6.6 Controlling the hours of use

Substantial savings can be made by controlling the hours of use for lighting. The lighting may be on unnecessarily whilst the building or part of it is unoccupied and it's not uncommon to find buildings fully illuminated at night whilst a small band of cleaners do their rounds.

6.6.1 Switch control

If the building receives adequate daylight for part of the day, then artificial lighting is not required and could be switched off. The simplest method of control is to encourage people to switch off unwanted lighting. This is not usually very effective but self-adhesive labels, reminding people to turn off unwanted lights, are always a worthwhile investment. You can rely upon occupants to turn lights on, but not to turn them off.

6.6.2 Daylighting and switching

Figure 6.2 shows how the probability that lights will be switched on varies with window design and time of day. The vertical scale is daylight factor. This is simply the ratio of how much daylight illumination is achieved at a point in a room compared to that in the open.

In a shallow office with lots of window area, the daylight factor at a desk could be 1% to 2%. What does this mean?

Suppose that the daylight factor at the desks in an office is about 1% to 2%. If the workers come in at about 8 a.m. then, on average they will tend to turn on 50% to 60% of the lights, and they will stay on through the day.

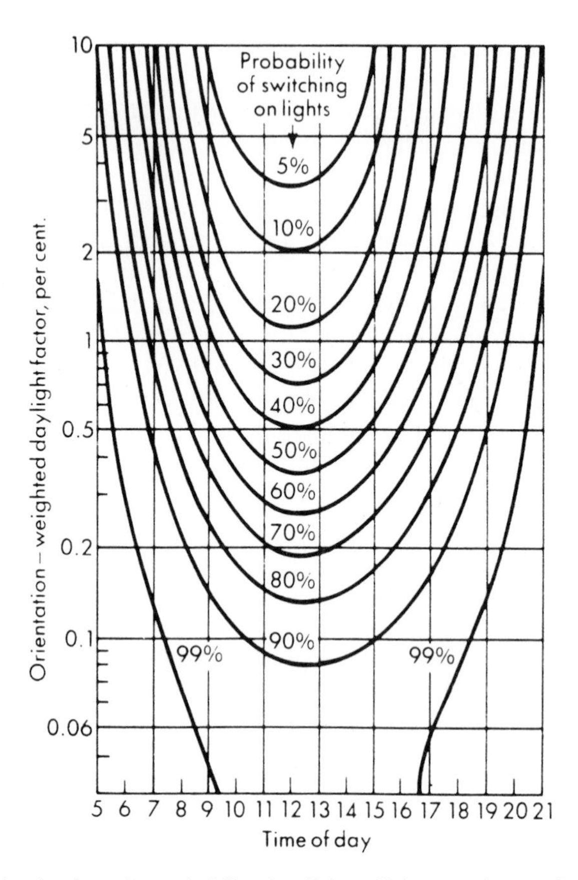

Fig. 6.2 Graph showing how the probability that lights will be turned on varies with window design and time of day.

If, however, we could trigger the lights off at lunchtime, then only about 20% would be switched back on because there is enough daylight. More lighting will be turned on as the daylight fades, but there will be a saving. Of course we could keep on turning the lights off through the day, but this would be annoying and save little extra, so a lunchtime switch-off is all that we would normally recommend during the working day until going-home time. Infra-red, ultrasonic and other forms of remote switch can help to provide more flexibility in switching and are becoming popular. These allow individuals or small groups to control lighting in their part of the office.

Combined daylight and artificial lighting systems can be designed in which the inadequacies of daylight are supplemented by the electric lighting. These can be automatic or manual, with the rows of fittings near the windows being switched off or preferably dimmed when there is sufficient daylight.

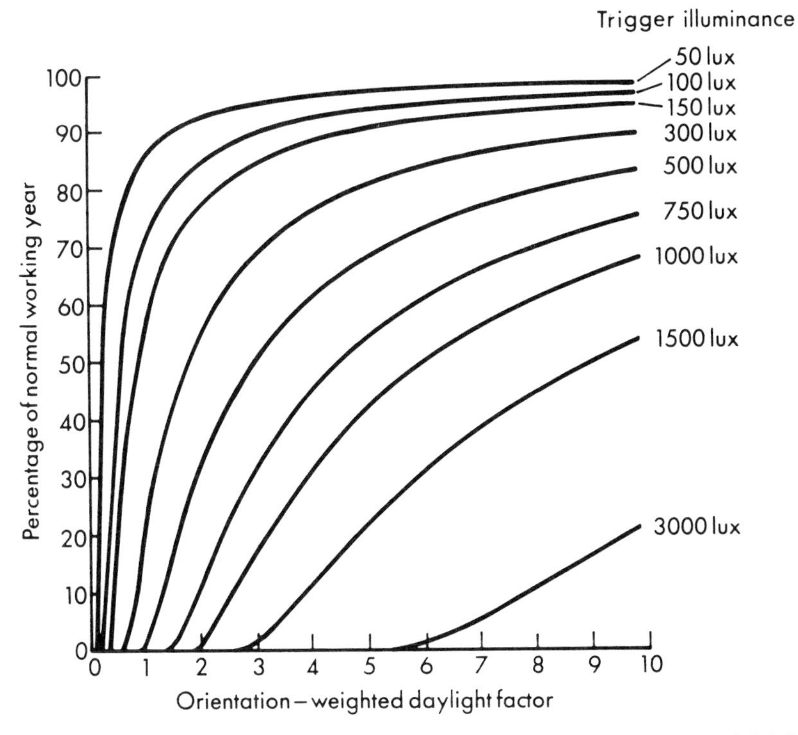

Fig. 6.3 Graph showing the savings made by dimming and the daylight factor, and lighting levels required.

6.6.3 Daylight and dimming

Figure 6.3 shows the relationship between the savings made by dimming and the daylight factor and lighting levels required. We will assume that we want to provide 500 lux on the desks. If the daylight factor at the desk is 2% and we simply want to maintain a level of 500 lux, then we can see that a dimming system linked to daylight levels could save 60% of the lighting energy during the working day. But what happens in practice? Figure 6.4 shows the energy profile of two identical floors in a building – one with conventional wiring and the other with local switch control, photocells and a timed off, at lunchtime and in the evening. Even though dimming has not been used, the saving is dramatic, but no one is receiving less light when they need it.

6.6.4 C-VAS

C-VAS – the controlled visual amenity system – is a logical extension of this. In this system, controllable high frequency light fittings are adjusted in response to management commands. The lighting adjusts to match daylight availability. This

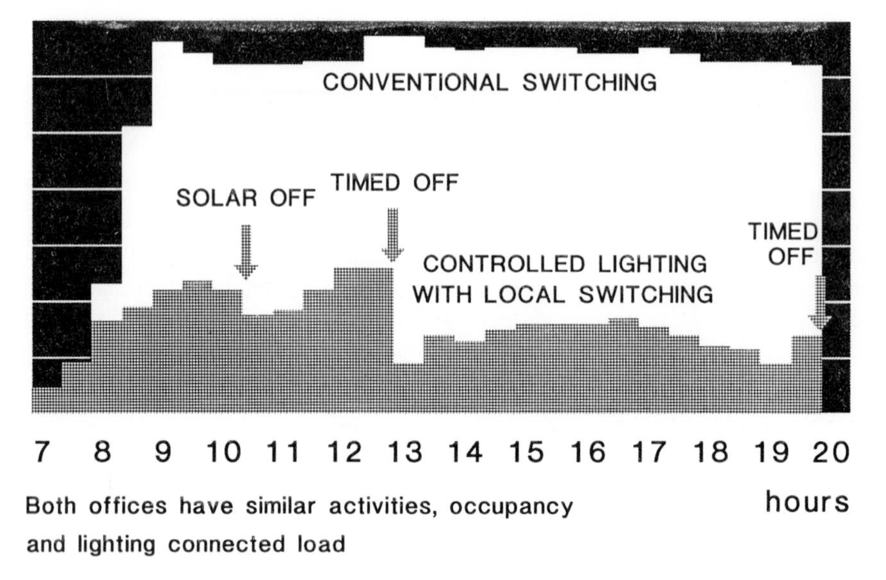

Fig. 6.4 Annual variation of lighting energy use.

saves energy and creates natural variety and change of modelling. The lighting also adjusts to respond to the occupancy pattern, so that lighting is on when it is needed and is off or dim when it isn't. It can be programmed to change lighting levels to match the different activities that occur during a normal day. These range from normal office work, to cleaning and security.

In any real building, the type of work can alter with time. This can be on a short scale when people need to increase or decrease their lighting, or on a longer term as the work changes. The lighting can be adjusted to reflect this. These days offices get reconfigured regularly, and the lighting can be reprogrammed to match the new use for the space. In modern buildings it is often difficult to predict the intended uses of different spaces, and flexibility of control can ensure that all uses are catered for.

Lastly, and most importantly, giving flexibility and control to the occupants increases their efficiency and effectiveness. There are sound physiological and psychological reasons why people work much better when they like their environment and can control it.

6.6.5 Building management

A final approach is to link the lighting to the other services by using a central processor to control the whole system according to a defined but flexible program. This type of approach is still expensive, but such systems are becoming attractive.

6.7 Maintenance

Many lighting installations are wasteful and inefficient, not because of poor design but because of neglect. The abysmal standards of maintenance that many organizations adopt without realising often result in a 60% or 70% reduction in efficiency. No one would buy a car and not service or clean it – yet we do with lighting.

For example, one site manager couldn't understand why night theft and vandalism, which had been eliminated by his security lighting five years ago, was now again a problem. The answer shocked him. Poor maintenance had allowed lamp failures to go uncorrected. A new high pressure sodium floodlighting system paid for itself completely in three months and reduced maintenance and security problems to a manageable level.

6.8 Conclusion

The performance of light sources and luminaires has been steadily improving over the years and will continue to make advances. The new generation of fluorescent and high pressure sodium lamps and the new localized lighting systems for offices are recent examples. Despite this, most lighting installations in use today are inefficient. This is mainly due to:

- Inadequate briefing.
- Too much emphasis on capital cost.
- Lack of thought during design and installation.
- A frequent failure to realize the importance of a sensible maintenance programme.

Efficient and effective lighting does use energy wisely, but that is only a minor part of its true value. People function better and more happily if the lighting is right. The benefits in this area far outweigh the rest. A better office environment for example, means less absenteeism and each reduction of three hours absenteeism per person per year saves as much money as the most energy efficient lighting scheme. A reduction of one day per person per year will pay for the lighting completely. Similarly, each annual 0.2% improvement in worker productivity is worth as much as the most energy efficient scheme, and 0.6% will pay for the lighting.

These may seem difficult things to measure accurately, but they do occur and they are worth much more to the organization than mere energy savings. However, this isn't a trade-off. You can design lighting to provide an efficient and effective environment and you can make it use energy efficiently and effectively. All it takes is a willingness to use good design.

The Building Envelope: Materials and Glazing

Chapter 7

Real low-energy buildings: the energy costs of materials

John N. Connaughton
BSc(Hons), ARICS
Davis Langdon & Everest Consultancy Group

In order to maximize energy saving in building designers must look not only to the traditional factors included in energy auditing methods but must also look at the hidden energy costs of the materials of which buildings are constructed. These include the embodied energy in materials – the energy expended in their production and manufacture, the energy used in construction, for periodic refurbishment, and the life cycle period of the building from which one can assess the annual cost of production of the building. The energy costs of materials can significantly alter the energy efficiency performance of a building and should be recognized by all designers.

7.1 Background

Yesterday's concerns about energy use and profligacy in the developed world focused on the implications of depleting the Earth's stock of fuel resources. Today's interest in energy conservation and energy efficiency is spurred by a different and considerably more compelling concern.

There is still uncertainty over the link between the build-up of greenhouse gases in the Earth's atmosphere and long term climatic change (see, for example, Department of Environment, 1989). Furthermore, full implementation of the Montreal Protocol will limit future consumption of CFCs which currently account for about 20% of the effective warming from the greenhouse gases. However, a projection of future world energy demand was made at the World Energy Conference in Montreal in 1989. This indicated a 75% growth in energy demand and a consequent 70% rise in carbon dioxide levels by the year 2020. If this happens, and the expected resulting global warming occurs, the consequences for the planet and its population could be catastrophic.

Currently, CO_2 accounts for some 55% of the effective warming from the greenhouse gases (Wigley, 1989). In the UK almost half of CO_2 emissions from human activities come from energy used in buildings (Shorrock & Henderson, 1990). In addition, around 10% of all the UK industry's energy requirement is used to produce and manufacture building materials and products.

7.2 Introduction

Attempts at reducing energy use in buildings have, up to now, concentrated largely on the energy used by the building's occupants to heat, cool, ventilate,

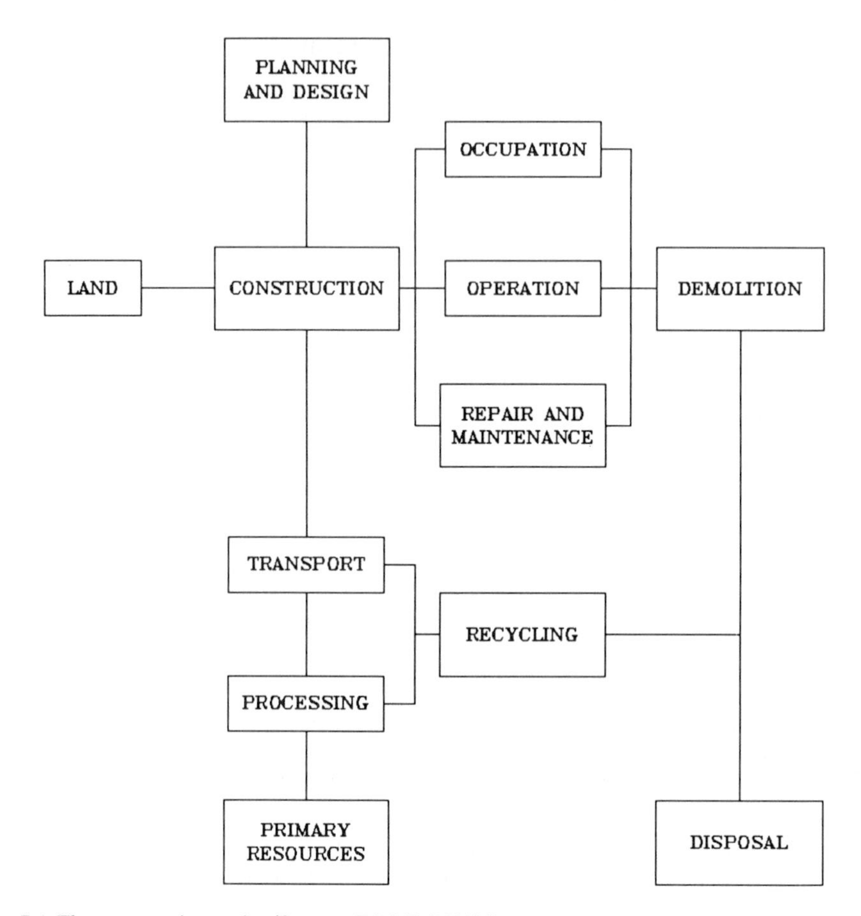

Fig. 7.1 The construction cycle. (Source: DL&E (1991).)

light, operate equipment and so on (see, for example, Energy Efficiency Office, 1985). Design strategies evolved have consequently focused on energy used after the building has been built. Such strategies have virtually ignored the energy required to produce the building's constituent parts, to get them to the site and to assemble them.

Although assessments have been made of energy saving potential in the processing and manufacture of building materials (see, for example, W. S. Atkins and Partners, 1983), little or no attempt has been made to include this as part of low-energy building design. Real low energy buildings use relatively small amounts of energy not only in use, but at other points in the building's life-cycle (see Fig. 7.1) and, particularly, during production. For a new building, the energy 'embodied' in the building's structure, its fabric and its services can be upwards of five times the amount of energy used by the occupants in the first year.

Techniques for the calculation of the embodied energy (or energy content) of materials (including building materials) are not new (see, for example, Chapman, 1974; 1975). However, these have not yet found their way into a low energy design approach aimed at achieving an optimum balance between energy use at all stages in the building's life cycle. There are many reasons for this: there is no generally agreed or accepted methodology for calculating energy content; techniques are time consuming; there are many materials which make up even the simplest building, and it is frequently argued that energy content is only a small proportion of the total energy used in a building over its whole life. This latter point is disputed here. Furthermore, although there is a lack of comprehensive data on the energy content of building materials, there is much that designers can do based on what is available now.

This chapter discusses the concept of embodied energy in the context of building design. It is intended to introduce building designers to the main issues so that more attention will be paid not only to the energy used by the occupiers of buildings, but also to the energy used in producing the building's constituent materials and components.

7.3 A life cycle costing approach

An appropriate concept for the consideration of a building's embodied energy and energy in use is that of life cycle costing. The rationale for life-cycle costing grew in part from the building failures and short life expectancies associated with many 1950s and 1960s buildings, where a concentration on capital cost economies spawned a legacy of high running costs, or 'costs-in-use'. By looking at initial capital costs and at other costs over the building's life, the design team could indicate a client's total financial commitment to the building and identify options which could be most cost effective in the long run. Capital costs, when expressed in terms of whole life costs, are frequently a relatively small proportion. This has led many designers to advocate spending more now to help offset higher future running costs.

In the area of low energy building design, however, the emphasis on initial capital costs has been reversed. Designers here concentrate on reducing energy in the use of the building. Just as in the 1950s and 1960s, little attention was paid to building maintenance and running costs, now many so-called energy efficient designs have taken little or no account of the energy used in materials production, transportation and fixing in place.

7.4 The significance of embodied energy

About 10% of industry's energy consumption (5% of the total UK energy consumption) is used in the production of building materials (Fig. 7.2).

Most of the energy used in buildings is consumed in old buildings. The annual rate of new building is only a very small proportion of the total existing stock; in the housing sector, for example, the annual new-build rate is usually no more than 1%. In some sectors, e.g. offices, growth rates of up to 3% have been recorded but this is high and other sectors are effectively declining – their new-build rate is less than their demolition rate. New buildings are not only a small proportion of total stock, but they are also built to significantly higher thermal

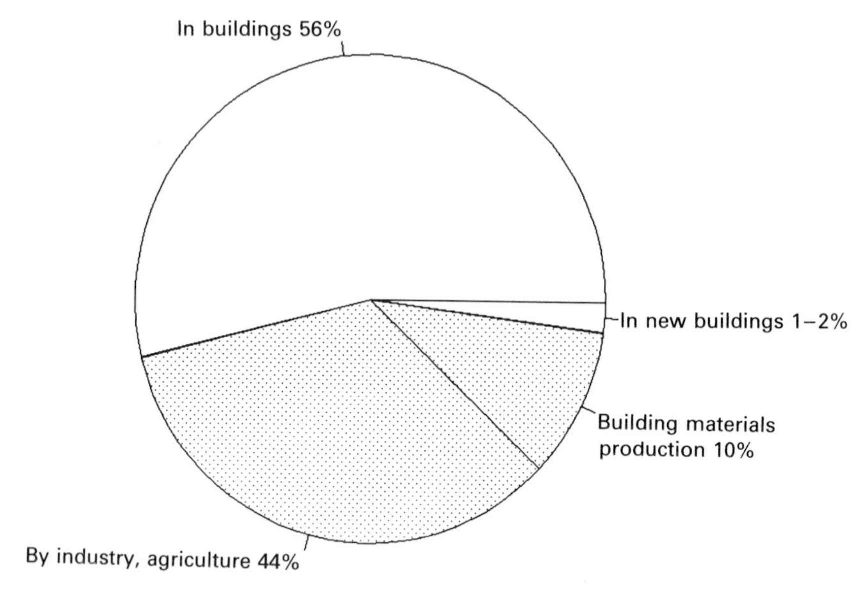

Fig. 7.2 UK annual energy consumption. (Source: *Digest of UK Energy Statistics.)*

standards and, as a result, in many cases are relatively low consumers of energy. This is particularly true in the housing sector but perhaps less so in the commercial sector where many new buildings are very highly serviced. Of the total amount of energy used in buildings in any one year, therefore, only about 1% to 1.5% is used in that year's 'new' buildings. But more than five times that amount of energy is used in the production of the materials which are incorporated in these new buildings.

However, a building's 'embodied' energy is incurred once, whereas energy in use accumulates over time. Assuming stable construction output and a rate of new additions to the existing stock of some 1%, the simple model in Fig. 7.3 shows that for a given (say 10%) energy saving, it will take about 10 years for the cumulative effects of the saving in energy-in-use in the new additions to the stock to outstrip the cumulative energy saved in producing these buildings in the first place. This is because a constant saving in energy for a relatively stable annual amount of materials production accumulates arithmetically (say 15 Petajoules in the first year, accumulating to 15 + 15 = 30 in the second year) whereas a saving on energy in use in new buildings accumulates at a faster rate as the stock of new buildings itself accumulates.

7.5 The importance of building 'lives'

The period over which the building is to be used and operated is therefore of key importance in determining the relative significance of the energy which goes into the production of a building's constituent materials and components, and the energy used in the building during occupation.

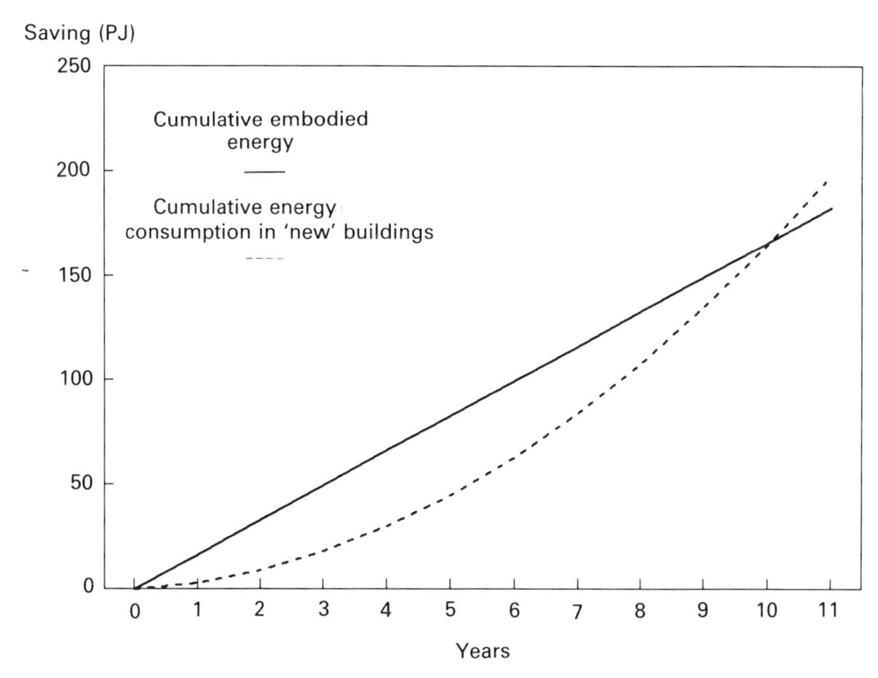

Fig. 7.3 Comparison of hypothetical energy savings in the UK–embodied energy and energy consumption in 'new' buildings. (Source: DL&E, (1990.)

Figure 7.4 illustrates the relative amounts of energy consumed in producing buildings ('energy content') and in using buildings ('energy-in-use') for four different dwelling types over an assumed building life of 25 years. It should be noted that since this data was prepared, changes in the Building Regulations thermal standards will have reduced the energy-in-use figures substantially. However, the embodied energy values will probably have increased as more energy intensive material is used to achieve improved thermal performance.

Moreover, for many building types, particularly in the commercial and retail sectors, major refurbishment (often involving substantial reconstruction) is undertaken at increasingly shorter intervals. The convention of relating building life to physical deterioration may no longer be appropriate. The effect of short 'real' building lives is to increase the relative significance of embodied energy costs compared to energy consumed in use.

Given that major refurbishment cycles are often as frequent as 10 to 15 years for some building types, the embodied energy value of these buildings may be almost as high as the energy used in occupying them over their real life-cycles.

7.6 Energy for building materials

In any given year, the energy required to produce sufficient building materials for one year's supply of new buildings is a significant proportion of total national

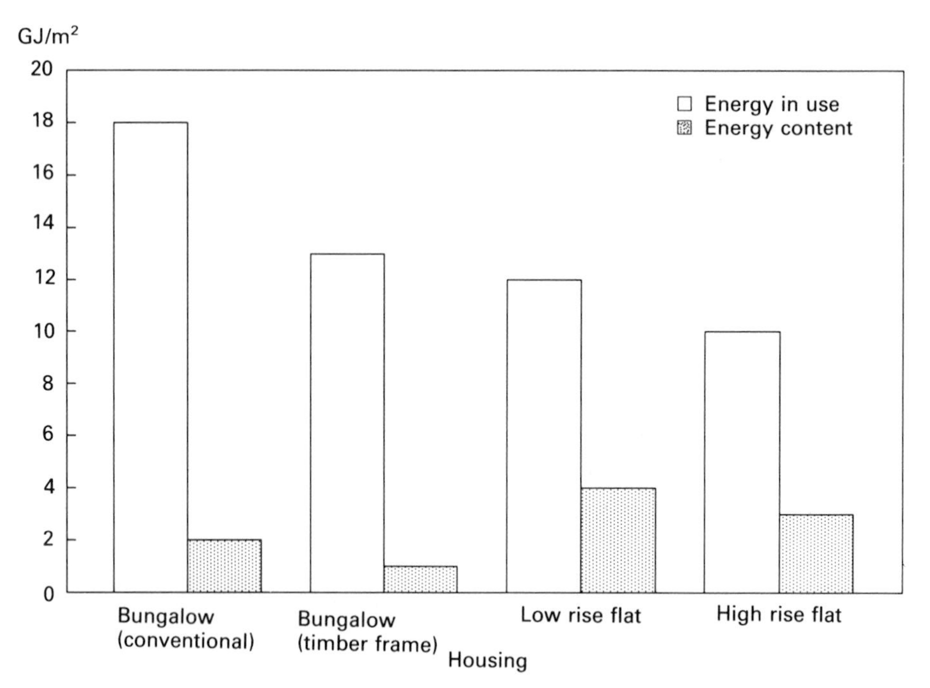

Fig. 7.4 Energy content and energy consumption over 25 year life. (Source: DL&E based on Barnes & Rankin, (1975).)

annual energy consumption – about 5% (or 10% of all industry's energy requirements).

The building materials industry is relatively energy intensive, second only to the iron and steel industry, of which some of the products are used in construction. This is because building materials are generally high-volume low-value products and their energy intensities are therefore relatively high, particularly when materials output is expressed in monetary terms. However, monetary value of output is perhaps the only common unit of measurement across the spectrum of all major industries (Fig. 7.5). This illustrates one of the problems encountered in comparing the energy content of different building materials, where common units of measurement do not exist (e.g. timber and glass).

For example, the energy consumed in the extraction of raw materials and the processing of these materials for use in construction may be calculated for different materials, but comparison between materials is potentially misleading. Flat glass, for example, is about four times more energy intensive than bricks per tonne (see Fig. 7.6). However, a square metre of one brick wall is about eight times more energy intensive than a square metre of 4 mm float glass. Before this kind of information can be applied usefully, therefore, it must be expressed in units appropriate to construction (tonnes of steel, cubic metres of concrete, square metre of glass), as part of a building assembly (a cladding system, for example) or a complete building type.

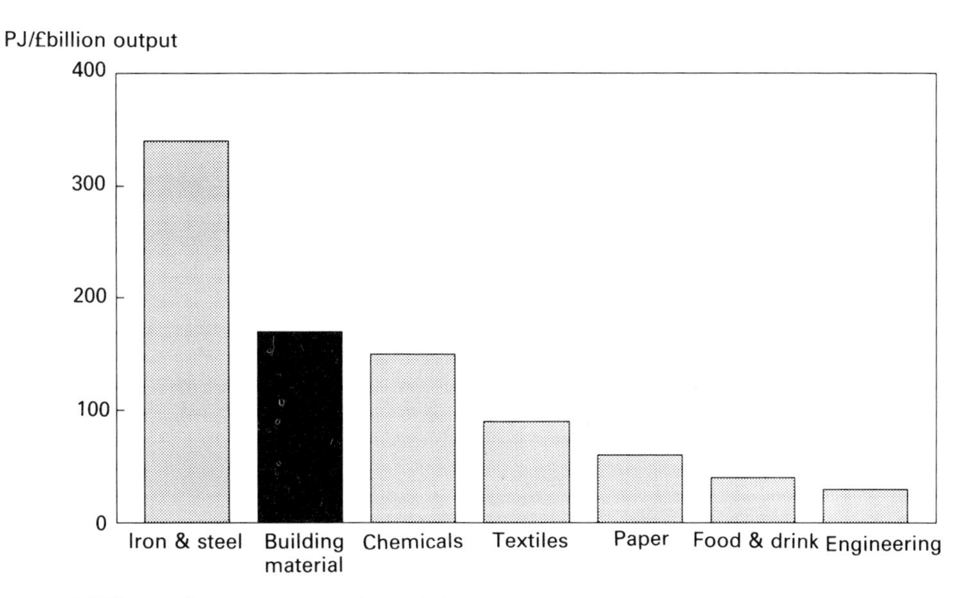

Fig. 7.5 UK annual energy consumption–relative energy intensities of different industries. (Source: Leach, (1979).)

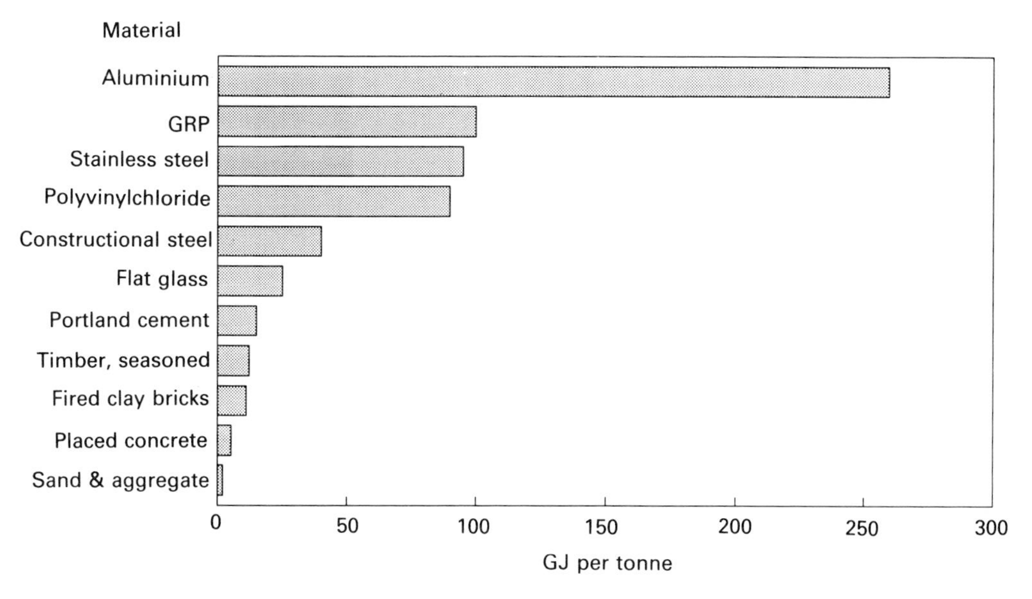

Fig. 7.6 Energy content of key building materials. (Source: Institute of Metals (1987).)

However, the variety of materials and components used in constructing even a relatively simple modern building is great and can range from primary industrial products such as aggregates and sand through traditional manufactured building materials such as cement, bricks, timber, etc., to precision engineered mechanical and electrical equipment. In the USA, Hannon *et al.* (1976) and Stein *et al.* (1981) have identified some 78 industries contributing 98% of all energy requirements for new construction. As few as 11 industries contribute more than 50% of all energy requirements for new construction, while 32 industries contribute more than 80% of requirements. Some 47 industries contribute individually less than 1% of requirements, and collectively just over 20%.

As far as is known, no systematic approach to the calculation of the energy intensities of all UK construction materials has yet been attempted. Considerable uncertainty prevails over the calculation methodology (in particular, over whether to include a proportion of 'indirect' energy, that is the energy used to provide materials and machinery for the process of manufacturing a building material or component), over what materials to include and the units with which to express both the material (e.g. mass, money, etc.) and its embodied energy value (e.g. primary, delivered, etc.).

However, taking a cue from existing literature and accepting that although construction uses a vast range of materials, a small number of materials tend to predominate, it is possible initially to identify a small number of key materials which account for a relatively large proportion of energy content.

This is significant for both designers at a project level and for policy makers at a national level: designer's attention may focus on opportunities for minimizing material use or substitution where the potential for energy saving is greatest; Government may concentrate on the key industries where energy savings may contribute significantly to savings in national energy consumption and to reducing the energy content of new buildings. From Stein *et al.* (1981), five key materials can be identified which may account for over 50% of the total embodied energy of new buildings (see Fig. 7.7). In more recent work on UK housing, Howard (1991) found that five key materials account for more than half of the embodied energy value of a three-bedroom detached house.

Key materials are those which by virtue of the energy used in their processing or manufacture *and* the quantity employed in construction represent the most energy consumptive inputs to construction. Materials which have a high embodied energy value per unit of output, such as aluminium, but a relatively low level of usage in construction may not account for a high proportion of construction industry energy requirements.

Large savings in the energy used to produce building materials can be achieved. An increase in energy efficiency in the UK building materials industry in the ten years after 1966 measured by a fall of 34% in the industry's energy intensity over the period has been identified by Leach *et al.* (1979). Considerable potential for further savings in energy required by the building materials sector exists, in particular in the brick and clay products industry, the china, earthware and glass industry, the cement industry and the iron and steel industry.

Energy savings as high as 30% of energy use per unit of output for the iron and steel industry are thought to be achievable. Taylor (1982) indicates an even higher energy saving potential of some 35% but claims that much of this may only be achieved in the long term following considerable capital investment and extensive research and development.

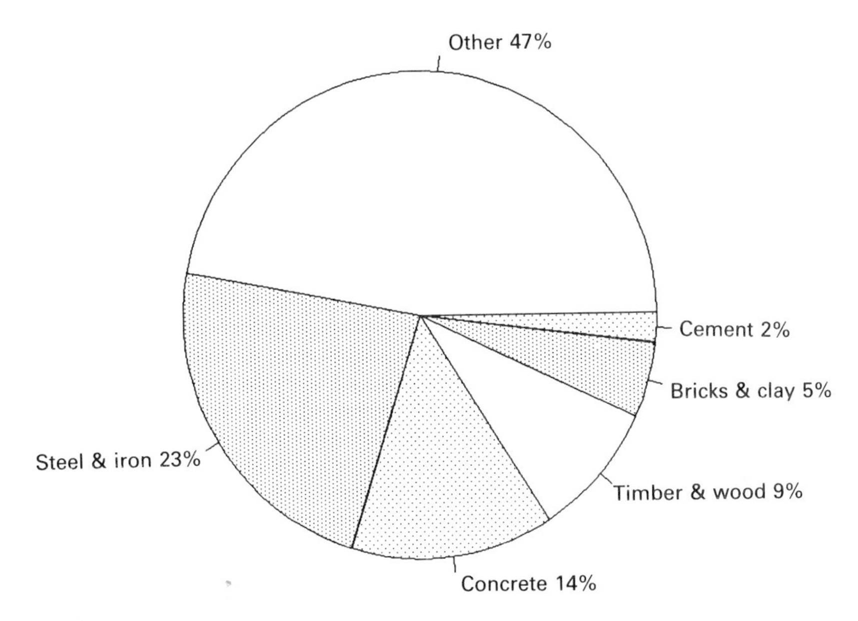

Fig. 7.7 Construction energy requirements. (Source: DL&E based on Stein, (1981).)

Similarly, Rose *et al.* (1977) and Malkin (1982) indicate potential savings of 50% and 20% to 47% for the brick and cement industries respectively. Although some of these savings may already have been realised, there remains considerable potential for further energy savings. For example, the UK cement industry is currently characterized by its predominant use of the energy intensive wet process. Dry process kilns such as those in operation in Germany use about half as much energy per tonne of cement as wet process equivalents. The opportunity thus exists for energy conservation as outdated plant may be replaced with more energy efficient equipment/processes.

7.7 Energy for building

The construction of new buildings is almost entirely a process of assembly of already manufactured or processed materials. The construction industry is characterized by labour intensive processes and energy requirements in the process of construction are relatively small. Of greater significance is the energy used to win the raw materials, to process and/or manufacture these raw materials into finished products and to transport them to the job site.

Estimates of direct energy consumed in the process of construction on site have been made for the UK (15% of all construction energy requirements: Casper *et al.*, 1975) and the USA (17%: Stein *et al.*, 1981). These relatively low proportions are reinforced by the view of construction as a relatively low 'valued-added' process. Construction purchases considerably more energy via materials

	1	2	3	4	5	6
Cement	5.7	6.1	6.1	6.8	7.8	7.8
Aggregates						0.1
Common bricks	3.4	2.8		6.3		3.1
Timber	6.4	2.3				2.7
Flat glass	11.9	15.0			22.2	12.4
Steel	24.0				37.3	31.0

Fig. 7.8 Energy content of key building materials (GJ/tonne). (Source: DL&E based on various sources, see references: Energy Audit Series; Barnes & Rankin (1975); Leach (1979); Stein (1981); Chapman (1975); Haseltine (1975).)

than it purchases directly. Howard (1991) has also found evidence of a direct relationship between energy intensity and monetary value of building materials and components.

There are opportunities for energy saving in the process of construction. Designers have more control, however, over savings which may be effected through material or component substitution. If energy saving through substituting energy intensive materials with less energy intensive alternatives is to be effective, however, designers need information, not only on the energy content of different materials and components, but on different assemblies, forms of construction and even building types.

Consider first of all the key material types identified earlier as being particularly significant in construction.

It is possible to derive estimates of the energy values of a number of construction materials from a variety of published sources and by an examination of the energy requirements of the industrial processes used in the production of building materials. Figure 7.8 presents coefficients of energy use (in Gigajoules per tonne of output) for a number of common building materials which have been drawn from a variety of sources. The data in this table are the result of extensive manipulation to bring them all to the common unit of GJ/tonne. The differences between values from different sources are probably a result of variations in definitions, errors in interpretation and actual differences in the production process involved. The presentation of energy intensity in terms of tonnes of output allows comparison between different materials. Figure 7.9 expresses the energy coefficients for materials (from the last column in Fig. 7.8) in units more common to the construction industry, for example thousands of bricks, or square metres of glass of a particular thickness.

The next step is to combine this data into common and alternative building assemblies, such as cladding systems, for example (Fig. 7.10). This is perhaps the kind of presentation most useful to designers. Here the relative energy intensities of different cladding systems can clearly be seen. Aluminium framed curtain walling costs about the same – in embodied energy content – as precast concrete spandrel units and aluminium windows. However, a brick cavity spandrel wall

	Unit of Measurement	GJ
Cement	cu m 1:2:4 concrete	2.460
Aggregates	cu m 1:2:4 concrete	0.100
Reinforcing steel	cu m of concrete	0.310
Common bricks	thousand	8.310
Common bricks	sq m one brick wall	0.960
Timber	cu m	1.280
Timber	linear m of 100 × 50 mm	0.006
Glass	sq m 4 mm thick	0.120

Fig. 7.9 Energy content of key building materials in common building units. (Source: DL&E based on Haseltine (1975).)

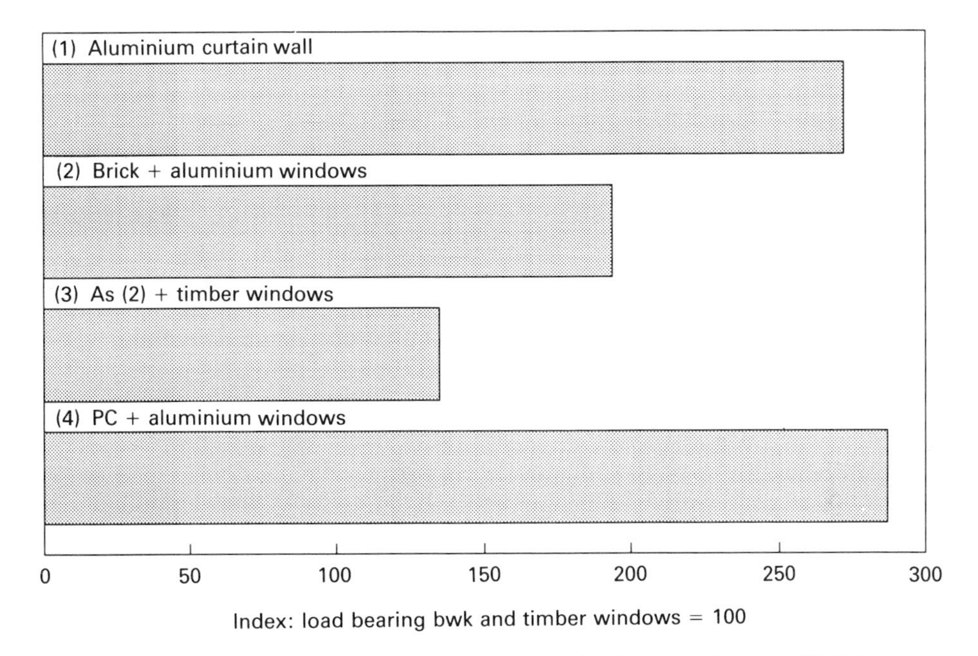

Fig. 7.10 Energy content of cladding solutions. (Source: DL&E based on Haseltine (1975).)

with timber windows costs about half as much in embodied energy terms. Similarly, a raised timber floor costs considerably less in embodied energy per square metre than a solid concrete floor.

A third step is to identify the embodied energy values of different forms of construction and types of building. This requires a good deal of information on the energy content of a variety of different material types which is not readily available.

For example, Haseltine (1975) calculates that a concrete encased steel frame with one row of internal columns and medium span beams is some 44% more energy intensive than a reinforced concrete framed equivalent. It should be noted that the costs of the two solutions are comparable, though steel may be preferred on the grounds of speed of erection (see, for example, Gray *et al.*, (1985); BDP, (1985)). In a comparison of structural floor framing systems, Stein *et al.* (1981) have calculated that a steel frame and concrete decking was some 48% to 65% more energy intensive than a reinforced concrete waffle slab.

In the domestic building sector, Howard (1991) has found that timber frame construction may be up to 20% less energy intensive than masonry walling, but much depends on whether the timber is imported (thereby incurring high transportation energy use).

7.8 Life cycle energy costs

For a low energy strategy concentrated on the energy used in the production and use of buildings to be effective, consideration must be given to the energy costs-in-use implications of materials and component selection at design stage. The future energy implications of design decisions need to be considered before an informed judgement can be made. It is conceivable that a particular building assembly costing less in energy terms than a proposed alternative could return an inferior thermal performance in operation.

It is important, therefore, to examine the total energy costs (embodied and in-use) before a lowest energy cost solution may be proposed. Similarly, the addition of fabric insulation to a building will incur initial energy costs but will save on energy costs-in-use. The energy cost-in-use savings must therefore be balanced against the initial energy costs to determine over what period of use the insulation is energy cost effective (see, for example, Howard, 1991).

A total or life cycle energy costing approach also involves consideration of the energy required regularly to maintain and periodically replace (where necessary) materials and components throughout the building life. Estimates of the life expectancies of building materials and components from published sources are known to vary considerably (NBA, 1985).

Additionally, the thermal performance of materials and components in use and energy requirements of buildings in use are climate specific and will depend very much on site locational characteristics. However, it is possible to make estimates of maintenance and replacement intervals for materials and components based on published information and experience. Energy use coefficients may be applied to these estimates over the lifetime of the building to determine the total or life-cycle energy costs of the materials and components.

The relative affects of the choice of different materials and components on total direct energy requirements by the building during its life may also be estimated, either manually or by using a thermal simulation computer model. It is therefore possible to estimate the embodied energy values and energy cost-in-use requirements of the constituent materials and components of various building types, and to estimate the direct energy requirements of these buildings during their lifetime.

Combining these and comparing the results for a variety of building types and forms allows the selection of the most energy efficient solutions.

7.9 Summary and conclusions

More than half the energy used in the UK annually is used in completed buildings. Government efforts to promote energy conservation have justifiably concentrated on this aspect of energy consumption. However, many of the measures which are thought to promote energy conservation in buildings (for example, fabric insulation and double glazing) involve energy expenditure in their production and incorporation into buildings. This aspect of energy consumption by buildings is frequently ignored.

The energy consumed in the building materials industry in providing materials input to one year's supply of new buildings is more than five times greater than the energy consumed by these building in the first year of use. Energy savings in the building materials industry in any year will therefore have a significantly greater initial impact than equivalent savings in the buildings built in that year. And although the savings in energy consumed in these buildings will accumulate, it may be some time before these savings outstrip the cumulative affect of comparable energy savings in the building materials industry.

A small number of building materials account for a large proportion of embodied energy in buildings by virtue of the high energy intensity of these materials and the relatively large quantities used in the construction of buildings. Steel and iron products, cement and concrete products, timber products and brick and ceramic products account for a considerable proportion of embodied energy in buildings, probably more than half. With the exception of timber products (of which a large proportion of UK construction industry's annual use is imported), considerable energy savings have been made in the production of these materials in recent years and there remains considerable potential for future energy savings, as high as 50% to 60% of total energy use in the case of the building brick industry, for example.

Further work is needed to identify the energy inputs to, and the use in construction of, a more comprehensive range of building materials so that a more significant proportion of the energy cost of construction may be covered by future energy saving strategies.

With short refurbishment cycles, particularly in the commercial and retail sectors (where US research has shown building types to have relatively high embodied energy values), the energy content of buildings may almost equal the total energy used in these buildings over their 'real' lives.

The most energy efficient building may become the least energy efficient one in the hands of a profligate occupant. Building designers have limited influence over what happens to the building after it is occupied. They have very much more control over what materials go to make up the building in the first place. Comprehensive information on the energy content of building materials and assemblies is now needed to add to the design vocabulary to help create real low energy buildings.

References

Atkins, W. S. & Partners (1983) *A Study of Energy Saving Opportunities in Brickmaking through Improved Process Control*, Energy Technology Support Unit (ETSU), Harwell, Oxfordshire.

Barnes, D. & Rankin, L. (1975) The Energy Economics of Building Construction *Build International* **8** 1975, pp. 31–42.

Building Design Partnership (BDP) (1985) *Frames for Multi-Storey Buildings. An economic comparison*. PCFA, Leicester.

Casper, *et al.* (1975) *Energy Analysis of the Report on the Census of Production, 1968* O.U. Energy Research Group Report ER G006.

Chapman, P. F. (1974) Energy Costs: A review of methods. *Energy Policy*, June 1974, pp. 91–103.

Chapman, P. F. (1975) The energy cost of materials. *Energy Policy*, March 1975, pp. 47–57.

Department of the Environment (1989) *Global Climate Change*, HMSO, London.

Energy Efficiency Office (1985) *Energy Efficiency in Buildings*, EEO.

Gray, B. A. & Walker, H. B. (1985) *Steel framed multi-storey buildings. The economics of construction in the UK*, Constrado, Croydon, Surrey.

Hannon, B. M. & Stein, R. G. *et al.* (1976) *Energy use for building construction: Final Report*, Energy Research Group, University of Illinois, Urbana, Illinois.

Haseltine, B. A. (1975) Comparison of energy requirements for building materials and structures, *The Structural Engineer*, **53**, September, No. 5, pp 357–365.

Howard, N. (1991) Energy in Balance, *Building Services*, May, pp. 36–8.

Leach, G. *et al.* (1979) *A low energy strategy for the United Kingdom*. The International Institute of Environment and Development, Science Reviews Ltd, London.

Malkin, L. S. (1980) *Energy Audit Series No. 11. The Cement Industry*. Department of Energy and the Department of Industry, London.

NBA (1985) *Maintenance cycles and life expectancies of building components and materials: a guide to data and sources*. NBA Construction Consultants Ltd, London.

Rose, K. S. B. *et al.* (c1977) *Energy Audit Series No. 2 Building Brick Industry*. Department of Energy and the Department of Industry, London.

Shorrock, L. D. & Henderson, G. (1990) *Energy use in buildings and carbon dioxide emissions*. Building Research Establishment, Garston.

Stein, R. G. *et al.* (1981) *Handbook of energy use for building construction*. US Department of Energy, Reassignment DC.

Taylor, P. B. (1982) *Energy Audit Series No. 16. The Iron and Steel Industry*. Department of Energy and the Department of Industry, London.

Wigley, I. M. L. (1989) Submission by University of East Anglia to House of Lords Select Committee on Science and Technology/*Inquiry into the Greenhouse Effect*. HL Paper 88–II.

Chapter 8
Smart glazing and its effect on design and energy

John Littler
BA, MA(Oxon), PhD, MCIBSE, MCIOB, MASHRAE
Professor of Building, Polytechnic of Central London

Professor Littler argues that glass and glazing are sources of delight and pleasure in building and that a reduction in glazed area is not the only, or most desirable, way to effect the energy savings made necessary by the global problems associated with overuse of energy. He reviews a range of 'smart' glazing (both that available at present and that close to production). The paper concludes by examining the use of atria and the contribution that glass structures can make to the energy efficiency of a building.

8.1 Introduction

Commentators have seized on the remarkable quality of the light in Venturi's Sainsbury Wing at the National Gallery. Shoppers are delighted by the airy feel to the Metro Centre at Gateshead, and the Ark and Broadgate rely for much of their eye catching nature, on large areas of glazing. The Züblin and Wiggins Teape office buildings (Stüttgart and Basingstoke respectively) owe much of their success to the use of core atria as circulation space. The indoor gardens in the John Deere offices (Illinois) and the Opreyland Hotel (Tennessee) serve as hugely enjoyable spaces for meetings with clients or friends.

Atria caught developers imaginations in the 1980s, to the extent that even countries such as Portugal where the lack of pavement cafés attests to the desire to escape the sun, have examples of office buildings with central atria which can easily reach 40°C: as Richard Saxon, the author of *Atrium Buildings* points out (Saxon, 1985; 1986) the architectural and social potential of atria are frequently left unrealized.

Glass delights; but it can provoke energy and comfort disasters. The following chapter suggests that this need not be so, and that, conversely, large areas of glazing can be energy savers and comfort providers. Composites technology and materials research is advancing faster than designers can grasp the new products. It seems very likely that 'smart' plastic glazing with the desirable properties described here, will arrive before we are prepared for its exploitation.

The scope for saving energy by better design of a buildings is enormous (Fig. 8.1).

Dramatic improvements have doubled the efficiency of automobiles in the last 15 years. Simple changes, widely adopted, such as aerodynamic shields on truck cabs, have greatly reduced the burden of pollution per person or per tonne carried. Further changes in the pipeline, to engines, controls and aerodynamics, will improve efficiency by an additional 40%, moves to alternative fuels may

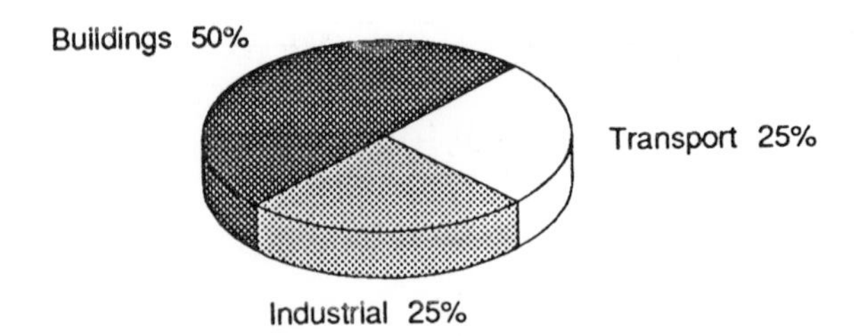

Fig. 8.1 Approximate division of primary energy use in developed countries.

further cut CO_2 emission. Trains such as the TGV have opened routes to those who would otherwise go by airplanes which cause more pollution per person/km.

In industry, enormous improvements have been made in the efficiency with which many operations are carried out, for example by using targeted microwave heating in drying processes, and by process heat recovery. Similar advances in energy conservation have occurred in isolated buildings; but widespread application of new techniques is woefully inadequate.

Notwithstanding the advances in vehicle technology, pollution levels are now so high in London that cyclists and police persons on point duty are commonly seen wearing masks. Firms are introducing air filters for buildings (both portable and fixed), to fight the internal pollution caused by chemicals from building products, but also chemicals arriving in the ventilation air (Humanair filter systems, Kessler Technology, Germany). Mounting pressures seem likely soon to force governments to follow California's example and introduce tough legislation concerning motor vehicle emissions, and one of the easiest (technologically) ways of reducing vehicle pollution is to use taxes to discriminate against inefficient vehicles. Governments show reluctance to do this, just as the UK government refuses so far, to make energy efficiency in buildings a high priority, in spite of advice from numerous Select Committees.

Planning was a dirty word in the Thatcher years, but as London's Docklands show, inadequate attention to public transport has dire consequences. Planners and architects have a role to play in assisting the movement towards teleworking (working from home using computer links) three or four days per week with only one or two days spent commuting thus reducing the burden on the transport systems. As the following figures suggest, designers also have a major role to play in reducing the burden of pollution by reducing the use of energy in our buildings.

Most of the technologies which have caught peoples' imaginations and have allowed improvements in efficiency, are highly *visible* (e.g. aerodynamic vehicles) or are highly exciting to the engineers and managers responsible for their introduction (e.g. the TGV – train grand vitesse). Measures to save energy in buildings have been the Cinderella partly because they have been invisible and technically unexciting (e.g. cavity wall insulation). When technologies lack a glamorous appeal to designers, or obvious economic appeal to the public, changes can only ensue if resort is made to sticks and carrots. The contention advanced here is that

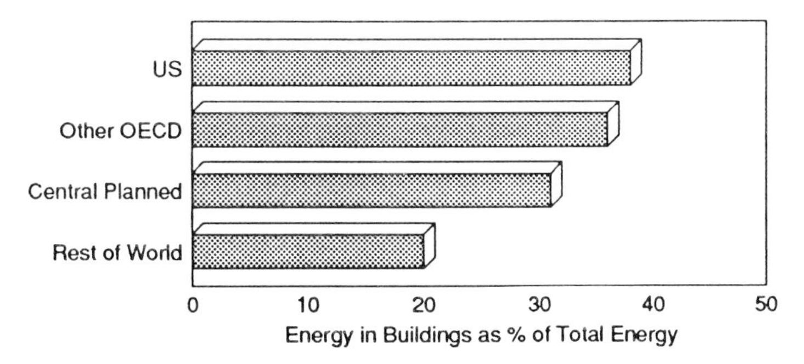

Fig. 8.2 Energy used in buildings as % of total energy for various stages of 'development'. (Source: H. Kelly, *Buildings of the 21st Century*.)

smart glass and *smart plastics* at last provide technologically glamorous and exciting products, which will also give great freedom to architects in designs for good bioclimatic buildings.

8.2 The Greenhouse Effect or why we need to reduce energy use

In the buildings sector low fuel prices have not provided the necessary incentive to the consumer to ask for higher levels of energy conservation (the carrot); in the UK, a combination of Government desire 'to lift the burden of controls' and the construction industry's immense resistance to change, conspire to avoid effective legislation (the stick). This laissez faire attitude is typified by a recent UK Government Report which suggests that CO_2 emission in the UK will rise by 73% by 2020 (Guardian, 1989); let us hope that the rest of the world does not share this disastrous 'head in the sand' attitude. Thus, at least in the UK, rather little progress has been made in reducing the burden of greenhouse gases caused by buildings. The effect is made worse by the reluctance of the electricity industry to add CO_2 scrubbers to its power stations.

The following data (Kelly, 1989) suggests that as countries develop, the fraction of their total energy devoted to buildings may increase. Since their *total* energy use will increase, the significance of buildings in the rising production of CO_2 is doubly important in less developed nations (Fig. 8.2).

Disaggregated US data illustrates the origins of CO_2 evolution as a result of energy used in residential and commercial buildings. The dominant sources are space heating in the domestic sector and lighting and space heating in the commercial sector (Fig. 8.3).

The data for CO_2 production may not yet be very reliable, (the UK probably does not approach the US in total CO_2 production as a result of building use) (Milbank, 1989); but the significant observation is that the shape of the two distributions is similar in both countries, namely space heating and lighting plus appliances, are the important areas to attack (Fig. 8.4).

Smart glazing will reduce space heating by allowing large areas of solar admission, without increasing cooling loads, and will reduce electric lighting by

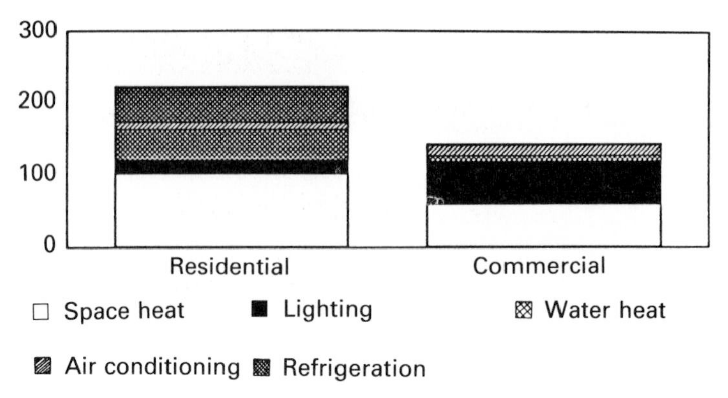

Fig. 8.3 CO_2 production resulting from buildings in USA (millions tonnes/year). (Source: H. Kelly, Buildings of the 21st Century.)

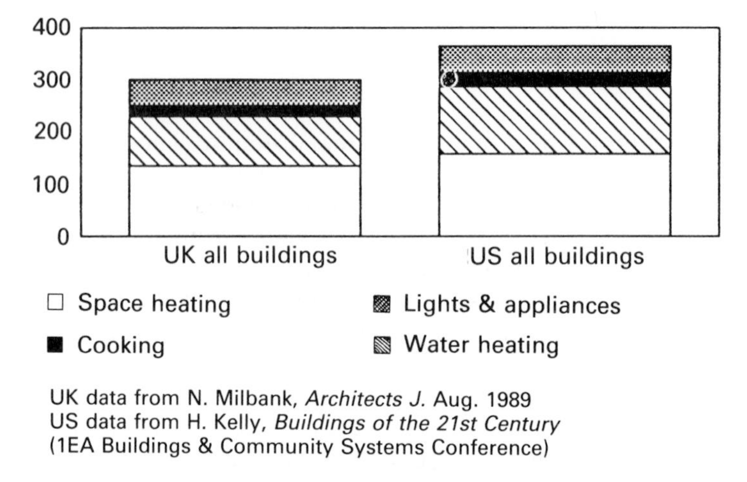

Fig. 8.4 CO_2 production resulting from buildings (millions tonnes/year).

increasing daylighting, also without increasing cooling loads. It thus has the possibility of attacking two of the most significant areas of energy use in buildings. Nonetheless, it will usually be the case that simple insulation, where it can be easily applied, will be the most cost-effective way of reducing space heating, but as the following sections suggest, this is only half the story.

8.3 Insulation does not equal *nirvana*

Figure 8.5 shows the total energy costs for a three bedroom house in the UK. Space heating, water heating and appliance use are included. The chart suggests that as the obvious remedies are applied, i.e. double glazing, higher fabric (wall, roof and floor) insulation and heat recovery ventilation, the effect on *total* energy is relatively modest. The real goal for designers should be the 'Energy Producer House'!

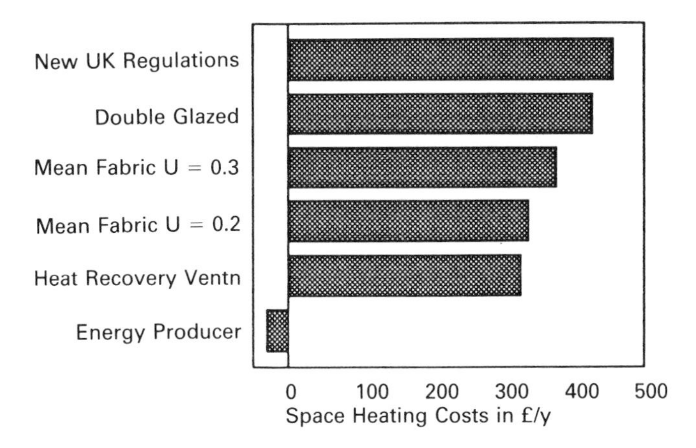

Fig. 8.5 Total energy costs for three bedroom house In UK.

Even with high levels of insulation and heat recovery, much of the energy cost and thus CO_2 emission, remains unaddressed. The superinsulated houses which the Polytechnic of Central London demonstrated for the European Commission in the UK (Littler & Ruyssevelt, 1987a; 1987b), are heated to very high comfort levels for the price of a cup of coffee per week, but the electrical energy for appliances and lighting were not reduced by similar fractions. The provision of energy has two sides: the supply side and the demand side. Many developed countries are increasingly choosing to reduce demand by conservation measures, rather than build new power stations. Surely it is this trend which conscientious designers should support?

As the efficiency of appliances in houses rises (Danish and UK studies suggest up to half the electricity used in homes could be avoided (Norgard, 1989; Doe, 1990)), the casual gains will fall. Frequently, solar gains are not useful because demand for heating has already been satisfied by casual gains – especially in the swing seasons. The opportunities to use more solar gain should thus rise as appliance efficiency rises. The following sections start with a description of smart glazing technology, and go on to suggest where smart glass can help in the reduction of energy use in the domestic sector.

8.4 Smart glass

A wide range of glasses which respond passively or actively to the environment are under development.

8.4.1 Thermochromic

Cloud gel invented by Chahroudi in his kitchen in 1971, has taken a long time to reach its present day effectiveness. It is a clear film which, when heated above room temperature reflects sunlight by turning an opaque white, turning clear again

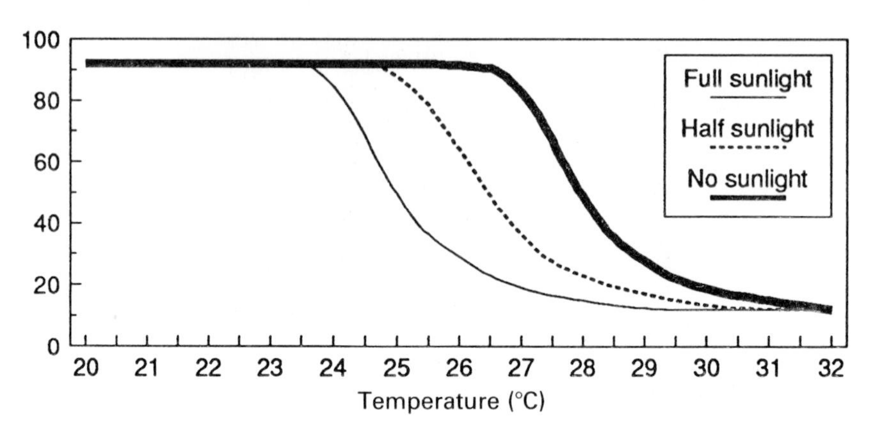

Fig. 8.6 Cloud gel opacity versus temperature and solar radiation. (Source: R. Chahroudi, Suntek Materials Inc., *The Climatic Envelope*.)

when cooled. The degree of opacity is also determined by the strength of solar radiation, so that a window in a cool room exposed to high sunlight levels remains clear; but one in a warm room turns cloudy, as illustrated in Fig. 8.6 (Suntek Materials Inc, USA).

An analagous product TALD has been developed in Europe (Boy & Meinhardt, 1988). In general, such organic gels contain water soluble polymers, which alter their chain shape or length dramatically over short temperature spans. Various oxides of vanadium which can be produced as coatings, also change visible transmission with temperatures near those required (Granqvist, 1989).

8.4.2 Photochromic

Large numbers of compounds undergo reversible change on illumination, and photochromic sunglasses provide a familiar example. Silver halides with small amounts of Cu^{+} are possibly the most common dopants used in glasses. For plastics, spirooxazines in cellulose acetate butyrate sheet is a strong contender (Granqvist, 1989).

8.4.3 Electrochromic

Certain compounds undergo reversible colour changes when a small voltage is applied across a thin layer, as illustrated in Fig. 8.7 adapted from Lampert, 1984.

The switching voltage is needed only to change the state. Most investigations have concentrated on transition metal compounds, notably WO_3, but oxidation/reduction reactions in some organics, for example the phthalocyanines, can

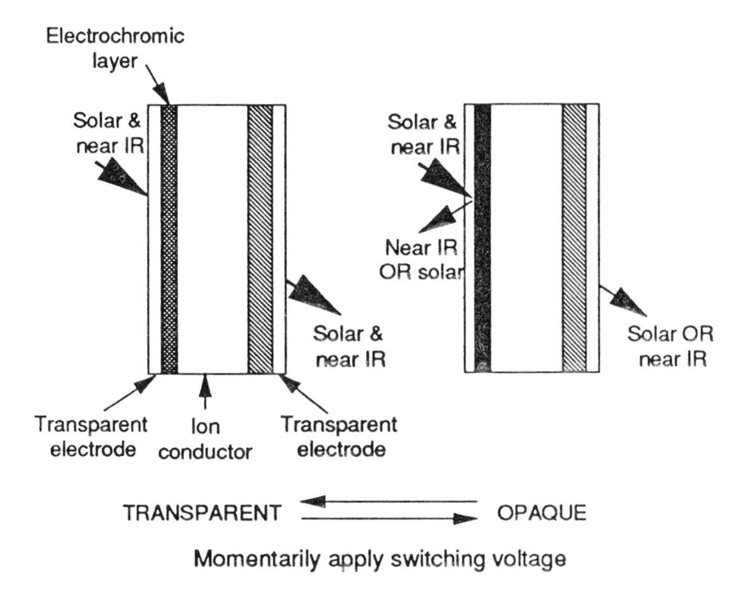

Fig. 8.7 Electrochromic switching glazing.

produce the desired effect. Asahi Glass has installed large numbers of prototype electrochromic windows in the Seto Bridge Museum and in the Daiwa House (Lampert, 1989). Liquid crystals, commonly seen in calculator displays, are now available in large areas from Taliq as switchable privacy screens in offices. Liquid crystals require continuous application of the switching voltage. Finally, some suspended particles can be aligned, so that transmission is increased, by the application of an electric field. Battery powered sunglasses using the technology are in the prototype stage (Research Frontiers, New York).

For architectural applications, a great merit of solid state electrochromics over passive switching such as thermo- or photochromics, is that of control. For example, in Hybrid Roofspace Collector Systems (Tindale & Littler, 1989; Hancock & Littler, 1986; 1987; Littler *et al.*, 1985) (see Fig. 8.8(a)), it is desirable in the winter to allow the roofspace to heat up as much as possible (below 60°C), so that warm air may be pumped into the house, but in summer it is desirable to keep out the sun at much lower temperatures. The picture (an illustration from an explanatory booklet about the houses handed to tenants of the London Borough of Newham), shows a south-facing twin wall polycarbonate roof. Warm air is blown into the downstairs living rooms and extracted back into the roof from the bedroom landing.

Indeed, the cost of extract fans and controls to avoid the risk of roofspace systems overheating the attics, greatly diminishes the cost effectiveness of the systems. This is particularly unfortunate because the simplicity of roofspace systems and theii relevance to the rehabilitation of houses with roofs in need of repair, makes them an attractive option. Technically such systems have the advantage that roofs are the least overshadowed surface of a building and fav-

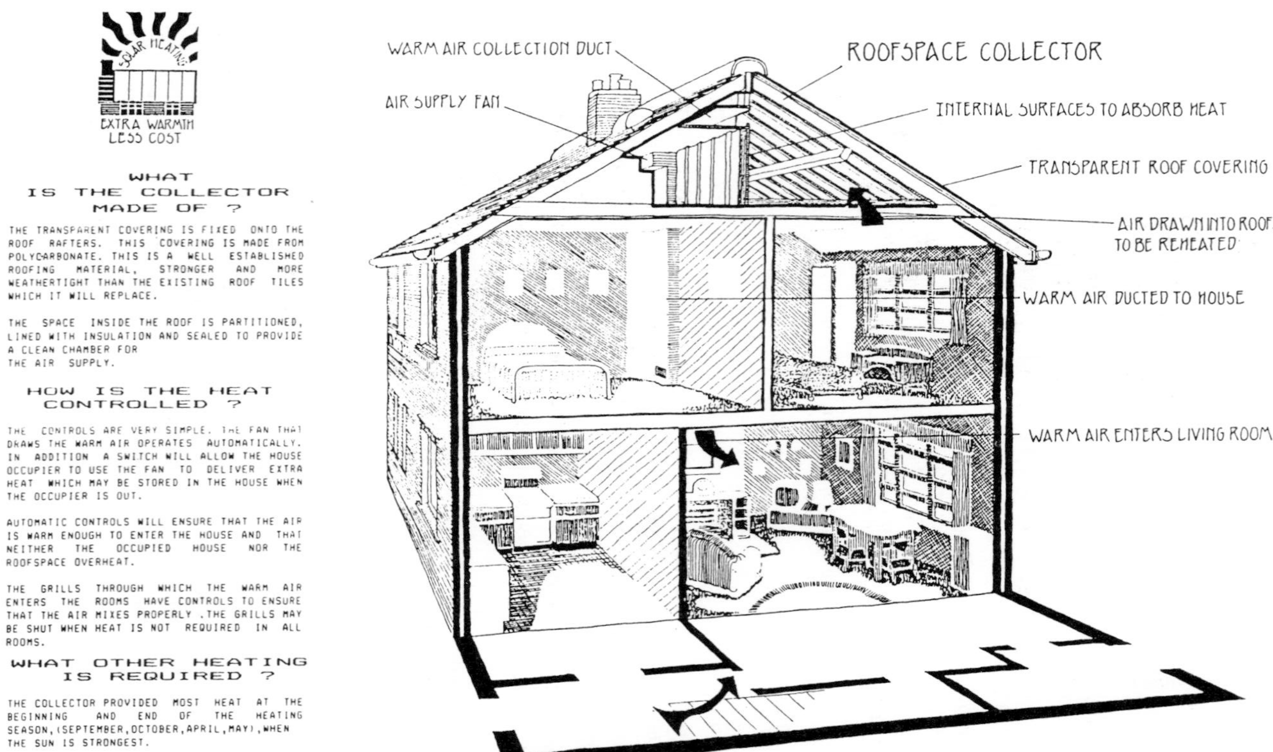

Fig. 8.8(a) Cross-section through a house showing roofspace collector and air distribution.

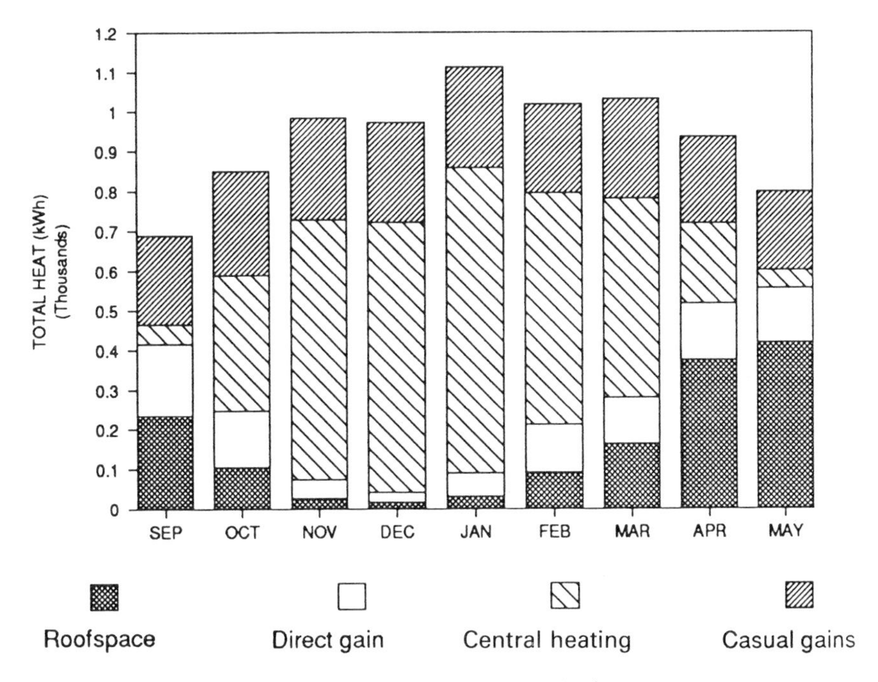

Fig. 8.8(b) Performance data of Hybrid roofspace collector at Newham.

oured by the diffuse nature of radiation in the UK. Figure 8.8(b) illustrates some performance data.

Control of solar radiation is even more desirable in airliner cockpits, and the PPG Glass Company recently announced at the Paris Air Show an electrochromic windscreen for jumbo jets. Unlike the electrochromic glass shown in Fig. 8.7, which controls *either* the visible or near infra-red, the PPG unit has multiple layers to control *all* the incoming solar radiation.

8.4.4 Holographic

Holographic films are currently used to allow drivers to see the road ahead, but yet also to reflect light at a different wavelength, say green, coming from the instruments below, thus providing a 'head up' display on the windscreen. Such holographic films may redirect beam radiation, throwing it to distant parts of a space for daylighting, and may be tailored to reflect or transmit any selected waveband, thus acting as selective filters reflecting longer wavelength infra-red (heat) back into rooms but transmitting visible wavelengths.

8.4.5 Other redirecting coatings

Prismatic glazing has been most effectively applied in the renovation of the Billingsgate Fish Market in London, by the Richard Rogers Partnership (see for example Architectural Record, 1989). Redirection of daylight may also be effected by obliquely evaporating chromium so that inclined cylinders 100–200 nm long by 20–40 nm wide are deposited on the glass (Mbise *et al.*, 1989).

8.4.6 Selective coatings

It has taken about ten years for selectively coated glazing to become commonplace, following its discussion in the technical press (Littler, 1979). The same article also pointed out the importance to the energy balance at a window (the energy *gain* arising from incoming solar and near infra-red radiation minus the thermal *loss*), of inert gas filling in the space between sealed glass units, and the value of anti-reflection coatings.

With the advent of cheap and powerful processors, researchers such as those at the Lawrence Berkeley Laboratories, have been able to produce detailed models of window performance, which outstrip our ability to test very advanced windows experimentally. (A more detailed discussion of the issue of testing versus modelling can be found in the LBL program which is public domain (Energy and Daylighting Group)). A sealed unit with two 3 mm panes currently provides a shading coefficient (SC – a measure of the solar heat gain relative to a single 3 mm pane of glass), of 0.88. If low iron glass is used the SC rises by about 8% to 0.96, and if the glass is anti-reflected on all surfaces, it rises by a further 11% to about 1.07. Since the thermal loss remains the same, the window now has the potential to collect an additional 19% of the available solar radiation.

In 1986 the Polytechnic of Central London designed superglazed houses as an European Community Demonstration, in which large areas of highly insulating glazing (U-value 1.0 W/m^2 K, and shading coefficient 0.6 (Robinson & Littler, 1988; Littler & Robinson 1987c; 1987d, 1988a; 1988b)) were installed in 6 houses (Section 8.5.1). In 1989 Lawrence Berkeley Laboratories built and installed in several test homes, sealed triple pane windows with two low emissivity coatings ($e = 0.08$) and gas fill consisting of a mixture of argon and krypton. The U-value is about 0.85 W/m^2 K and the shading coefficient about 0.7. LBL has developed units which are less heavy and thinner, by slimming down the centre pane to 2 mm and relying on the cushioning by the trapped gas layers – an idea related to Southwall's original suspended centre-film Heat-Mirror product.

The Solar Energy Research Institution (SERI) has a prototype evacuated double glazed unit with one low *e* coating. The glass is held apart by tiny glass beads, and the centre pane U-value is better than 0.57 W/m^2 K with a shading coefficient of about 0.8 (Fig. 8.9).

Figure 8.10 illustrates the case of an overcast winter sky in northern climates, with total daily radiation in the glazing plane of 0.9 kWh/m^2. The 24 hour heat balance is adverse for simple double glazing, but positive for several other combinations of U-value and shading coefficient. The best technology which could be assembled in a window at the present time from available components, (triple glazing, low iron glass, anti-reflecting treatment, two low emissivity coatings and Kr gas fill), would provide over 0.4 kWh/day of space heating per m^2 of glazing in the UK.

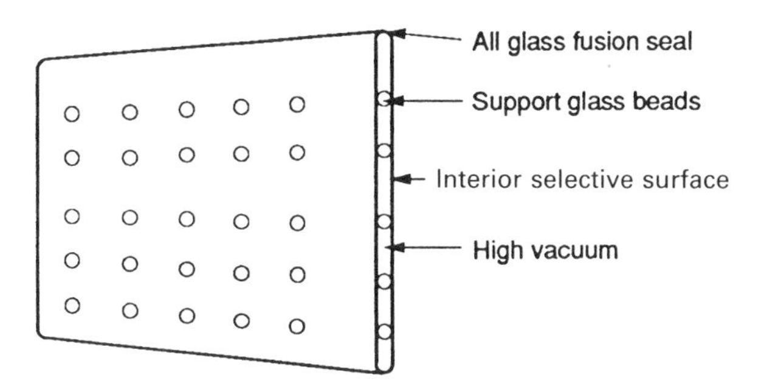

Fig. 8.9 SERI's evacuated window.

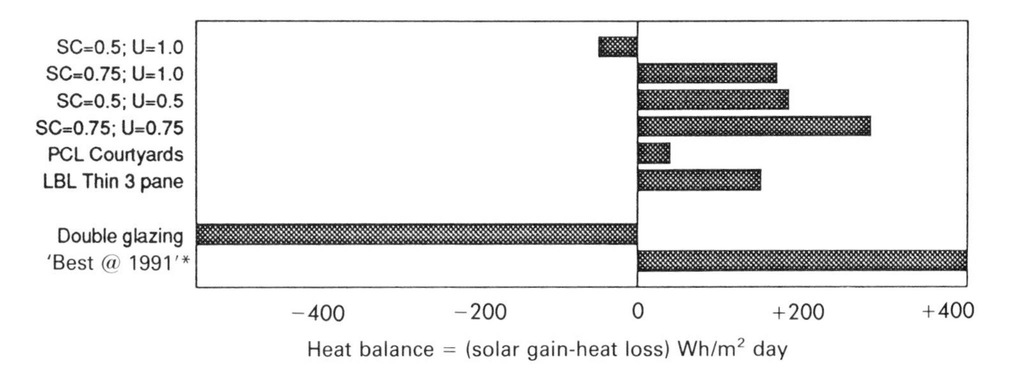

* 'Best @ 1991' calculated using WINDOWS – 3.1 from Windows & Daylighting Group, LBL, Cal. USA, triple, low iron, a – reflected, 2 low e, Kr, no Al.

Fig. 8.10 Heat balance at windows in winter outside 0°C, inside 20°C, daily incident radiation in vertical 0.9 kWh/m^2.

On overcast days in winter it is likely that all of this surplus can be usefully employed to offset space heating, but in the Spring and Autumn very high performance windows will often yield more solar heating than can be used. Thus it is more informative to use 'useful solar' as the guide to performance. Useful solar (and useful casual gains) is an output available from the simulation model SERI-RES. When this output is not available, designers using other models should perhaps concentrate on the residual space heating required rather than the amount of solar gain which arrives in each zone.

8.4.7 Transparent insulation

The Swedish company Airglass, has already made 600 × 600 mm panels of evacuated aerogel, sealed with a very thin stainless steel strip around the edge of the glass/aerogel/glass sandwich. A 20 mm thick slice of aerogel used in this way has a U-value of about 0.65 W/m^2 K and the solar transmittance of the aerogel layer is about 70%, whilst the enclosing glass is about 80%.

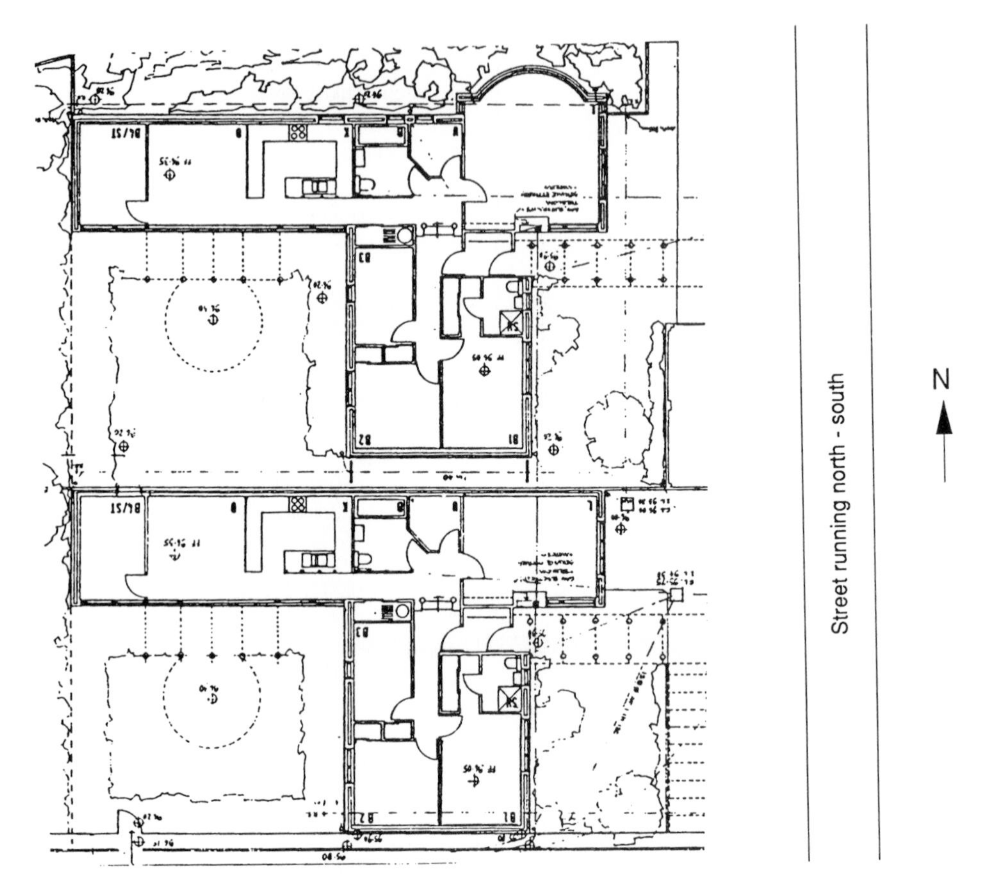

Fig. 8.11 Courtyard housing at Milton Keynes.

8.4.8 Air flow windows

Yuill (1988) has reported a 3 pane window in which outside air flows up through a 9 mm space. The *laminar* flow ensures low heat transfer between the glazings by suppressing convection cells yet picking up heat from the inner glass and transferring it into the building in the ventilation air. The U-value is about $1.0\,W/m^2\,K$ and the air flow about 3.5 l/s m^2 with a temperature rise of 8 K. (Outside temperature −18°C, roomside temperature 20°C.)

8.5 Space heating in dwellings

8.5.1 Provision of space heating

The rash of 'picture windows' in the 1960s and that of patio doors in the 1980s shows how easily northern Europeans take to large areas of glazing. The houses

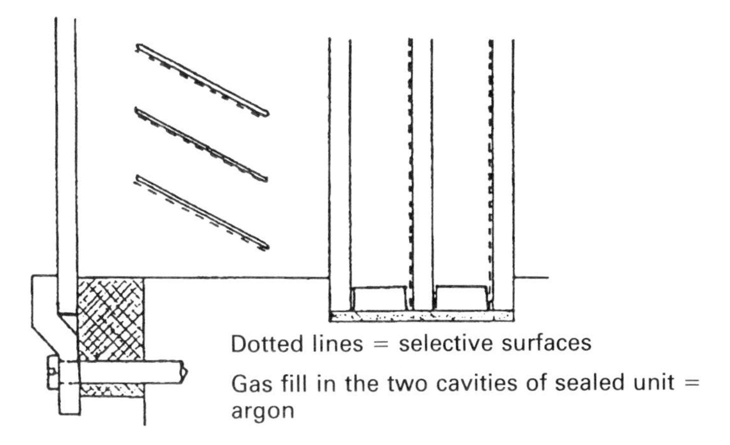

Fig. 8.12 Section through superglazed window.

with big windows built at Linford in Milton Keynes, showed large energy savings, when the occupants could be persuaded not to spoil the solar gains with net curtains. The Courtyard Houses (Feilden Clegg Design Bath) also built at Milton Keynes in the Energy Park (1985), (Robinson, 1991; Littler *et al.*, 1987), solved the problems of discomfort and cooling arising from large double glazed systems in adverse weather and at night, and the loss of performance caused by the net curtain syndrome, through the use of superglazing facing private courtyards. Figure 8.11 illustrates the site layout which allows neighbouring houses to enclose a private area facing south, without causing overshadowing.

The superglazing, illustrated in Fig. 8.12, was made up using off-the-shelf components and consisted of sealed triple units with two low emissivity coatings (for the opening lights). Fixed units possessed an additional outer layer of glass enclosing a ventilated cavity and an operable blind.

The U-value of the glazing was about $1\,W/m^2K$ or five times more insulating than single glazing.

Occupant surveys found that people would buy such highly glazed houses again (the glazing covered 100% of the south facade), glare and overheating were not major problems (the operable blinds helped in that regard) and energy was saved (see Fig. 8.13). Unfortunately, the transparency of the glazing was lower than expected, and the coatings were not of sufficiently low emissivity, thus failing to reduce the U-value as far as expected. The importance of this balancing exercise between insulation and transmission is addressed later.

One of the interesting features in Fig. 8.13 is the observation that the same amount of useful solar energy was collected in both the double glazed and superglazed houses. Roughly speaking the areas of double glazing were $\frac{2}{3}$ of the area of superglazing (the clerestories are unglazed); but the transmission of the double glazing is $\frac{3}{2}$ that of the superglazing. Thus the solar gain is about the same. The benefit of the superglazing is delivered by its reduced heat loss:

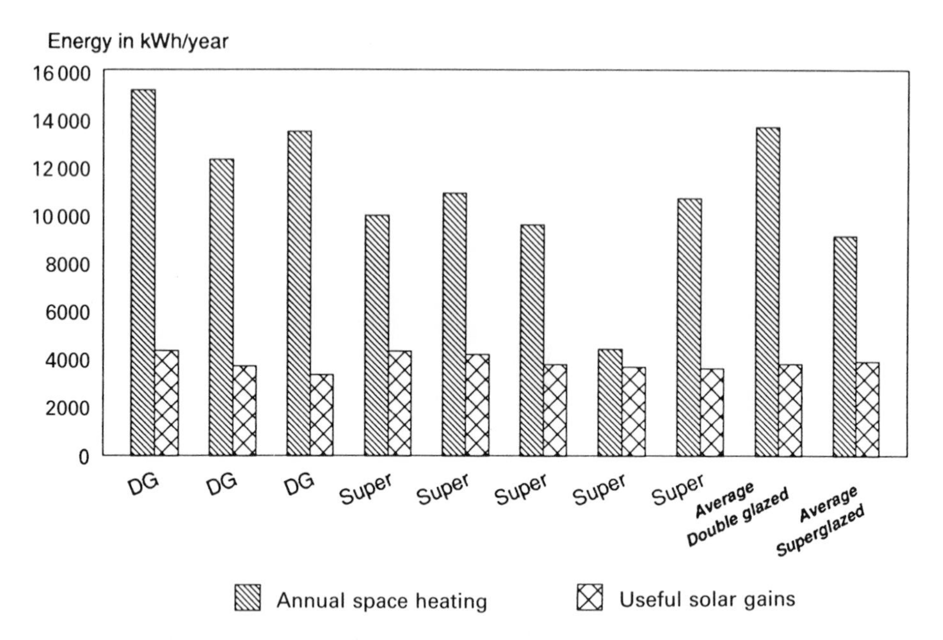

Fig. 8.13 Measured values from 8 courtyard houses: superglazed versus double glazed houses – annual space heating and useful solar gains.

$$(\mathrm{U}_{\mathrm{super}G}/\mathrm{U}_{\mathrm{double}G} \approx \tfrac{1}{3},\ \mathrm{area}_{\mathrm{super}G}/\mathrm{area}_{\mathrm{double}G} \approx \tfrac{3}{2},\ \text{heat loss}_{\mathrm{super}G}/\mathrm{U}_{\mathrm{double}G} \approx \tfrac{1}{2}) \quad (8.1)$$

Had the transmission of the superglazing been as good as expected (55% as opposed to 44%), a smaller area would have provided the same useful solar gain but with a heat loss reduced by 20%. Had the U-value been as low as expected (0.8 instead of 1 W/m^2K) the heat loss would have been reduced by a further 20%.

The following data was calculated using the simulation model SERI-RES (information about refinements to SERI-RES may be found in Haves & Littler, 1987). Figure 8.14 shows the variation of space heating which occurs when the area of the south-facing window is increased. Various kinds of glazing with high and low thermal resistance and varying degrees of transmission are illustrated.

Figure 8.14 shows the value to passive solar heating, of switching from single glazed windows (the top curve A), where the less window, the lower the annual space heating energy, to better windows. The next curve down B, for example, is for double glazing, where the optimum area as a fraction of floor area, is about 17%. Curve E illustrates a window which could readily be built today, and which reduces the space heating in the example house, from 40 to 50 GJ for single glazing, or 32 GJ for double glazing, to about 12 GJ/year.

To reach the dizzy goal of a zero energy house set in the Energy Producer House of an earlier Figure, requires glazing better than we can presently produce. The curve F represents the bounding case of a window with the transparency of single glazing and no heat loss. However, it would be quite realistic to assume

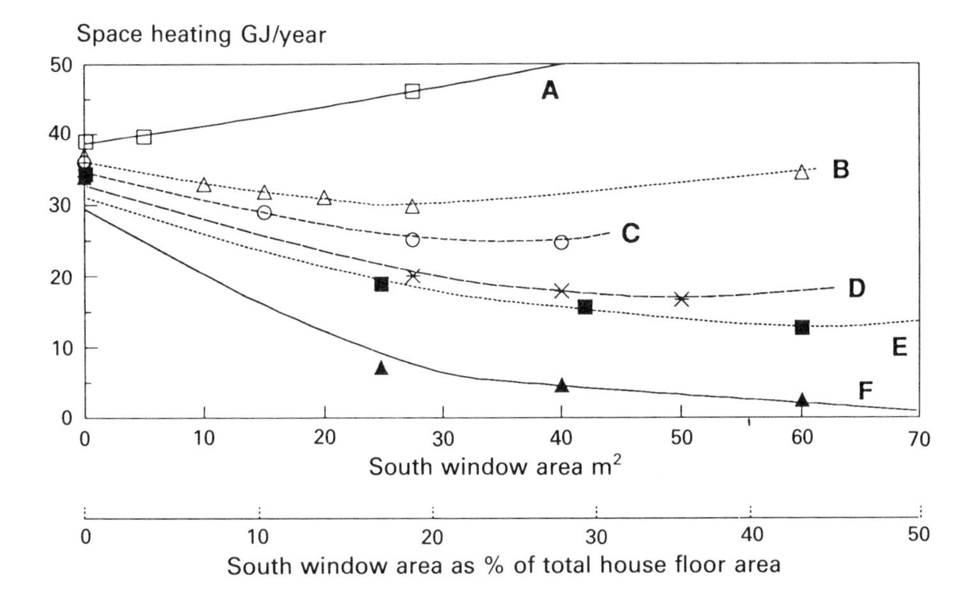

Data is calculated using SERI-RES and the Courtyard Direct Gain Houses.

	A	B	C	D	E	F
U-value	6.0	3.0	2.0	1.0	0.5	0.0
Shading coefficient	1.0	0.8	0.7	0.6	0.5	1.0

Fig. 8.14 Optimum window area for minimum space heating for a variety of window types facing south: types defined by U-value & shading coefficient.

that glazing between cases E and F *will* appear, allowing houses with large areas of glass, to use direct passive solar gain to reduce the space heating to negligible amounts.

The Courtyard Houses referred to above, do not overheat, even though they expose very large areas of glass to the south. At high summer solar altitudes, the shading coefficient is low and the solar gain modest, due to the high rejection of beam radiation by the multiple surfaces of triple glazing. In very advanced glazing of the sort which approaches case F, the shading coefficient must however be high in the heating season, and blinds or switchable glazing may be needed.

The relationship between shading coefficient SC, U-value U ($W/m^2 K$), south window area A (m^2) and space heating load E (GJ/year), for the Courtyard Houses is given, in round terms by:

$$E = 35.2 + (4.27)U - (14.15)SC - (0.275)A \tag{8.2}$$

The implication is that given a choice between trying to reduce the U-value below say $0.7\,W/m^2 K$, and attaining a high transmittance, it is the *latter* which is more significant, for highly driven passive solar dwellings.

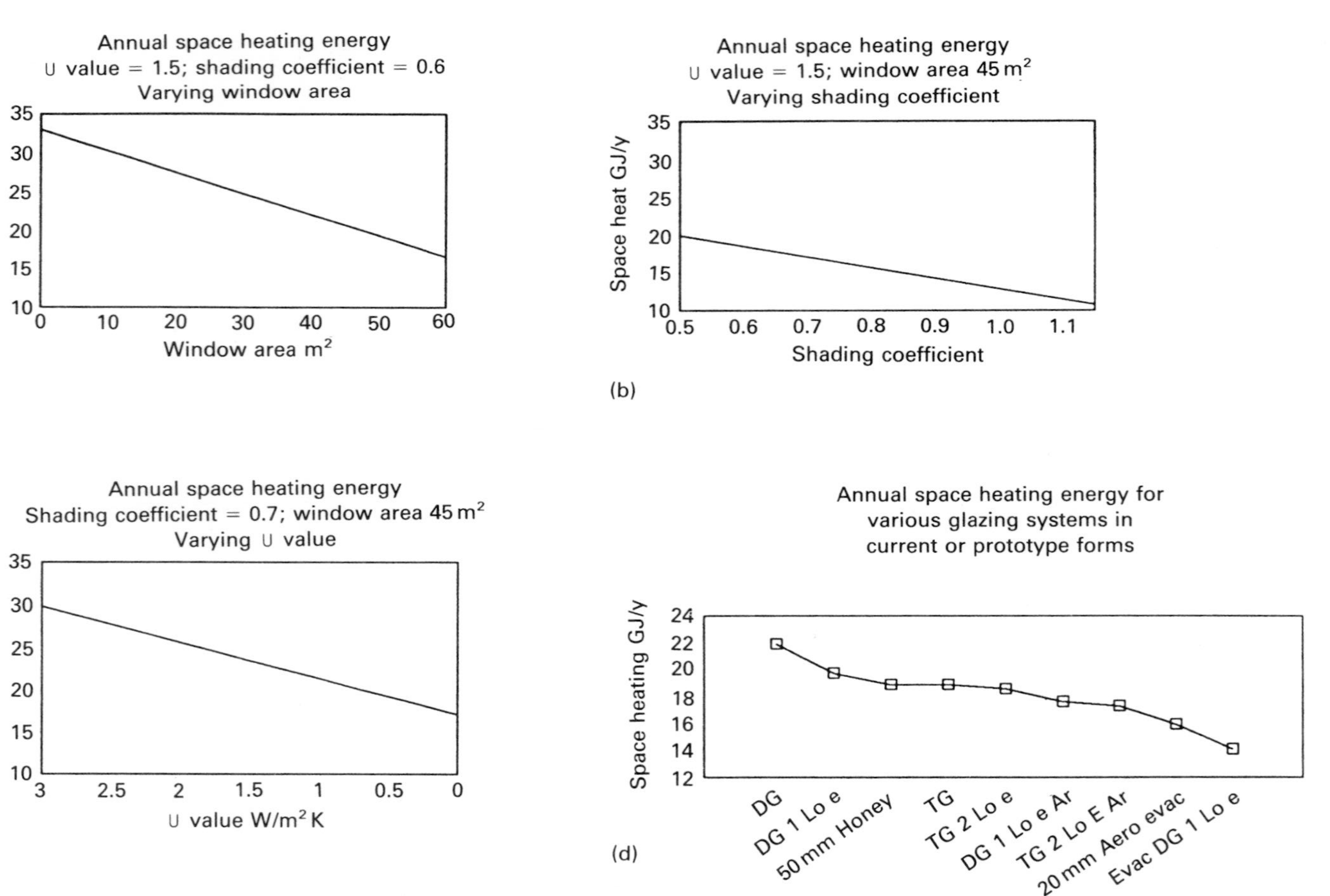

Fig. 8.15 Annual space heating energy for various glazing systems.

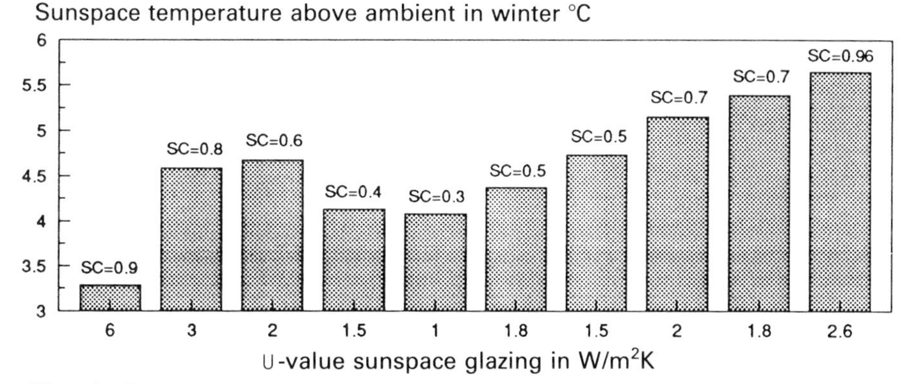

Fig. 8.16 Sunspace temperatures excess above ambient as function of glass U-value & shading coefficient–sunspace giving ventilation preheating to house.

The graphs in Fig. 8.15 embrace ranges of U-values, shading coefficients and glazed areas which can be achieved. Single glazing is rejected as an unlikely choice for energy conservative buildings. The range of U-values corresponds to a range of space heating energy from 30 to 20 GJ/year. The range of shading coefficient corresponds to a scale of space heating in houses such as the Courtyard Houses of 10 GJ/year and the variation in glazed area provides a range of 15 GJ/year.

The prototype or current window systems shown in Fig. 8.15(d) are all available or soon to be available. The abbreviations stand for:

- DG = double glazing with 4 mm uncoated glass.
- DG 1 loe = DG with one low emissivity coating.
- 50 mm honey = 50 mm thick honeycomb of the Arel type.
- TG = triple glazing.
- TG 2 loe = triple glazing with two low emissivity coatings.
- Ar implies argon gas fill.
- 20 mm aero evac = 20 mm thick aerogel between two layers of glass in a sealed and evacuated unit.
- Evac DG stands for evacuated DG of the SERI type, described above.

In the case of sunspaces, the rise in temperature above ambient, Δt, (which controls the value of ventilation preheating and the habitability of the sunspace), is given by a relationship of the form:

$$\Delta t = 3.60 - (0.68)^{*}\mathrm{U} + (4.25)^{*}SC \qquad (8.3)$$

The results for a specific case are illustrated in Fig. 8.16. The bar graph was calculated for a sunspace providing preheated ventilation air to the house, at a house air change rate of 0.75/hour. The height of the bars suggests that upgrading U-values from 6 to 3 W/m^2K has a potent effect on the temperature jump in the sunspace; but that further reductions in U-values are less worthwhile than increases in transparency of the glazing.

Selkowitz has provided related information for houses in the USA, where he shows for a fixed window area of 6.1 m^2 facing east, the relationship between *SC* and U-value, to achieve a positive annual contribution to space heating. For windows worse than those defined by the heavy line, more heat is lost than gained in the heating season (Fig. 8.17).

8.5.2 Prevention of overheating

All these cases suggest that highly passively driven solar houses with very low space heating can be achieved even in northern and adverse climates; but although overheating is not a problem in the Courtyard Houses, with the improved transmission of future glazing, the penalty will be overheating unless adjustable means are available to reduce solar gain under controlled conditions. It is here that *smart glazing* has one major role in dwellings. One product which goes part of the way towards meeting this demand is a cloud gel laminate (Fig. 8.18).

Large areas of the climatic envelope could be used to replace stretched fabric roofs such as that over the atrium of Herron's beautiful Imagination building in London (ABCD, 1989), and as the covering for Jeffrey Cook's Solar Oases in hot dry climates.

8.5.3 Privacy and preservation

Smart glass allows a particular part of a room to remain in shade as the sun moves round the building. Sections of a window will darken to keep objects of value out of the direct sun. Similarly in the smart homes of the future, it will be possible to program sections of a window facing the street, to become opaque or shaded, for reasons of privacy. Such provision may persuade people not to use net curtains which, in passive solar houses in the UK, have frequently removed much of the benefit of direct gain. Martin has carried out measurements in test cells of the disbenefits occasioned by net curtains (Martin, 1987).

8.5.4 Daylighting during the day and the electronic wall at night

With very large areas of glass on the south facade, excess daylight may be redirected from the upper sections of the facade by holographic glass and distributed to adjacent rooms by smart walls in which for example upper sections become clear during the day. A smart wall, consisting of active glass panels, may also become the large area video screen, the display area for functions within the

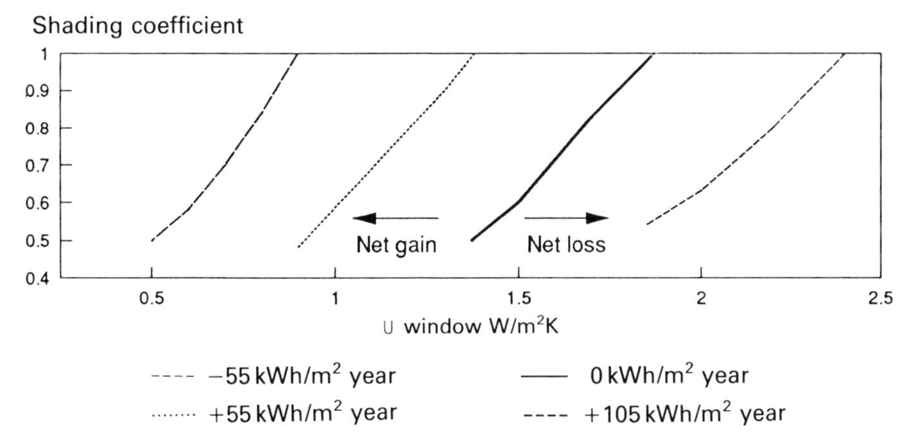

Fig. 8.17 Net annual useful energy flux/m² – south window in a house in Madison, Wisconsin, USA.

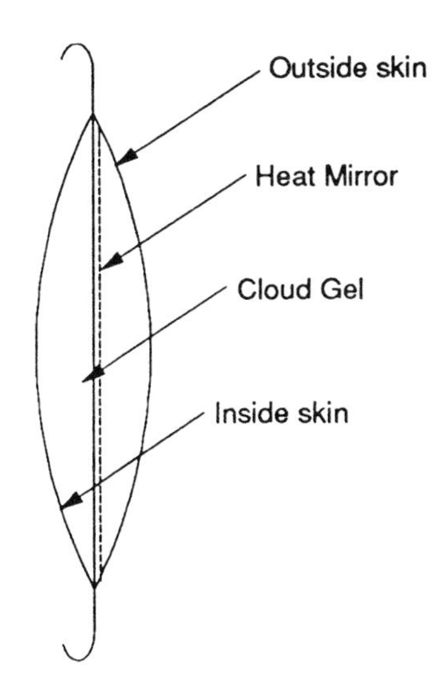

Fig. 8.18 Climatic envelope by Suntek Inc.

smart home and the entry-phone screen. It may be programmed as a wall sized light show.

In cases where sufficient light is not available in north rooms it is possible to envision the collection of diffuse daylight from the roof, with redistribution to the rest of the house. Fluorescent edge collectors have been proposed for concentrating light admitted to photovoltaics. Daylight is absorbed by a dye contained in a thin sandwich. Fluorescence at a *different wavelength* is totally internally

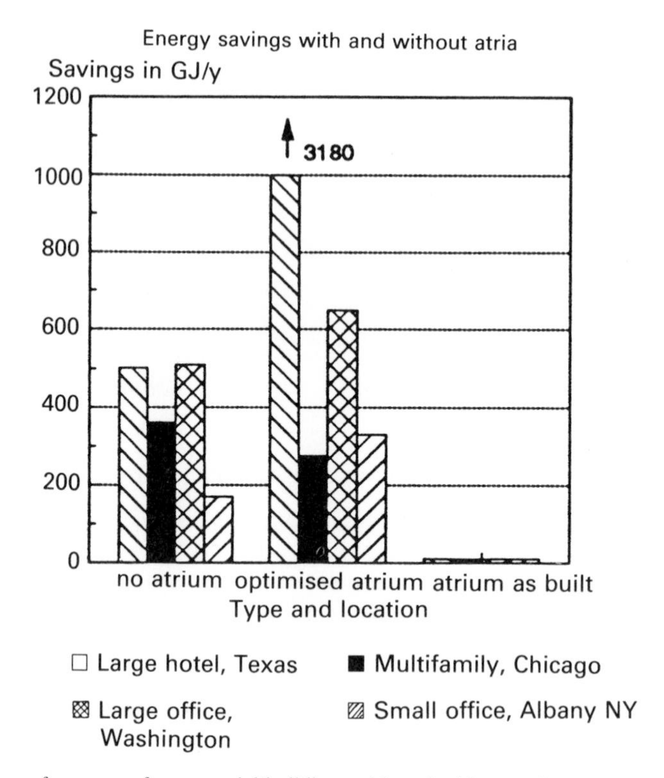

Fig. 8.19 Energy performance of commercial buildings with and without atria.

reflected at three of the edges of the sandwich and escapes as a line source at the fourth edge. From there it could be ducted by the 3M light pipe system which is already available.

8.6 Commercial buildings

So far this chapter has concentrated on smart glass in housing; but clearly after applications in planes and cars, the first use is likely to be in prestige office buildings and art galleries, to control daylighting on pictures and reduce glare, particularly adjacent, for example, to computer screens.

8.6.1 Atria

In the USA, with its southerly latitudes, the desire for atria and very highly glazed spaces often overrides all other considerations, and heroic methods are then required to maintain comfort conditions for both people and plants. These heroic measures cause increased energy consumption, and one of the very rare systematic studies of atrium performance (Landsberger *et al.*, 1988) suggests that few of the examples studied in the USA *save* energy by comparison with the same building without an atrium, but that optimizing the design can reverse this situation (Fig. 8.19).

8.6.2 Advantages of atria

The *advantages* of atrium spaces over open streets or courtyards, are much more obvious in northern latitudes – hence their prevalence in Scandinavia, Germany and the UK, and their rarity in Italy, Portugal and Spain. The following advantages are often quoted for such atria:

- The provision of semi-outdoor space with protection from cold and wet weather (for example the circulation spaces in the Züblin Construction Company, Stüttgart and Wiggins Teape offices, UK).
- The conversion of open courts to daylit and protected space, which can be used for circulation, restaurants, play areas in hospitals etc. (Erwine *et al.*, 1988).
- The possibility of using the atrium as a useful sink for warm extract air, for example the Suncourt housing rehabilitation in Stockholm (see Suncourt in references) or as a preheater for ventilation air.
- The reduction of heat loss from building surfaces which would otherwise be exposed to winter weather.
- The reduction of maintenance costs to facades otherwise exposed to the weather.
- The enhanced use of daylighting so that for the majority of the year no electric lighting is required during office hours (e.g. the Mount Airey Library, N. Carolina).
- The provision of vast and entertaining interior gardens (e.g. the John Deere offices and the Opreyland hotel).
- The provision of links, both within one building and between streets, (for example, The Galleria, Milan by Mengoni).

8.6.3 Atrium problems

The *problems* however, involved in designing atria, are also serious:

- Added fire and smoke risks.
- The provision of ventilation to spaces which would otherwise be open to ambient air.
- The loss of daylight to rooms adjacent to courtyards, caused by roof structures.
- The risk of cross contamination in hospital atria, and the spread of odours in glazed malls.
- The costs of glazing.
- The risk of overheating.

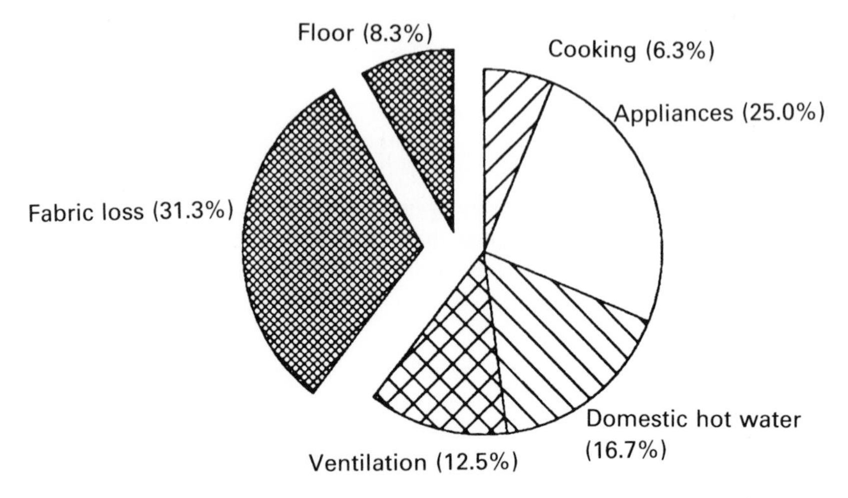

Fig. 8.20 Breakdown of domestic energy use with energy costs as % of consumer's total energy bill. (From Hancock *et al.*, 1989.)

Richard Saxon examines many of these issues and Bednar provides further examples (Bednar, 1986), the CEC aids to the competition *Working in the City* (O'Toole, S. & Lewis, J., 1990) are also a very helpful source of information about the performance of highly glazed spaces.

8.6.4 Atria as collectors of useful heat

8.6.4.1 Solar collection

Commercial buildings are often said to possess small heating loads – they possess large volume to surface area and high internal gains. Figures 8.20 and 8.21 show typical values for domestic and non-domestic buildings. However, the UK commercial buildings shown in Fig. 8.21 (which were all designed as low energy examples), show heating loads usually exceeding lighting loads. The *costs* of meeting those loads will however be reversed, since electricity costs about three times as much as gas. Similarly the pollution load generated by the provision of heating and lighting will reflect the costs. Thus the designer or client who wishes to reduce costs in use of a commercial building is unwise to concentrate too much attention on the space heating.

Added glazing will only save heating energy if the *useful* solar gain captured, exceeds the added thermal loss through the glass, over say a 24 hour period. Excessive solar gain which merely has to be vented cannot be regarded as useful. The 'LT' (Lighting & Thermal) method from the EC document '*Working in the City*' (Suncourt), illustrates the typical impact of added glazing on space heating energy for a variety of atrium forms and locations in the Community countries.

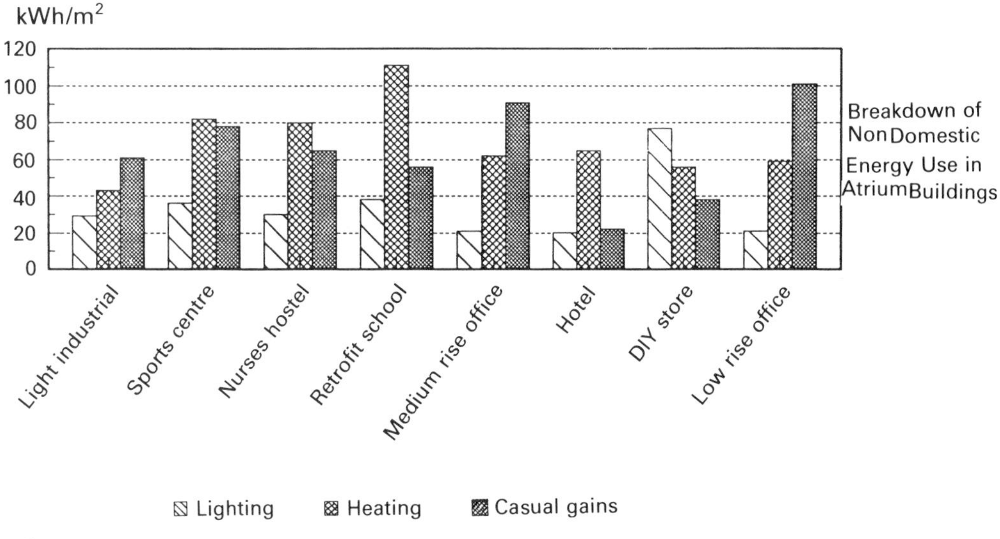

(From Ove Arup, 1988.)

Fig. 8.21 Heating and lighting energy in a variety of atrium buildings. Breakdown of non-domestic energy use in atrium buildings.

When a building is in a climate which imposes a heating load, then increased north glazing increases that load, but decreases the lighting load. The detrimental effect on the heating is less for east and west glazing, and for south or horizontal glazing the heat load may be reduced. However, the gains by reducing the heating load are almost always overtaken by the gains made by reducing electric lighting and may be outweighed by the loss due to added cooling induced by excessive glazed areas.

8.6.4.2 Overheating

8.6.4.2.1 Prediction

Most of the dynamic simulation models will give the designer a good measure of overheating problems. None yet embraces sufficient convection modelling to provide accurate data about stratification. Examples such as the State of Illinois Building in Chicago, which exposes a huge glass cylinder truncated at the top and facing south, to high ambient temperatures and solar radiation and which severely overheats (Selkowitz, 1988), illustrate the problems but offer no solution.

8.6.4.2.2 Remedies

External shading is far more effective than internal shading, and movable devices prevent the loss of useful daylight and heat when these might be obscured by fixed devices.

Natural ventilation (described in chapter 3.3) can readily achieve 20 air changes per hour.

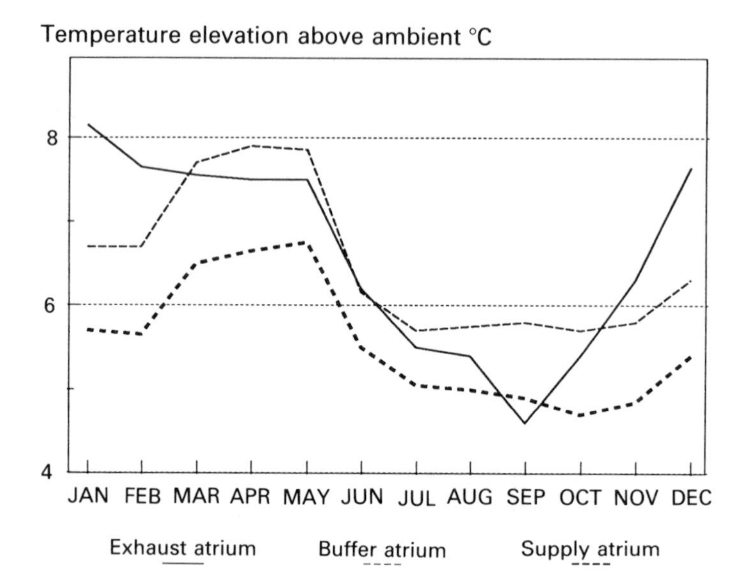

Fig. 8.22 Temperature elevations in various atrium designs (in °C above ambient).

8.6.4.3 Heat recovery from adjacent spaces

Enclosing an open court channels heat escaping from the enclosed walls, into the atrium, and clearly the effect is at its greatest when the area of glass to walls is at a minimum, but the amount of heat is always relatively small. Waste air from adjacent zones may be vented through an atrium, providing some cooling in the summer and heating in the winter.

8.6.4.4 Supply air atrium

Make-up air, for mechanical ventilation systems, or for natural ventilation, to adjacent offices, may be drawn through the atrium from outside, collecting heat on its way; however, such a mode cools the atrium in winter and may overheat offices in summer, thus for large atrium buildings the path noted in 3.1.5 is more common.

8.6.4.5 Exhaust air atrium

Using the atrium as an outlet for ventilation air, does ensure slightly higher temperatures without heating in winter, as illustrated in Fig. 8.22.

In January for example, the elevation of atrium temperature above ambient is highest for the exhaust air design (~8K) and lowest for the supply air atrium (~5.6K). In both cases the savings are small. Figure 8.23 shows data from a

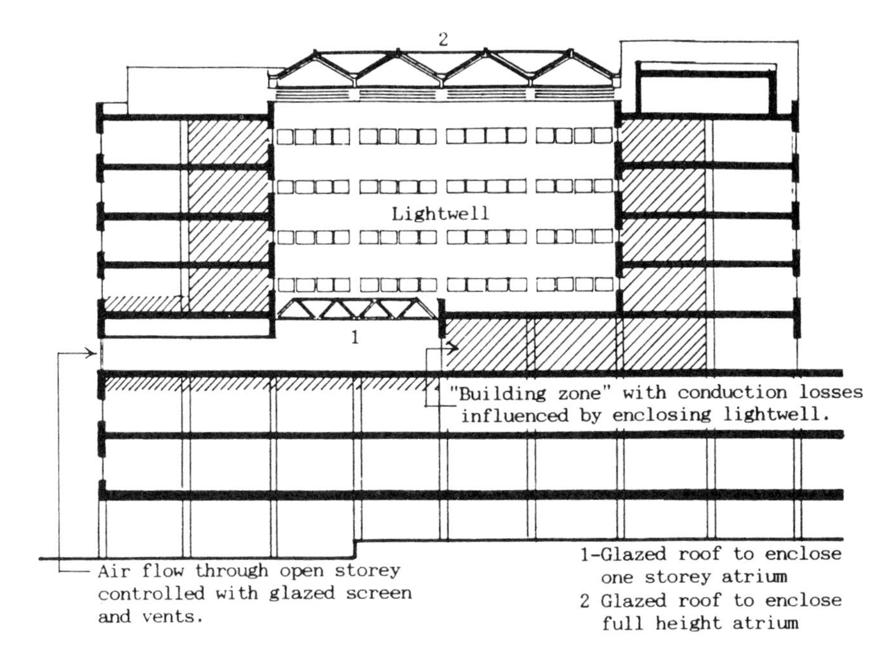

	No atrium: open courtyard	Exhaust air atrium	Supply air atrium	Buffer atrium: no ventilation
Heat loss from building thousands of kWh/year	640	410	396	437
% reduction in heat loss	0	36	38	32

Fig. 8.23 Energy savings from options in the retrofit overglazing of a hospital courtyard.

hospital case study where the open courtyard shown in the diagram was to be glazed (Erwine *et al.*, 1987).

8.6.5 Lighting the atrium

Tasks requiring high lighting levels are not normally carried out in atria, and where they are, local lighting is possible (for example at security desks). However, plants need very high light levels and heroic methods are often needed to keep them alive. It is essential to choose suitable plants and examples such as the Standard Charter Bank in London where magnolia trees have been sustained by very high levels of electric light are not a good guide to energy conscious design! The Landeszentral Bank, Wiggins Teape and Züblin Offices all provide high levels of pleasing daylight to the atria, and electric lighting is not required during most office hours. Similarly the spectacularly successful Mount Airey Library is fully daylit by shaded windows and extensive skylights. Numerous studies show the savings in electricity (Mazria, 1979).

8.6.5.1 Daylighting adjacent spaces

The opacity of the atrium glazing system does affect the amount of light which reaches the rooms adjacent to the atrium. None of the daylight models readily available handles specular reflection well and designers may find it easier to build scale models to examine the intensity of the daylighting and the effects of surface reflectivities. For example, see the Building 2000 Project at the Newbury Library (Hancock & Littler, 1989).

The Züblin Building in Stuttgart, the Wiggins Teape HQ in Basingstoke and the Landeszentral Bank in Frankfurt all provide good examples of large office buildings where attention was given to maximizing daylight gain to the offices. In the Züblin building occupants of the offices looking into the atrium claim that the borrowed daylighting is quite sufficient for office tasks and indeed very successful for working with VDUs. On the other hand the beautiful daylighting to the Economics School near Copenhagen (Larsens, 1989) whilst it provides excellent illumination to stairwells and passages, provides none to adjacent offices.

Daylighting is the area where the *designer* can make the greatest savings of energy – especially when primary fuel is considered.

References

Architectural Record (1989) January, p. 121.

Bednar, M. J. (1986) *The New Atrium*, McGraw-Hill.

Boy, E. & Meinhardt, S. (1988) TALD a Temperature Controlled Variable Transparent Glass, *CIB Journal*, **4**.

DOE (1990) *Energy Efficiency in Domestic Electric Appliances*, Number 13 in the Energy Efficiency Series, Department of Energy, HMSO, London.

Energy and Daylighting Group, Lawrence Berkeley Laboratories, Berkeley, California.

Erwine, B., Hancock, C. & Littler, J. (1987) Lightwell to Atrium Conversions, *Energy in Buildings*, September, pp. 24–35.

Erwine, B., Hancock C. & Littler, J. (1988) Daylighting and Thermal Gains – a case study of a retrofit atrium, National Lighting Conference, CIBSE.

Granqvist, C. G. (1989) Chromogenic Materials for Transmittance Control of Large Area Windows, Internal Report from the Physics Dept., Chalmers University of Technology and University of Gothenburg, Sweden.

The Guardian (1989) Figures from planners at the Department of Energy, reported in *The Guardian* 17 Nov.

Hancock, C. & Littler, J. (1986) Retrofit and Monitoring of Six Local Authority Houses at Newham, with Hybrid Solar Roof Space Systems, ASES Conference, Boulder, Colorado, June.

Hancock, C., Littler, J. & Tindale, A. (1987) Construction and Performance of Retrofit Hybrid Solar Roofs to Low Cost Housing, European Conference on Architecture, Munich, April.
Hancock, C., Littler, J., Oreszczyn, T. & Robinson, P. (1989) The Proposed Building Regulations: Real Progress or a Shabby Compromise?, *Atrium Journal*, July.
Hancock, C. & Littler, J. (1989) Newbury Library – Building 2000 – an example of a daylight design study. ISES Conference on Daylighting Buildings, London.
Haves, P. & Littler, J. (1987) Refinements to SERI-RES, Final Report to the Department of Energy, ETSU S-1130.
Herron Associates (1989) *Imagination Building* Architect, Builder, Contractor and Developer (ABCD), May, p. 25.
Kelly, H. (1989) US Office of Technology Assessment, in *Buildings of the 21st Century*, International Energy Agency Conference Proceedings of the Buildings and Community Systems Agreement, May.
Lampert, C. (1984) Electrochromic Materials and Devices for Energy Efficient Windows, *Solar Energy Materials* **11**, p. 1.
Lampert, C. (1989) Advances in Optical Switching Technology for Smart Windows, ISES Solar World Congress, Kobe, Japan, September.
Landsberger, R., Misuriello, H. P. & Moreno, S. (1988) Design Strategies for Energy Efficient Atrium Spaces, ASHRAE 2996 (Technical Committee 9.8 Research Project-315).
Larsens, Henning (1989) Copenhagen School of Economics and Business Administration.
Littler, J., Hancock, C. & Ruyssevelt, P. (1985) Passive Solar Hybrid Roof Space Collector Systems: Results and Simulations at Woodbridge and Newham, PLEA, Venice, December.
Littler, J. (1979) Thermal Balance at Heat Reflecting Windows, J Energy Research, **3**, p. 173.
Littler, J. & Ruyssevelt, P. (1987a) Superinsulation, *Passive Solar Journal*, **4**, p. 2.
Littler, J. & Ruyssevelt, P. (1987b) Results of Monitoring in Superinsulated Houses at Milton Keynes, European Conference on Architecture, Munich, April.
Littler, J. & Robinson, P. (1987c) 100% South Facing Superglazing to Courtyard Houses, Proceedings of the Third International Congress on Building Energy Management, ICBEM87, Conference, Lausanne, Switzerland, November, p. 189.
Littler, J. & Robinson, P. (1987d) Monitored Performance of R7-Superglazing at the Courtyard Houses, Energy World Milton Keynes, European Conference on Architecture, Munich, April.
Littler, J. & Robinson, P. (1988a) Superglazing U-value 0.9, Procs Canadian Solar Energy Society, Ottawa, June.
Littler, J. & Robinson, P. (1988b) Methodology of Window Thermal Assessment Techniques in the UK and the US, Report to the Department of Energy, London.
Martin, C. J. (1987) Test Cell Studies 1: Measurement of Parameters Describing the Effect of Curtains, Net Curtains and Venetian Blinds, Report to Dept. Energy via ETSU.
Mazria, E. (1979) The Passive Solar Energy Book, Rodale Press, and a sequence of papers by B. Andersson *et al.* from the Lawrence Berkeley Laboratories, Berkeley, California.
Mbise, G., Smith, G. B., Niklasson, G. A. & Granqvist, C. G. (1989) Proc. Soc. Photo. Opt. Instrum. Engr. 1049.
Milbank, N. (1989) Response to Climate Change, *Architects Journal*, August 2, London.
Nørgård, J. (1989) Low Electricity Appliances – Options for the Future, from Electricity: Efficient End Use and New Generation Technologies and their Planning Implications (Eds T. Johansson, B. Bodlund & R. Williams) Lund University Press, April.

O'Toole, S. & Lewis, J. (1990) *Working in the City*, CEC EUR 12919 EN, Eblana Editions. Copies are available from J. Owen Lewis at the University College, Dublin, Eire.
Ove Arup Partners (1988) Passive Solar Design Studies for Non Domestic Buildings, UK Dept. of Energy ETSU 1157-2.
Robinson, P. & Littler, J. (1988) Methodology of Window Thermal Performance Assessment in the UK and US. Report to the UK Dept. of Energy/ETSU.
Robinson, P. (1991) Advanced Glazing for Passive Solar Applications: Assessment of Glazing and Whole House Energy Performance by Measurement and Prediction, PhD Thesis, Polytechnic of Central London.
Saxon, R. (1985) Building Study, *Architects Journal*, 13 February.
Saxon, R. (1986) Atrium Buildings – Development and Design, Architectural Press.
Selkowitz, S. (1988) At the Workshop on Atria, held by the American Solar Energy Society, Boston Annual Meeting, July.
Suncourt Low Energy Multifamily Housing, Swedish Council for Building Research, Svensk Byggtjanst, Box 7853, S-103 99, Stockholm, Sweden.
Tindale, A. & Littler, J. (1989) Passive Roofspace Systems: Performance Evaluations, Second European Conference on Architecture, Paris, December.
Yuill, G. K. (1988) Laminar Air Flow Superwindows, Proceedings Annual Conference Solar Energy Soc., Canada, Ottawa, June.

Chapter 9
A case study of atria

Graeme Robertson,
BArch, ANZIA, MNZIA, FRSA
Department of Architecture, University of Auckland, New Zealand

While the source of the modern commercial building atrium can be traced back to buildings such as the Reform club in London (1837), the Pension building in Washington DC (1882) and more recently the hotels of Portman (Hyatt, Atlanta 1967, etc.), it is suggested that only in quite recent years has the atrium space been used so dominantly. It is now widely recognized that atria can provide wonderful buffer zones between the internal spaces and the external environment; however, it is argued that their real potential lies with climate change and worker efficiency benefits. Often the aesthetic benefits appear to outweigh the thermal or energy saving considerations, especially when the climatic extremes are relatively minimal. Too often the thermal considerations have been ignored at the design stage and quite inadequate buildings have resulted. This chapter briefly explores the development of atria from Mesopotamian times and discusses their thermal characteristics as part of the energy efficient model for commercial building design. The 20th century atrium is examined in more detail and conclusions are drawn as to important design considerations. The justification of atria in a variety of climate zones is discussed in terms of sustainable development, climate change phenomena and worker efficiency benefits.

9.1 Introduction

The first energy crisis of 1973 produced a new interest in energy efficient building design as it was recognized that the energy use in buildings was a significant proportion of a developed country's energy consumption. The proportion of total energy use attributable to buildings generally ranges from 40% in the USA, 30–35% in OECD countries to as low as 10–15% in undeveloped countries. With the production of hydrocarbons often directly related to energy production and usage, concerns regarding the impact of the greenhouse effect and climate change has further enhanced the importance of energy conservation in buildings, including commercial buildings.

In the design of commercial buildings there are two basic models available:

(1) Climate adapting buildings where the purpose is to use the climate to advantage to serve the needs of the building occupants. The positive and negative climatic influences are selectively filtered and balanced at the building's boundary to provide internal environment control.
(2) Climate rejecting buildings where the purpose is to insulate the building from the environment with the form and envelope serving solely as barriers between the exterior and the artificially conditioned space (Fig. 9.1).

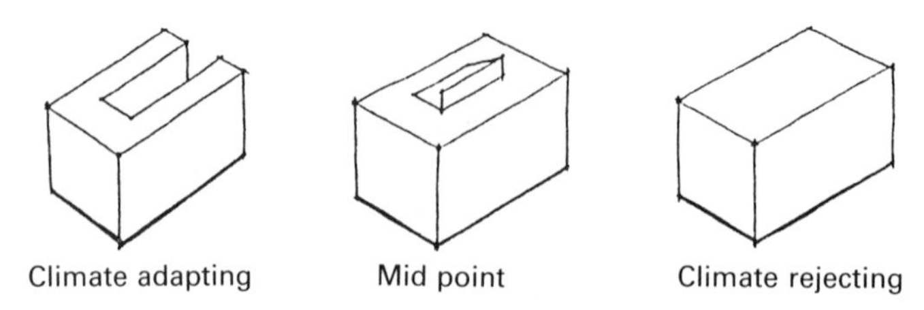

Fig. 9.1 Building forms.

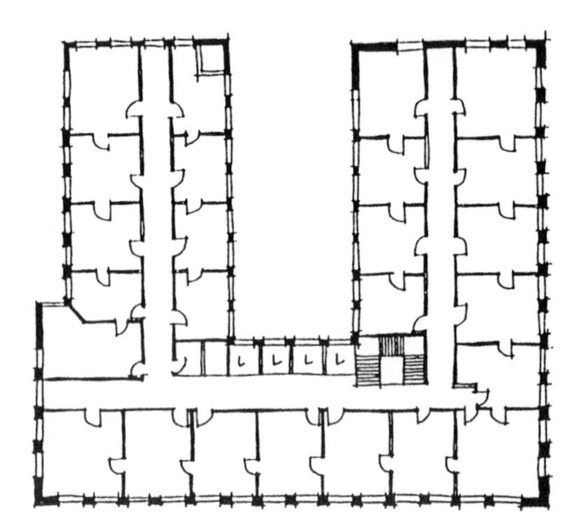

Fig. 9.2 Wainwright building.

Climate adapting buildings provide a more interesting and stimulating internal environment with energy costs potentially lower than the climate rejecting building type. Typical climate adapting commercial buildings, depending on climate zone and actual usage, tend toward:

- Increased perimeter area.
- A thick 'friendly' wall – the envelope being a filter not a barrier.
- Efficient daylighting.
- Good orientation.
- Efficient solar radiation control – glare, overheating.
- Natural ventilation.
- More efficient artificial lighting systems.
- Clear zoning of different activities.

All of the above factors can obviously be satisfied particularly easily by employing an atrium form.

Climate adapting commercial buildings using atria or lightwells were in evidence last century, as mechanical cooling or humidity control systems were not yet available and incandescent lights had just arrived. Louis Sullivan's Wainwright building (1891) in St Louis, Missouri, is a good example (Fig. 9.2). The U shaped plan form was selected for optimum natural ventilation and illumination. Openable windows, awnings and blinds were utilized to control the natural environment. Sullivan placed all corridors, stairs and lifts in the central area where access to natural light and ventilation was limited. Floor to floor heights of 3.9 m and associated high ceilings allowed better circulation of air and deeper penetration of natural light. A 'cube' would be more efficient to heat, but difficult to cool and light naturally.

In every city of the world there are good examples of climate rejecting buildings. Typically they suffer from internal gains with light, air temperature, humidity and ventilation control all provided mechanically. The new technologies of temperature and humidity control (developed during the 1920s) and the fluorescent light (commercially available in the 1930s), provided design solutions using low cost energy. Buildings such as the Seagram building in New York (1958), typical of the climate rejecting type, have an external skin to floor area ratio of just 0.3 with the Wainwright building, by comparison, having a ratio of 0.8. The trademark of the climate rejecting building type is a compact plan form where the impact of the external environment is reduced to a minimum. Atria have no place in this form of building.

9.2 Atria origins

The thick wall characteristic of the climate adapting building is exemplified in buildings incorporating a courtyard or atrium. The intermediate space so created produces the perfect buffer zone to modify the external climate and hence produce a more habitable interior. Natural light, ventilation and temperature control are all assisted by the atrium or the more basic courtyard form.

The earliest courtyard buildings are the houses of Mesopotamia dating from 3000 BC. Here the correct application of courtyard width, to length, to height produced a relatively sophisticated modified climate inside the buildings. The Middle Eastern pattern of building development, courtyard with introverted buildings, continues today and is, of course, highly appropriate.

Mediterranean cultures generally exploited the thermal advantages of the courtyard. Michael Bednar (1983) summarized this development thus:

> 'In hot dry climates with modest summer temperature (North Africa), shallow atria, one storey high, serve as collectors of cool night air and a source of shade in the daytime. In regions with higher summer temperatures (southern Spain), atria with deeper cross sections are utilized for the same reasons but for greater efficiency. In temperate climates with moderate to severe winter seasons (Rome), shallow atria serve both as passive solar collectors and as wind shelters. In warm-humid climates, shallow atria serve to generate wind-induced ventilation, which is necessary for cooling effects.'

The 'modern' atrium developed late last century with the advent of cast iron and wrought iron and the associated structural freedoms. As so often happens,

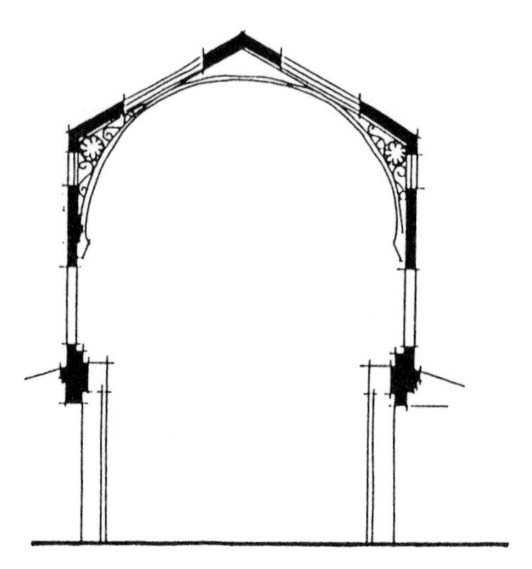

Fig. 9.3 Royal Arcade, Melbourne.

the arrival of a new technology developed a new architecture. While John Sloane played with a lantern of glass and iron in the Bank of England consuls office (1792), the origin of the modern atrium is generally attributed to Sir Charles Berry with his Reform club in London (1837). Here the court has become a grand internal room housed under a wonderful glass vaulted metal form.

The atrium form, proliferating from the 1840s, was suited to a range of building types with train stations, exhibition halls, shopping arcades, conservatories and other public buildings all incorporating atria. The atrium form had become a common architectural response throughout the world (Fig. 9.3).

Perhaps the most spectacularly ambitious building, in terms of the thermal environment created, was in North America where the new structural freedoms were shown to good effect by Montgomery Meigs in the Pension building in Washington DC (1882). This was an incredible building in terms of the thermal environment created. With all vertical top glazing, high mass, double glazing, good insulation and an excellent natural ventilation system providing an air change every two minutes into the offices, the Pension building must remain a model for today's architects (Fig. 9.4).

The advent of the skyscraper and planning ordinances, along with a change in public perceptions, meant that atrium buildings became rare during the first half of the 20th century.

The atrium space, while being utilized occasionally in office buildings and department stores, eventually became synonymous with hotel design. John Portman with his Hyatt hotel in Atlanta (1967) is often recognized as the originator of this application. His aim to design a building 'to explode the hotel, to open it up, let the elevations become a giant kinetic sculpture' is still being followed by hotel designers worldwide almost to the point of becoming a cliche. Energy efficiency aspects do not appear to be among the reasons for the use of

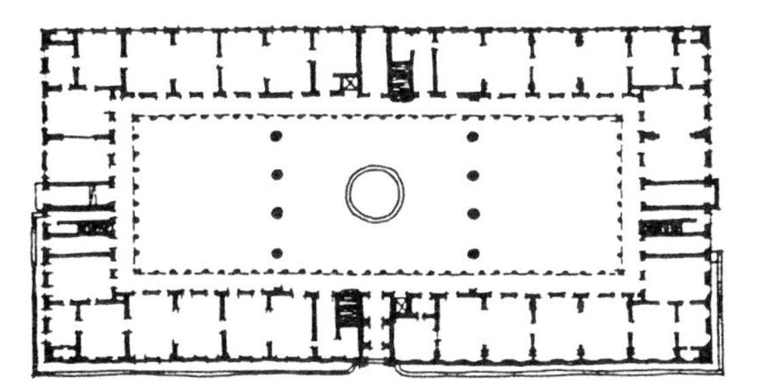

Fig. 9.4 Pension building.

the atrium form in this building type. The work of Chicago architect Frank Edbrooke to produce the Brown Palace (1892) in Denver, Colorado, should be recognized as the first hotel designed to incorporate an atrium. Here the connection between the main lobby, atrium and guest room set the precedent for the modern hotel – and out of necessity was an energy saving form.

9.3 Post-energy crisis

While Portman and Edbrooke had developed the hotel atrium for essentially social and spatial regions, the first energy crisis of 1973 gave architects and clients an incentive to develop energy efficient atrium buildings. The age of the skyscrapers and other climate rejecting building forms should have been over at that time. Examples of responsible climate adapting buildings, often utilizing an atrium, evolved slowly, usually with government incentives. The Bateson building in Sacramento, California (1981) designed by Sim Van der Ryan incorporates an atrium serving to create a humane work environment and an energy efficient building (Fig. 9.5). The building is a relative success on both counts. The atrium is enclosed by south-facing angled clerestories with automatically operated vertical louvres to minimize excessive solar radiation in summer. The thermal mass of the building, including the rock bed, is used to reduce daytime temperatures by using cool night air to flush the excessive heat from the building. Possible air stratification in the atrium is reduced by the use of large canvas tubes with low velocity fans inside which, in themselves, are a dominant architectural feature.

Other buildings, many in the USA, deserve mention as important in the evolution of the modern atrium. The Lockheed Missiles and Space Company in Sunnyvale, California (1983) by Leo Daly is critical, as it carries the pursuit of energy efficiency to unprecedented levels (Fig. 9.6). This larger-than-life building incorporates a central atrium to collect and distribute available daylight. In this building the 'thick wall' concept is carried to an extreme by utilising very efficient light shelves.

In Chattanooga, the Tennessee Valley Authority office building (1984) again concentrates on daylighting as the major energy saving possibility and incor-

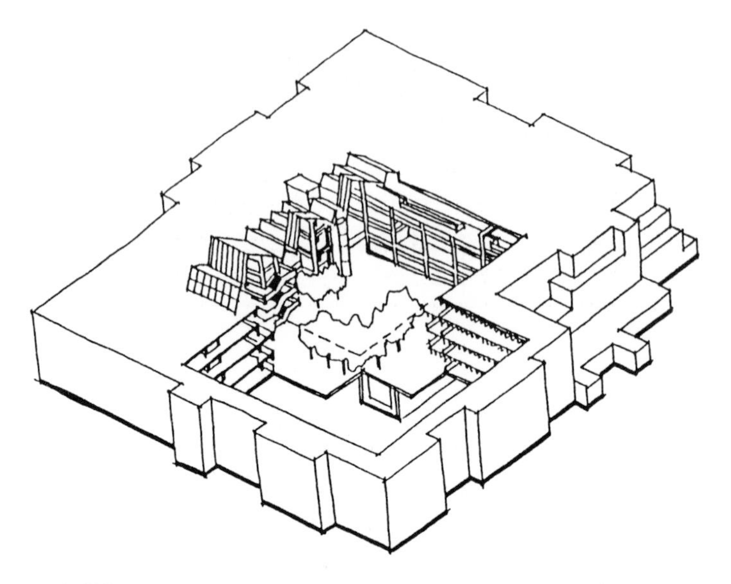

Fig. 9.5 Bateson building.

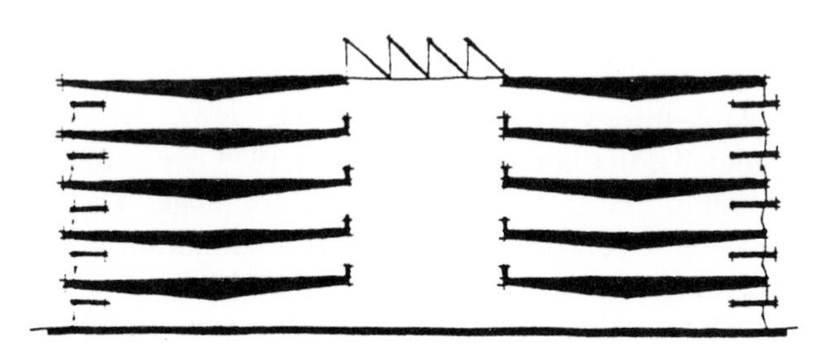

Fig. 9.6 Lockheed Corporation building.

porates east/west linear atria (Fig. 9.7) which become both a sophisticated architectural space and a good example of energy conservation. These sophisticated atrium spaces use light scoops on each projecting floor to reflect daylight into the 3.6 m high offices.

An interesting and important but little recognized reconstruction project was carried out in London in 1978 by Sir Frederick Gibberd and Partners (Fig. 9.8). Gibberd designed for Coutts bank a central atrium behind the existing Regency facades of John Nash. This scheme illustrates well the potential of atria to assist the natural light penetration on large sites and the reuse of existing resources – surely a necessary feature of buildings as part of a sustainable society?

The concept of stacking several atria vertically in an office building has been explored by Skidmore, Owings and Merrill in the 875 Third Avenue building in New York and the tower at 33 West Monroe Street in Chicago. By modifying several floors and incorporating an atrium it is possible to achieve daylighting and

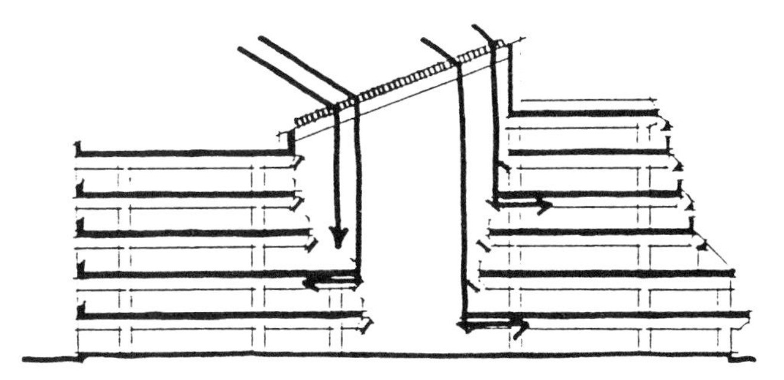

Fig. 9.7 Tennessee Valley Authority building.

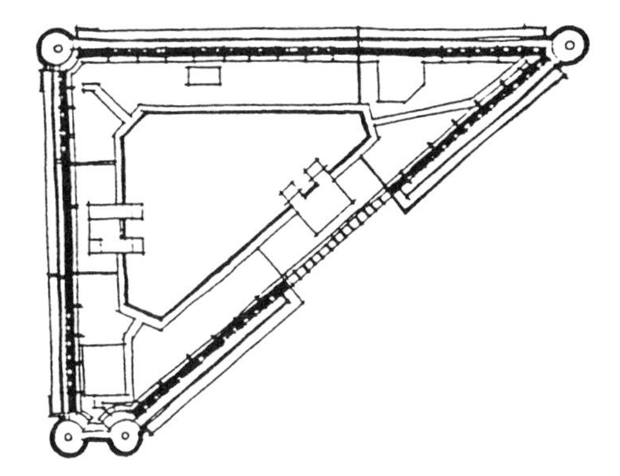

Fig. 9.8 Coutts bank.

energy saving plus views and an improved ambience. In a multi-tenant building this technique provides great opportunities to better identify each individual tenant. The deeply recessed atria with fragile connections to the exterior, as in the West Munroe building, perhaps have more social significance than energy conservation potential? Atria in this type of application – the token atrium – do little more than slightly modify what is increasingly being seen as an unsustainable building form.

Sir Norman Foster's Hong Kong and Shanghai bank offers another quite sophisticated adoption of the modern atrium that is only challenged by Richard Rogers' Lloyd's Insurance building in London. Both atria provide a strong social focus combined with energy efficiency considerations and are certainly climate adapting in harsh environments. They do not compromise the 'towers' concept that is so much part of the present city form and yet do offer a more environment friendly building form. Buildings such as the Lockheed Corporation have none of these constraints and the form reflects this relative freedom.

The work of Helmut Jahn of Murphy/Jahn must be seen as leading in the development of modern atrium form in a range of downtown commercial developments particularly in Chicago and New York. Here is a designer who has stepped beyond the constraints and fragmentation of 'late modernism' and 'post

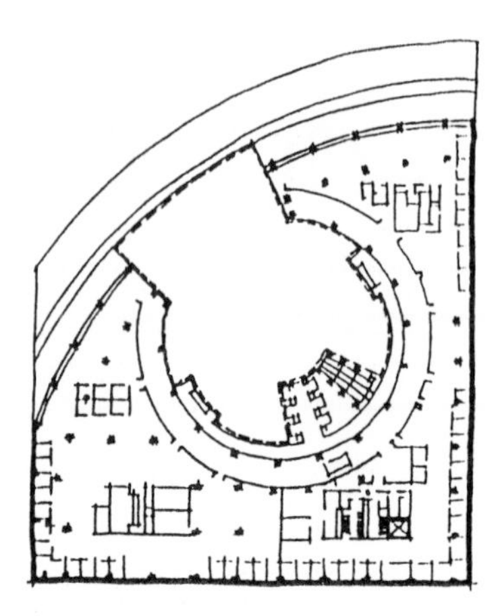

Fig. 9.9 State of Illinois centre.

modernism' by extending the modern movement into a 'new synthesis'. His philosophy

> 'not of an abstract nature, not of a technological Utopia, nor looking back to borrow from history, but a recomposition of classical and modern elements of the building arts leading to an historical continuum based on conceptual relationships,'
>
> (Jahn 1986)

certainly allows for new forms to appear and rediscovers more traditional forms such as atria. His recent work certainly includes a large proportion of atria buildings which are generally climate adapting.

Jahn's Board of Trade extension in Chicago (1982) exemplifies this compassion for old and new and exploits the possibilities of the atrium as a connector and a focus architecturally. This 1930 Holabird & Root Tower is connected to an eleven storey atrium adding reflections and daylight to the new office space.

Also in Chicago is the State of Illinois centre (1984) where Jahn, in attempting to produce a new form for the Downtown office building, has given the city an amazing rotunda rising through the entire building to project through the roof in a grand sloping skylight (Fig. 9.9). Here is a building incorporating an atrium that is clear and precise, efficient and exciting, dynamic and public. Here the open space of the atrium is both a positive element and functional in terms of usage and thermal environment modification. In terms of one of Jahn's stated aims, 'buildings as a realization of fantasies', it is an impressive building.

The Northwestern Atrium centre in Chicago (1986) again by Murphy Jahn incorporates an atrium, or series of atria which seem completely part of the building, not self-conscious add-ons. Hence in the lower levels the delights of natural daylight in large public spaces are exploited to the fullest. Unfortunately the office tower above is somewhat less climate responsive. This train station, along with the Illinois centre (which has two transport connections) and the United Airlines terminal at O'Hare airport, provide wonderful examples of grand public spaces which respond to and use the external environment to its fullest. Whilst still not meeting the full energy saving potential of the atrium form, these examples can be seen as the first step in the evolution process of the true environment-friendly commercial building.

9.4 Design options

The atrium is an addition to the architectural palette of designers allowing better connection of interior spaces to the external environment as well as providing the opportunity to create interior volumes of significance and public use. From an energy conservation viewpoint the atrium, by providing a buffer zone between interior and exterior climates, especially in extreme climate regions, offers real possibilities. While some possible advantages of an atrium are very much dependent upon climate zone and building use (internal loads), daylighting remains a feature of atria of universal advantage.

To be able to control direct solar radiation, which may provide excess heat or glare, but still allow diffuse daylight into the interior of the building is particularly easy to achieve with atria. The most basic factors which determine any daylighting scheme's efficiency are the nature of the top skin to the sky, the height and widths of the atrium space, the orientation, the reflectivity of the wall surfaces and the characteristics of the connection to the other interior spaces in the building. The geometry, orientation and construction of the glazed surfaces are normally the most important factors.

Atria have a second main function of providing passive cooling or heating, again depending upon the climate zone and the building use. In the cooling mode a well designed atria can provide shading at the right time of the day and year to improve the internal environments with associated natural ventilation. Where large diurnal temperature ranges are experienced the use of shading, building mass and night time flushing can reduce day time temperatures significantly. Where passive heating is relevant then the aspects of atria orientation and configuration become important by allowing solar radiation to penetrate, without glare problems, and connect with the available thermal mass. The heat loss through the atrium skin should be reduced by double or triple glazing or moveable insulation systems.

Natural ventilation is the third thermal design possibility of atria. The vertical height of the atrium space can provide a thermal chimney effect which can be enhanced by providing cool air intake at the lowest levels. A Trombe wall incorporating some thermal mass may also be used to assist air movement. Atria provide air movements without the typical draughts prevalent with openable windows in non-atria buildings.

The introduction of water and plants into the atrium space can further enhance the climate modifying possibilities particularly in low humidity regions.

Each possible form of atrium must be environmentally assessed in terms of

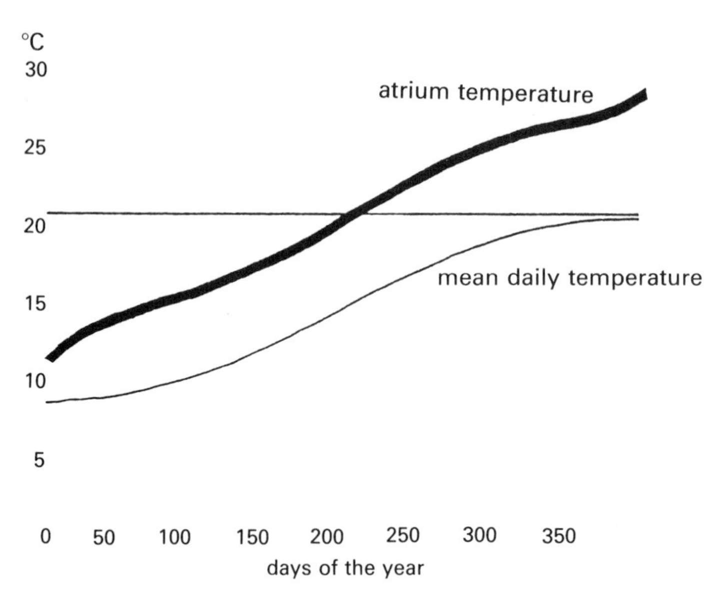

Fig. 9.10 Allowable temperature range.

building use, location, capital and on-going costs, quality of comfort required, lighting levels expected, transmission gains/losses to the parent building and to the exterior, ventilation losses/gains, mass, surfaces (colour, reflectivity) and insulation possibilities.

Obviously the allowable comfort range within the atrium space will have a major bearing on these decisions along with the on-going economics. If the temperature in the atrium is allowed to run free the auxiliary heating/cooling demand will be zero whereas if the atrium temperatures are to be kept at precise levels equating to those in air-conditioned spaces then auxiliary loads and associated costs increase (Fig. 9.10). Here is the main opportunity for the energy efficient atrium form as part of a sustainable future (Robertson, 1991).

Careful analysis of each design possibility and comparison with a 'base' building will assist in these decisions, although many items will be interconnected in their performance. Designers must avoid the incorporation of strategies which solve one problem but create others. Generally the simpler the atria design the more thermally effective it is likely to be.

9.5 Justifications

The growing concerns regarding the possible impact of climate change (IPCC, 1990) may well create even better opportunity for the application of atria to a range of building types and climate zones. Whilst the comfort and energy cost justifications will continue to be used in the more severe climate regions, the

Component	%
Personnel costs (salaries & direct benefits)	80–87
Leasing costs (or capital replacement in the case of owner occupied space)	1.5–8
Air conditioning/heating	0–5.5
Lifts	3–4
Cleaning	2–5
Building maintenance	0.5–2.5
Energy costs	0.3–2.2
Other (insurances, taxes etc.)	1.5–8.5

Fig. 9.11 Breakdown of the costs of running an office building.

savings in overall energy use with associated carbon dioxide reduction, will be of increasing importance in even temperate climates.

The efficient use of atria will reduce the artificial lighting component of energy use through daylighting possibilities. Natural cooling, ventilation and, perhaps, heating as appropriate will continue to be attractions. The general movement toward a more environmentally sensitive 'green' architecture will also ensure that the atrium form increases in popularity.

The other area of increased justification of atria is the improved worker efficiency coming from better work environments. Recent studies (Robertson, 1990a,b) and confirmed in North America (Wineman, 1986) suggest that savings to be had from increased staff efficiency outweigh all other possible economic justifications in temperate climate commercial buildings (Fig. 9.11).

It has been shown (Hedge, 1986) that from a survey of 896 office staff, that even though the ambient environment was being maintained around an 'optimum' level in physical terms, adverse reactions were still quite pronounced (Fig. 9.12).

The preliminary results from the Auckland study (Robertson, 1990c) of in excess of 560 office workers suggest a similarly very high level of dissatisfaction with the quality of the office environment with clear contradictions arising. The conclusion, because of the high level of dissatisfaction and contradiction, is that office staff desire to be able to control their own environment better. By providing a better work environment, involving a more natural quality of space ideally using

Ambient conditions		% Agreement
Temperature	– Too hot	48
	– Too cold	21
Ventilation	– Desire to open widows	77
	– Too stuffy	61
Lighting	– Prefer more light	76
	– Lighting too bright	33

Fig. 9.12 Factors contributing to comfort of office buildings.

atria, staff dissatisfaction can be reduced with immediate economic advantages (Samuels, 1990).

Buildings must become more efficient both in terms of energy efficiency and worker friendliness if sustainable, quality built environments are to evolve. A well designed atrium properly connected thermally to the adjacent building can satisfy these requirements and prove the 'delight' so necessary in architectural terms.

9.6 Conclusions

Sustainable forms of commercial building are likely to become increasingly important as environmental awareness, generated by the climate change phenomena, increases within the building industry and the wider public. Forms of building utilizing climate adapting techniques will become essential – the style of the typical commercial building is likely to change dramatically. The atrium form will, because of these developments, become even more prevalent. If the full energy saving potential of the atrium form is to be utilized then their design must be more carefully considered. To achieve energy savings – or more correctly CO_2 reductions – of in excess of 50%, will require atria that become the perfect filter between the exterior and the interior heart of the building. The non-air-conditioned atrium, with temperatures running free over a slightly greater range, is likely to become the norm. With this situation all will benefit: the building occupier through better work conditions; the building owner through higher rentals and lower energy usage; and the wider community through reduced greenhouse gas emissions.

9.7 References

Baker, N. (1988) The Atrium Environment, *Building Technical File No. 21*, UK.

Bednar, M. J. (1983) *The New Atrium*, McGraw-Hill.

BOMA (1989) *Optimising Energy Conservation*. Seminar Proceedings, Auckland.

Breuer, D. (1988) *Energy Efficient Passive Solar Commercial Building Design*. NZERDC 173, Wellington.

Bryn, I. Atria IEA *Monitoring and Evaluation of 16 Buildings* SINTES, Technical University of Norway, Trondhein.

Gentilli, J. (1971) Climates of Australia and New Zealand, *World Survey of Climatology*. **13**, Amsterdam.

Gillette, G. (1984) The Energy Performance of Atria in Commercial Buildings. American Institute of Architects, *Building Redesign and Energy Challenges*. Boston.

Hedge, A. (1986) The Impact of Design on Employee Reactions to their Offices, *Behavioural Issues in Office Design*, Van Nostrand Reinhold.

Herzberg, F. (1976) One More Time: How Do You Motivate Employees? *Concepts and Controversy in Organisational Behaviour*. Goodyear Publishing.

IPCC (1990) Working Group 1 Report, *Scientific Assessment of Climate Change*. UN Geneva.

Jahn, H. (1986) The First 20 Years. *A + U June 1986 Extra*. Tokyo.

Ministry for the Environment (1990) *Responding to Climate Change – A Discussion of Options for New Zealand*. Wellington.

Robertson, G. (1990a) Energy Efficient Commercial Buildings – A Realistic Market Objective. *Proceedings ANZScA/ADTRA Conference*, Sydney.

Robertson, G. (1990b) The Marketing of Energy Efficient Commercial Building Design. *Proceedings C I B Symposium Property Maintenance, Management and Modernisation*. Singapore.

Robertson, G. (1990c) Survey of 150 Auckland CBD Buildings, *Report, University of Auckland*, Auckland.

Robins, C. D. (1986) Designing Atria, Light Courts, Reentrants, and Sun Control. *Daylighting, Design and Analysis*, Van Nostrand Reinhold.

Samuels, R. (1990) Solar Efficient Architecture and Quality of Life. *Proceedings 1st World Renewable Energy Congress*, Pergamon, Oxford.

Saxon, R. (1983) *Atrium Buildings – Development and Design*. Architectural Press, London.

Wineman, J. D. (1986) The Importance of Office Design to Organisational Effectiveness and Productivity. *Behavioural Issues in Office Design*, Van Nostrand Reinhold.

The Building Envelope: Insulation

Chapter 10
Designing effective domestic insulation

Anthony E. Mould
Dipl Arch, RIBA
Energy consultant

A basic requirement of an energy efficient house is good insulation and yet high standards of thermal performance of building insulation are difficult to achieve in practice due to poor standards of workmanship on site and lack of site supervision, inadequate attention paid to thermal leakage through the envelope, ignorance of the impact of the moisture content of materials, drying problems, and thermal bridging. If designers are to seriously attempt to conform to current Building Regulations standards many of these problems will have to be tackled by both designers and the insulation industry.

10.1 Energy targets

Only in recent years have the Building Regulations imposed what are now considered reasonable minimum standards of thermal insulation for housing. It follows, therefore, that the overwhelming proportion of the national housing stock is poorly insulated. Even though only about one per cent is added to this stock each year by new building, for simplicity this chapter will refer to new construction only, as the principles involved are applicable generally, including improvements to existing houses.

The enclosing envelope of a house needs to fulfil complex requirements. These include the ability to reduce significantly the rate of heat loss from the interior to the exterior during the heating season. The rate of heat loss is dependent upon the temperature difference between the inner and outer surfaces of the envelope, its porosity and its thermal transmittance, or U-value.

U-value targets, which have been made successively more stringent, now include ground floors (Fig. 10.1). Trade-off options permit an increase in some U-values provided that compensatory improvements are made elsewhere. These statutory requirements are minima, but are generally perceived as maxima in order to minimize construction costs. However, local authorities and housing associations have obligations to their tenants in respect of their heating costs and these considerations are leading to the specification of enhanced U-values in order to reduce running costs. The Building Regulations prescribe U-values but they do not include requirements for air-tightness. Other bodies such as Milton Keynes and the Electricity Association have their own overriding specifications which are much more comprehensive. The latter's 'House 2000' specification, for example, includes not only lower U-values than those in the Building Regulations, but also includes maximum air leakage rates for the whole house.

The purpose of regulating U-values is to restrict heat loss. This in turn will limit

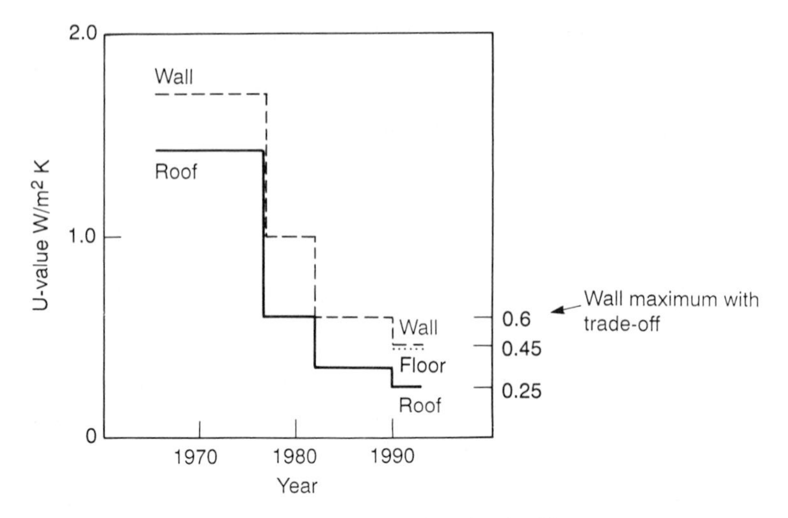

Fig. 10.1 Recent history of mandatory maximum U-values for dwellings.

space heating costs and CO_2 emissions, improve comfort and reduce the risk of surface condensation. The converse is that it may have detrimental effects, such as reducing the durability of the fabric, for example by increasing the likelihood of frost damage. In addition to heat loss calculations, analysis of the design might be carried out to obtain an energy rating. This may be a requirement of the owner or be used as a design aid by the designer.

Some speculative builders affirm that their houses reach specific National Home Energy Ratings. This is a rating scheme created and administered by the National Energy Foundation and based on the Milton Keynes Energy Index described in chapter 14. This rating (NHER) is a level of energy economy ranked on a scale of one to ten. Ten is the most efficient and one the least efficient. Houses complying with the thermal standards of the present Building Regulations would score about seven. In addition to the energy rating, the program displays the 'Building Energy Performance Index' (BEPI), which shows the degree of compliance with the Building Regulations. A design which precisely meets their requirements scores 100.

The NHER is a suite of comprehensive computer analyses which calculate the total yearly energy costs per square metre of floor area for a standard occupation pattern, using data about the particular house, including exact location, its degree of exposure and its orientation. The computer programs incorporate their own data for air change rates and U-values, depending on the requested input. In some simplified versions of the programs default values are used.

10.2 Designer's liabilities

As well as any complaint about the material aspects of the house envelope, such as cracking or rain penetration, complaints might be levied about inadequate or expensive space heating or condensation. Surface condensation could be due to injudicious thermal design, described in the following chapter, or faults in construction. Complaints about heating could be due to faults in the heating system itself but are more likely to be due to heat loss through the fabric exceeding heat input from the heating system. It would then be necessary to check the calculations for error and then to assess whether the design U-values have in fact been met.

There are three simple relationships between the heat energy liberated within the building and the thermal performance of the envelope enclosing it, as follows:

(1) The rate of heat energy flowing out through the envelope is equal to or less than its designed value.
There are no grounds for complaint.
(2) The heat energy flowing out through the envelope is greater than the designed rate of heat loss and is more than the combined rate of energy output from the space heating system plus heat from other sources. This will result in a lower than planned air temperature.
Complaint is likely.
(3) The rate of heat energy loss is greater than designed but the space heating system capacity, together with the heat output from other sources, is able to match it and so to maintain the required internal air temperature. In this case a higher than planned heating cost will result.
Again complaint is likely, unless the householder has a low energy house where the heating cost, although higher than it should be, is still moderate.

Where an energy rating has been carried out on a house or design, giving predicted internal temperatures, energy consumption and running costs, the householder will later be able to compare these predictions with actual figures. It follows then that the designer should give particular attention to really achieving his target U-values, and to take active measures to restrict air leakage to the assumed background air change rates. In short he will need to reassure himself that his design, when built and occupied, will meet his predicted performance.

The designer should be aware also that in the case of dispute, his responsibility is overriding. It is no defence to claim compliance, say, with the Building Regulations, if the conditions they require lead to building faults. An example of this could be that the amount of thermal insulation needed in a wall results in insufficient heat energy reaching the external brickwork to promote drying, as was customary, so leading to frost damage. The designer should have recognized the danger and issued an appropriate specification.

People vary greatly in the way in which they operate houses. Similar families in similar houses may have energy consumptions which may vary by a factor of three to one. Their expectations differ; some will operate their space heating systems to give continuous whole house heating at temperatures significantly above those used in design calculations. Others will opt for economy and lower temperatures.

It has been found that, given good levels of thermal insulation, which result in moderate or low space heating requirements, people tend to opt for higher air temperatures, rather than take the benefit as reduced running costs. It is for these reasons that the NHER programme gives three levels of heating in its running cost estimates, the middle one being for standard conditions. Householders with energy bills resulting from higher room temperatures may assume that they are operating in the standard mode and blame their higher than expected costs on inadequate thermal insulation. The only way to refute positively such a suggestion is by making a thermal check on the dwelling, or at least recording internal and external air temperatures over part of the heating season, together with metering of energy input.

In the local authority and housing association sphere, tenants are at present subject to three running costs relating to their dwelling – community charge, rent and energy costs. The first is a fixed cost while the other two are interrelated if a heat with rent policy has been implemented. Basically tenants have a right to reductions in rent if space heating costs are greater than agreed. It follows then that the landlord, in turn, has a keen interest in the continued effective thermal insulation level of his property.

The thermal performance of the house envelope itself is likely to be lower than the design on paper would suggest depending on the type of construction. Those constructions requiring the maximum of site assembly being most subject to shortfall, while predominantly factory-assembled structures are more likely to function as calculations suggest. There are a number of possible reasons for substandard thermal performance, namely:

(1) Higher than specified density of materials used.
(2) Higher than standard moisture content of materials.
(3) Convective air movement through or around the materials.
(4) Subsequent damage during building by following trades.
(5) Deterioration of materials.

Several of these reasons may be jointly implicated in one particular instance but for clarity possible causes will be examined here individually.

10.3 Thermal insulation

Thermal insulation is positioned between a warm material and a cooler one and functions by slowing the rate of heat energy transfer from the warm to the cooler material. In a house wall insulation is flanked by other materials but essentially it is interposed between warm room air and cold outside air during the heating season.

Throughout the building industry there is a widespread ignorance about thermal insulation, how it functions, the way in which it should be installed, and the ways in which its effect can be degraded. This lack of understanding operates from labourer to top management. It therefore follows that supervision of work in progress, to ensure that the installation of the insulation is effective, is likely to be cursory and lacking in understanding. Management and workforce, as well as the professions, need to have a basic knowledge about the subject, including installation procedures, in order to build with a reasonable likelihood of effective thermal performance which lasts throughout the life of the building.

It should not be forgotten that house building, certainly with traditional construction, proceeds in the open, with no shelter from inclement weather and quite often in mud, with the construction materials dumped on the ground nearby. Add to this the fact that the site agent needs to finish within the contract period and within the contract price, using itinerant labour, then the exact positioning of building components becomes difficult.

A vital characteristic of any thermally insulating material is its thermal conductivity or 'lambda' (λ). The lower the value the better the insulating properties of the material. The values usually quoted are measured under standard conditions in a laboratory, in which the insulation is tested in a sealed box. If the insulant is used in similar conditions these values are valid. Manufacturers of insulating materials quote the thermal conductivity (λ) value of their materials as tested in accordance with BS 874, which is the laboratory test. They can also publish a copy of their Agreement Certificate, which is another test. In practice the insulation might well be fixed where different conditions exist which change its performance.

The designer should look carefully at test certificates and compare only like with like. In their Investigation Report (Eurisol, 1991) Eurisol have drawn attention to variables of density and moisture content. These are discussed in the sections which follow. The most instructive test is to measure in situ the U-values of a variety of constructions. Some different types of construction have been measured and the U-value results obtained are disturbingly high.

10.3.1 Density

In simple terms, good thermal insulation is a durable, dry, low density material with a high proportion of small void to solid material, in which the voids do not interconnect. The lower the density of insulation the better it performs thermally but the weaker it is structurally. Insulation is therefore usually sandwiched between other materials which function as the load-bearing structure, such as the inner and outer leaves of a cavity wall. These other elements, in this case inner and outer leaves, also contribute to the overall thermal resistance of the wall, but are optimised for characteristics other than thermal insulation.

Until recently, for masonry construction, an inner cavity wall leaf of 100 mm thick aircrete block of appropriate density satisfied the U-value requirements of the Building Regulations. Now, with lower stipulated U-values, either thicker blocks or additional insulation is needed. The designer could seek reassurance from the supplier that the insulating blocks comply with their stated nominal density. If the blocks are heavier a higher U-value will result.

The surest way to check is to weigh random blocks on site and then calculate their weight. This will give only an approximate density because the moisture content of the chosen blocks will be uncertain. The standard moisture content assumed for the inner leaf is 3% moisture content by volume. The actual amount of moisture present may well be significantly greater, particularly if the blocks have been exposed to rain. Aircrete blocks are made of lightweight concrete, with small voids induced in the mix before setting. The mix can be varied to produce a

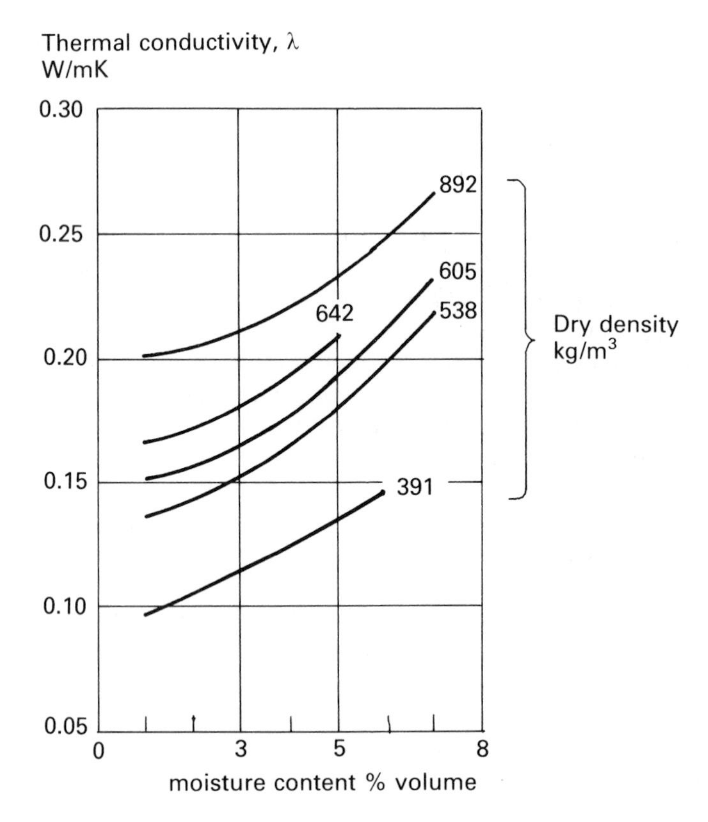

See Stukes, A.D. & Simpson, P. The effect of moisture on the thermal conductivity of aerated concrete, BSERT **6** No. 2, 1985, pp 49–53.

Fig. 10.2 The effect of moisture on the thermal conductivity of aerated concrete blocks of differing density.

range of densities with correspondingly differing strengths. They therefore have differing thermal conductivity values (Fig. 10.2).

10.3.2 Moisture content

Moisture absorption by materials will increase their thermal conductivity and so reduce their insulating qualities. The moisture content of materials forming the envelope of a dwelling will vary with the seasons and be affected by factors such as moisture generation within the dwelling, rainfall and windspeed. Moisture may be trapped or condensed within the materials within the envelope, such as those forming flat roofs, in certain wall constructions and where cold bridges occur. The designer should of course seek to eliminate such occurrences by judicious detailing (BRE Report 143, 1989).

In order to produce standard thermal transmittance values, standard moisture content levels have to be assumed. That for brickwork exposed to rain is 5% and for concrete internal wall leaves is 3%; both by volume. Insulation materials are assumed to be dry. Standard thermal transmittance values are used in U-value calculations, therefore if moisture content values are higher in practice, U-values will also increase, and so the rate of heat loss.

The moisture content of materials in a new, traditionally built, house is likely to be high, until it has dried out, and particularly if it is in winter and the house was rain soaked during construction. About 20 tonnes of water is used in the construction of this type of three bedroomed house, with its wet finishes. Later, when occupied, a family of five will generate about 14 litres of water vapour a day within the house.

Due to higher internal temperatures during winter and the likelihood of closed windows, there will probably be a higher moisture pressure within the house than outside it, so tending to push moisture vapour out through the walls, roof and floor, possibly leading to interstitial condensation.

10.3.3 Porosity and air movement

Probably the most widespread reason for the inadequate performance of the insulation in the house envelope is that of excessive air movement, either through it or by perverse convection within it. The usual air change rate used in calculating heat loss is one air change an hour in the house itself, although this is modified if, for example, a balanced ventilation with heat recovery system is installed. Ventilation rate is dependent on the degree of airtightness achieved in designing and building the house shell, the vagaries of the weather and the habits of the occupants.

The house can be calibrated on completion, or at any subsequent time, by pressure testing. The test will give the overall leakage rate at a particular pressure, usually 50 Pa. The contribution of particular leaks to the total can be found by selectively sealing them with tape. Leaks through a loft hatch or through patio doors, for example, could be quantified. Otherwise a smoke puffer can be used very effectively to identify the sources of leakage. Smoke velocity and direction will give a good indication of their importance. Figure 10.3 shows the average aggregate void area found in 200 m^3 houses, compared with the void area in similar sized houses conforming to the 7 ac/h at 50 Pa standard.

Not all air movement that affects the performance of thermal insulation connects with the inside of the house, so it will not be detected by pressure testing. Within the fabric itself very narrow cracks and very low air velocities can be important in increasing heat loss. Examples of air movement leading to additional heat loss are included in sections 10.3.3.1 to 10.3.3.5.

10.3.3.1 Mortar joints

It is a common experience to see numerous lengths of vertical joint omitted in the inner leaves of cavity walls. In ventilation terms it is thus potentially a honeycomb wall, unless the wall is then plastered (Fig. 10.4).

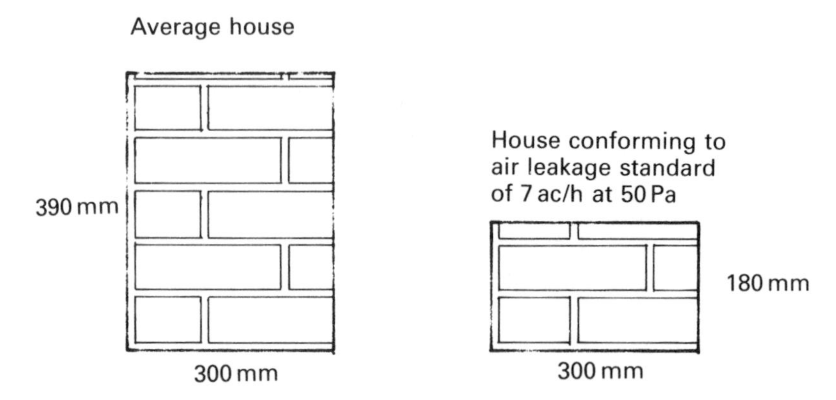

Fig. 10.3 Total void area found in the internal surfaces of a three bedroomed house of 200 m³/volume.

Fig. 10.4 Voids in mortar and around joist ends.

A particular case occurs in the walling within the thickness of upper floors, particularly where there are timber joists. During building, first the joist ends are placed onto the wall, then beam infilling butted up to the joists. After that timber drying and shrinkage produce an open joint (Fig. 10.5).

In a 7.0 × 7.0 m house plan, there are likely to be 24 joist ends bearing on the wall. The joist depth of 175 mm, not being a brick or block coursing height,

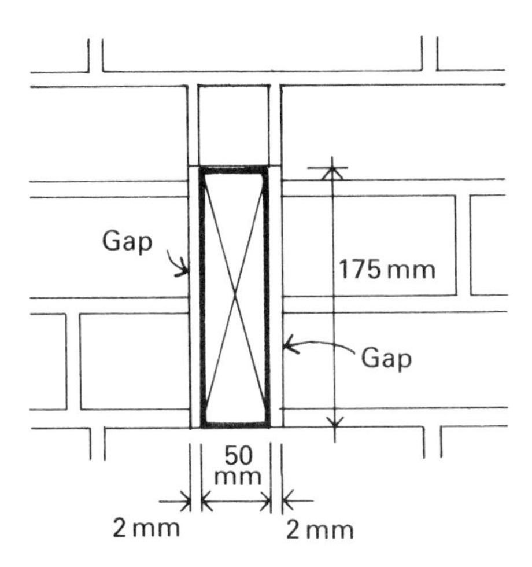

Fig. 10.5 End of floor joist bearing on inner leaf of a cavity wall.

means that beam filling between the ends of the joists is likely to be part blocks or bricks, imperfectly jointed. Blockwork and the timber will abut with dry joints and the joists themselves will dry and shrink, leaving gaps at their sides. Assuming that these gaps are 2 mm wide, then the equivalent of nearly a 100 × 100 mm hole is likely to result on opposite sides of the floor. It is then likely that with wind pressure on one side of the house and wind suction on the other, the floor void will be well ventilated by cold air. This is in addition to the effect of omitted vertical mortar joints in the blockwork as a whole. An alternative construction, potentially better draught sealed, would be to support the joists by joist hangers built into full course height and ensure that the blockwork is sealed.

10.3.3.2 Partial cavity insulation

Partial cavity fill is adopted for two reasons. One is to economize on insulation. The other is to preserve an open cavity. The open cavity will inevitably be vented to outside air by weep holes and voids in the mortar. The insulation will be a board or a stiff fibrous material, held in place by the wall ties or by pinning. To be effective the insulation needs to be firmly fixed against the inner wall face and to be continuous, without breaks.

The usual process is that the inner leaf is first partially built, with incomplete vertical mortar joints and bed joints squeezed out by the weight of the block above, creating continuous mortar snots. The insulation boards are then offered up to the inner leaf, creating a sizeable cavity between the insulation and the face of the blocks (Fig. 10.6). This allows air to circulate freely between them. The vertical joints between boards might be butted together in straight walling, but generally they do not meet at internal and external corners. Furthermore, the boards are often fixed in bonded formation, so that fractional boards occur at

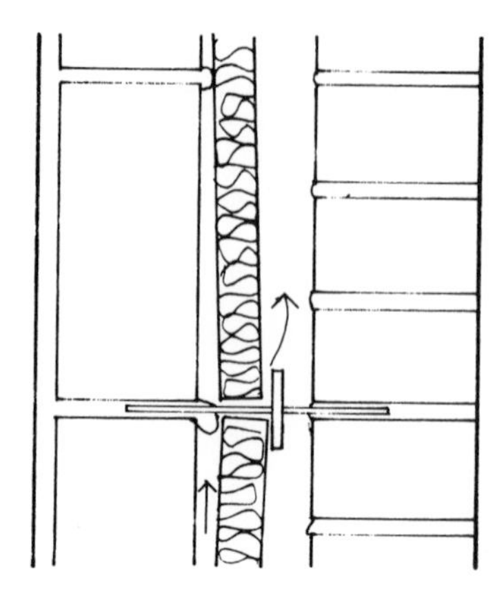

Fig. 10.6 Air circulation behind partial cavity insulation.

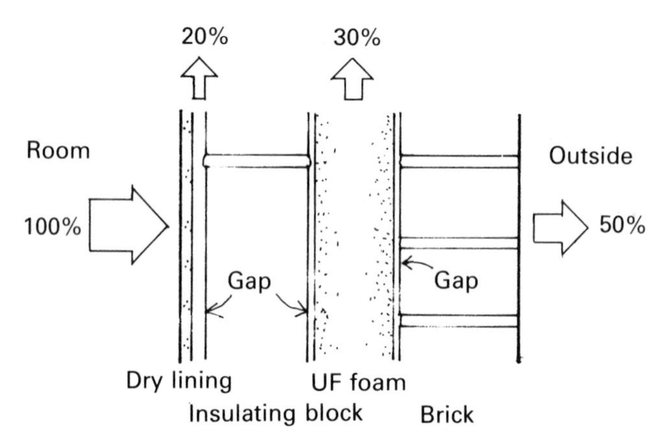

Fig. 10.7 Heat loss destination through a filled cavity wall.

corners, which are not properly supported. The whole process can be made worse by 'cutting' the insulation with a trowel.

This standard of workmanship is common and probably ensures that in the worst cases, little thermal benefit is gained from the insulation material. An indication of the deleterious influence of even a very narrow cavity is that the fraction of the measured heat loss due to the narrow shrinkage fissures found in a foam-filled cavity wall was 30% of the total loss through the wall (Siviour, 1982) (Fig. 10.7) (see section 10.3.3.4). UF foam has an ultimate volumetric shrinkage of about 8%, giving a 2 mm shrinkage gap, on each face of the cavity filling.

Two aims need somehow to be achieved: to fix the insulation hard against the block face, and to make the insulation continuous. Much better control during site installation would improve matters, but further development is needed to ensure the close fit of the insulation to its backing. One alternative is to use foam faced blocks, where the insulation is fixed in the factory. Unfortunately more than

7% of the finished wall face will be an uninsulated mortar joint (see section 10.5.1). Foam-filled hollow blocks are another alternative, but here again thermal bridging by the mortar is extended by the concrete bridging in the blocks themselves although manufacturers make allowances for this in their thermal conductivity data.

10.3.3.3 Timber

Timber is an organic material which, in its dried state, is subject to moisture movement. The amount of movement depends upon ambient humidity and temperature.

Structural timbers, when delivered to site, may well have a moisture content of about 22% by weight, particularly if they have been preservative treated. When built into a house the moisture content will fall to about 16%. Internal joinery is drier and will attain an equilibrium moisture content of about 12%, or lower if high internal air temperatures are sustained in winter. Drying shrinkage across the grain in softwoods is about 1% for every 5% drop in moisture content. Thus a 175 mm deep joist is likely to lose 2 mm in height by natural drying after being built in. This will, for example, further increase the draught past the joist ends in Fig. 10.4 by 15%.

Suspended ground floor timber joists rest on their supports, so that shrinkage will cause the top of the joist to drop and a tendency for gaps to open around the floor perimeter. This will occur between the floor surface and the skirting board, which itself shrinks upwards, thus expanding the gap. A solution here is to fix a draught seal, such as a continuous strip of a flexible sheet material, as illustrated in Fig. 10.8.

Window and door frames, even if built in during construction, will probably have air leakage occurring around their perimeter. Externally, at jambs and head, a mastic bead is often applied, to seal frame to brickwork, but the characteristic needs of a successful mastic joint design are not generally understood. The joint often fails by parting.

An important source of draught is likely to be below timber window sills (Fig. 10.9), for a number of reasons:

(1) It is rare to stoop and examine under a sill. The frame is generally dry mounted onto brickwork.
(2) Inside, a wood-based window board is quite likely not to have an architrave.
(3) Dry assembly. Timber drying shrinkage, and often the additional drying affect of a radiator, cause the sill assembly to open up so that daylight can be seen below it. This is where the stoop and look test is worthwhile.

Be suspicious of timber bay windows and dormer windows. There is plenty of scope for omitting thermal insulation. Usually the framing size is insufficient to allow full insulation. More importantly there is potentially a very long footage of crack between adjacent timbers and frame to wall.

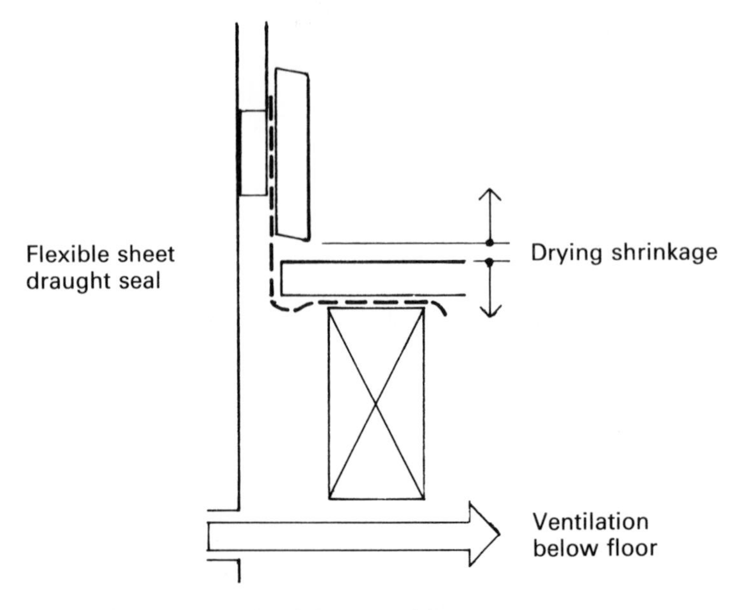

Fig. 10.8 Draught sealing a suspended timber ground floor.

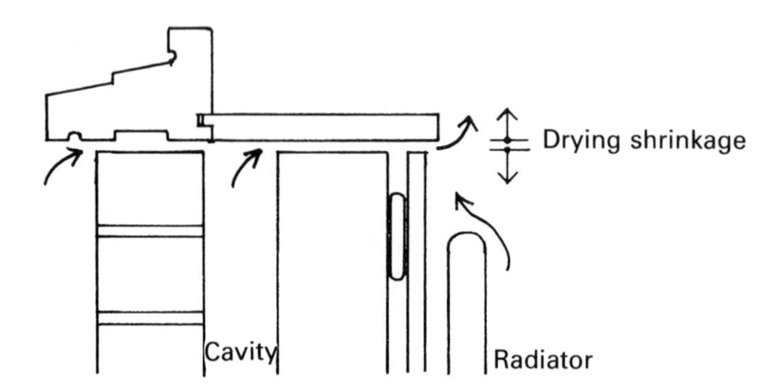

Fig. 10.9 Draught routes below a window sill.

The cracks which open up around the perimeter of window and door frames may be only 1 mm wide, but the total perimeter length around them all together may exceed 50 m. This aggregates to 5000 mm^2, or a hole 70 × 70 mm – not likely to be welcomed on a frosty day! Perhaps we should consider the establishment of a new trade, that of the draught sealer.

Timber is extensively used in house construction. Traditionally built houses may be viewed as a timber framed house in which the timber framed walls are replaced by masonry, but there are notable thermal advantages in having a completely timber framed house. They are:

(1) The whole structure is to a large extent factory made. Maximizing the amount of construction in the factory offers good control on its quality.
(2) The wall insulation is completely encapsulated and therefore performs efficiently.
(3) The walls and roof need to be lined with continuous vapour barrier to block the passage of moisture from the house interior into the interior of the structure where it could be trapped as interstitial condensation. The vapour barrier also acts as an air barrier, giving potentially excellent draught proofing as well.
(4) Timber, while acting as a thermal bridge, does possess significant thermal resistance, so that condensation is unlikely. Bridging is described in section 10.4.4.

10.3.3.4 Dry lining

A dry lining of plasterboard is justly popular. It takes the place of wet plaster and removes the burden of drying out from the house. However, the two materials have other differing characteristics. Wet plaster keys into a wall and draught seals it (except for the unplastered perimeter strip at first floor thickness where fissures and draughts persist).

Consider the building process where there are masonry walls. The blockwork has a rough surface and is a little irregular, so as a result there is an average gap of, say, 3 mm between the edge of the ceiling plasterboard and the external walls, giving an equivalent total void area of 300 mm × 300 mm at both ground and upper floor ceiling level.

Adhesive mortar dabs are trowelled onto the wall and the plasterboard offered up, to fit snugly with the ceiling. The wall board size being of necessity less than the room height, the plasterboard has a generous gap at the bottom, not effectively sealed later by the skirting. The omission of some vertical mortar joints in the inner leaf of the wall is described in section 10.3.3.1 and the ventilated cavity behind any partial cavity insulation in section 10.3.3.2. This setting of a perforated wall with a ventilated cavity on each side, easily able to extract warm room air and to supply cold replacement air results in uncontrollable passive ventilation. The ventilating air will travel by complex routes, driven by buoyancy and wind forces (Fig. 10.10).

In windless conditions, such as a foggy day, warm room air will drift upwards through any opening, while denser outside air flows in to replace it. Wind pressure on one face of the house and corresponding suction on the other drives air horizontally through the house. This occurs, for example, through the first floor joist void, where it can readily divert up or down into the rooms, in the loft and below any suspended ground floor (Fig. 10.11).

Wind driven convection, however, may be confined to the wall thickness, not entering the rooms. Imported cold air can still chill the back of the dry lining plasterboard, drawing heat from it and in turn from the room resulting in increased heat losses (Fig. 10.12).

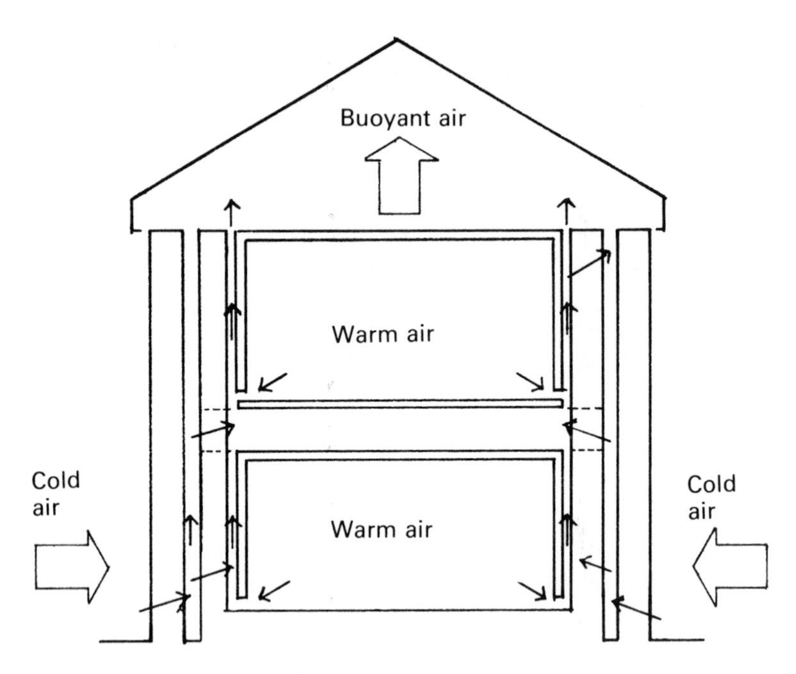

Fig. 10.10 Ventilation by warm air displacement.

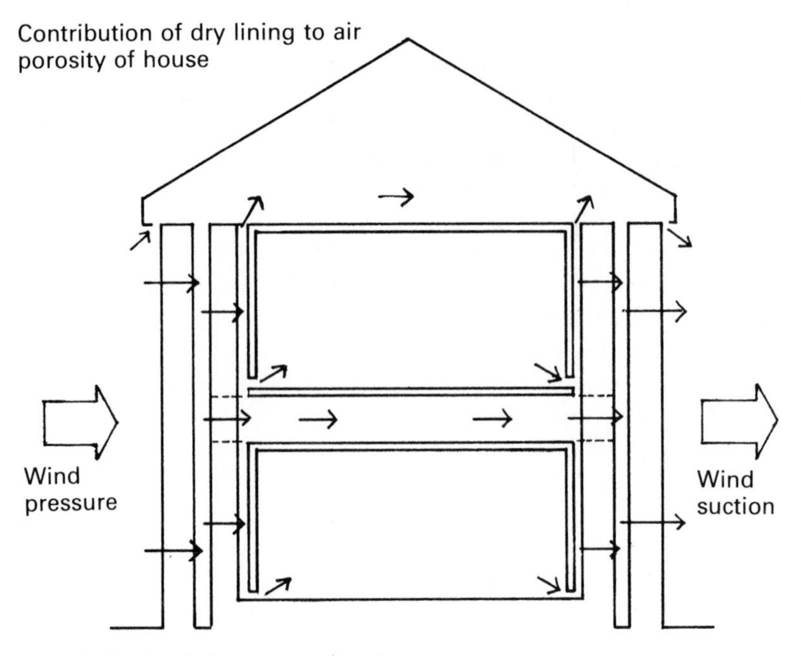

Fig. 10.11 Ventilation by wind pressure and suction.

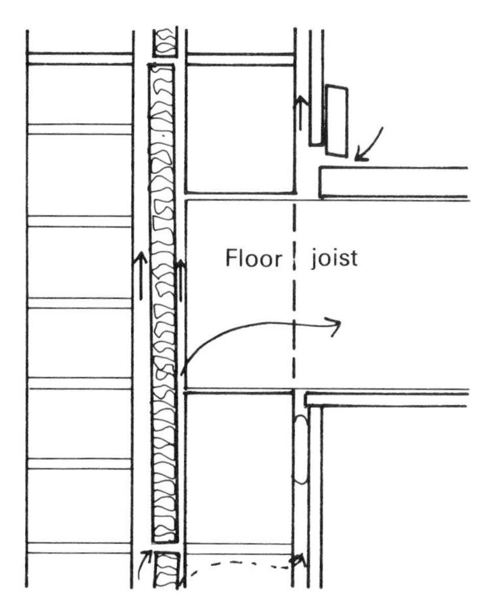

Fig. 10.12 Draught routes associated with dry lining at first floor level.

These and previous examples show that air can readily enter and leave cavities associated with a wide range of building constructions, particularly those associated with dry lining. Obviously such uncontrolled ventilation can vary in amount, but standard allowances are included in energy rating procedures for a variety of building elements, such as an unfilled cavity, the height of the house and various types of window and draught sealing.

Investigations suggest (Siviour, 1989) that 0.2 W/m^{2}K should be added to the U-value of a traditional cavity wall which incorporates dry lining. One manufacturer responded to fieldwork on the thermal penalty of dry lining by developing installation techniques for edge sealing the cavity behind their plasterboard dry lining. Their revised technical literature now needs to be assimilated by the trade and designers alike. However, the plasterboard manufacturer cannot be responsible for open mortar joints in the wall, which will still lead to cooling of the hidden face of the plasterboard.

One clear message from this section is that no cavity should be created on the warm side of any thermal insulation, unless it is completely and dependably sealed against ventilation.

10.3.3.5 Services

There are numerous services which penetrate the surfaces of rooms and so potentially lead to added ventilation. They are installed by tradesmen, subcontractors or by public utilities, none of whom are normally concerned with considerations of energy conservation or even making good the hole they drive through the fabric of the building. A bricklayer does not go back onto site to make good the results of a bolster and lump hammer attack on one of his walls. The damage is done long after he has left the site and only visible damage is covered over. Normally only the plaster surface is made good, and that may not be a durable repair, as it may shrink and fall out.

The following service installations both leave and re-enter room surfaces:

- Water pipes.
- Gas pipes.
- Electricity cables.
- TV aerials.
- Telephone cables.
- Security cables.
- Heating controls.

The following exit only from room surfaces:

- Flues and chimneys.
- Plumbing wastes and soil vent pipe.
- Fan ducts.
- Air bricks.

Prefabricated meter boxes built into outside walls occupy much of the wall thickness. Severe air leakage tends to be found where they connect with an internal consumer unit or pipework as well as into wall cavities. An additional factor is that they act as cold bridges, since they often reduce significantly wall insulation thickness.

Experience has shown that there are particular problem areas associated with passive air leakage around services. They are ones which recur frequently and ones which builders are reluctant to change. One area is that of sink units. They often contain the incoming water service pipe and stop tap. This sometimes rises through a suspended ground floor, bringing a cold air supply with it, so that cold air fills the cupboard plinth.

The sink waste normally exits through the external wall. Here again holes in the structure, connecting with the wall cavity are not made good, usually because the sink unit has a back which is slotted over the waste and fixed into position. Thereafter there is fresh air leakage under and around the sink unit which is impossible to seal in any reasonable way.

In the bathroom a similar situation prevails. Here though the plumbing holes connect the floor void with the bathroom, via the bath panel. The bath is sealed around its top rim but below this level the void under the bath sometimes connects with a horizontal duct carrying the waste from the basin and the WC. There is thus likely to be a direct connection between the bathroom itself and the duct, the underside of the bath, the floor void, the cavity in the external wall and so to the outside air.

A greater and more far reaching effect is associated with an internal soil vent

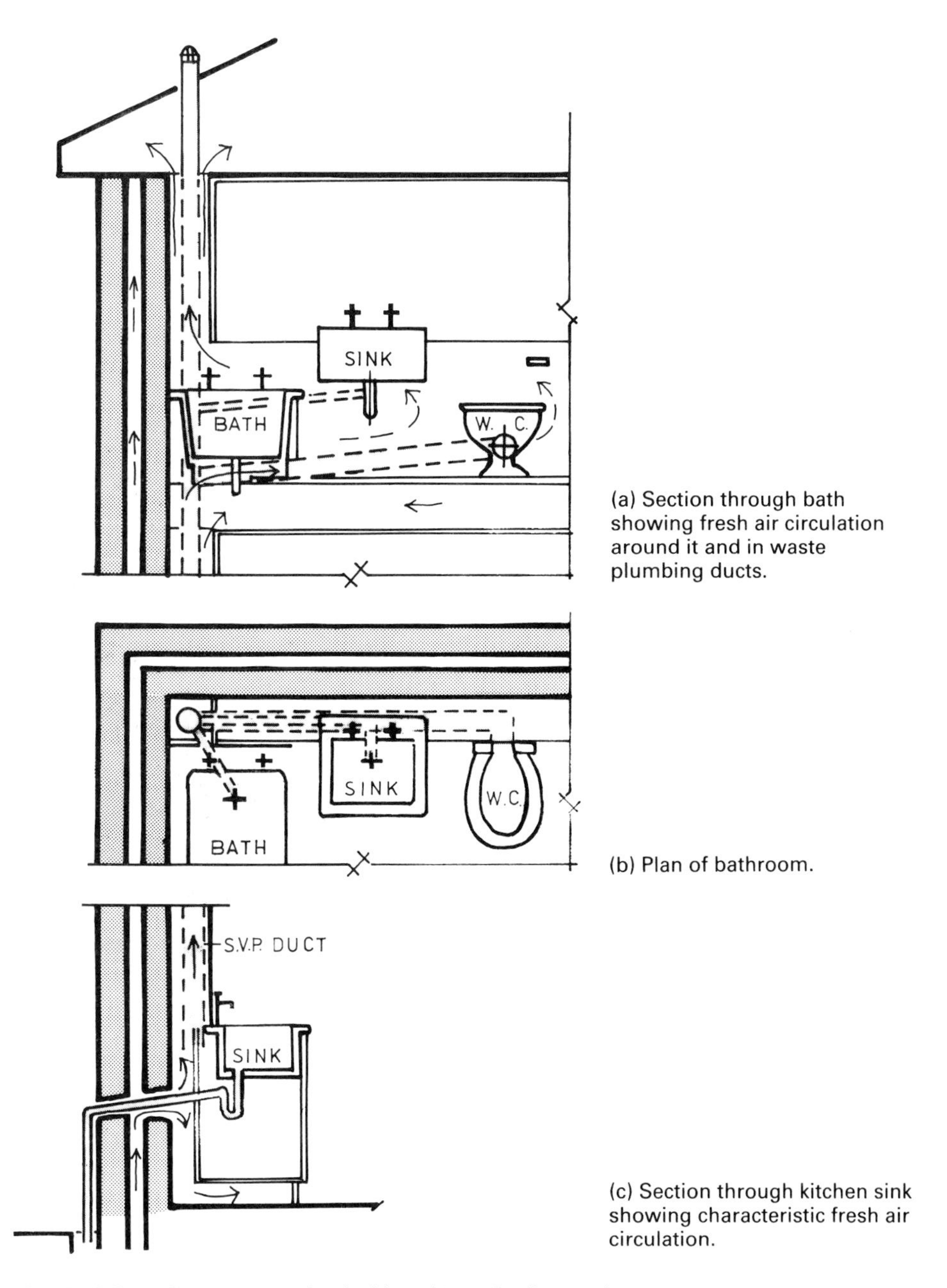

Fig. 10.13 Draught routes associated with an internal soil vent pipe.

pipe (Fig. 10.13). This is usually installed after the ceiling plasterboarding has been fixed. Consequently generous holes are swiftly made into the loft by the plumber. Later, where it is visible, the soil pipe is boxed in.

The air volumes thus interconnected are likely to be the cupboard units in the kitchen, the first floor void, the bathroom, via the waste branches, and the loft. If there is a suspended ground floor the void under that will be added, too. What is then established is a really efficient way of continuously ventilating the house by means of a vertical open chimney.

The air around a soil pipe tends to connect with the space behind any dry lining to the walls, as this is likely to be fixed after the soil pipe is in position, so that the plasterboard stops short and is hidden when the pipe is boxed in. A further variant is where the soil pipe descends through an internal garage, where there is another set of air leakage conditions.

The air leakage examples described here which are associated with house services are very difficult to draught seal satisfactorily after the services have been installed. A great deal of time and cost can therefore be involved, with little chance of complete success. New techniques need to be evolved to overcome such air sealant, but it is costly and will certainly not overcome the main leakage areas at their source. Awareness of the problem is the first step in education. Air leakage can be demonstrated vividly by using a smoke source in conjunction with a pressure testing kit. This is described in section 10.6.2.

10.3.3.6 Radon

In some areas of the United Kingdom radon is now a matter for serious concern. The problem is that of airborne radioactive particles which rise through the subsoil from igneous rocks. There are many small fissures into and within the external walls of a house, as there are also through the various forms of suspended ground floor. Concrete floor slabs cast onto the soil will have a perimeter shrinkage crack where they abut load bearing walls and probably have small air paths through the slabs themselves as well. All such cracks could act as routes for air to enter the dwelling, entraining the minute radon particles within them.

There are a number of possible strategies to arrest this airflow up from the soil. The main need is to create a durable and effective air barrier at ground floor level which extends to the outside of the external wall. This might be augmented by a thoroughly ventilated air space under the air barrier. Any necessary penetrations through the barrier need to be sealed, the most difficult ones being the services. It should not be too difficult to design and test a range of solutions. New techniques are needed. A plastic drain or soil pipe, for example, is subject to frequent thermal movement, while water pipes must be protected from frost.

As with any ventilation barrier the difficulty lies in developing simple and reliable test procedures. On completion, tests can be carried out to search for radon particles, but the tests might take the length of a heating season to complete. The designer cannot know in the short term whether his design and the builder's endeavour have been successful. A robust and perhaps duplicated barrier is needed.

In the parts of North America where natural radon levels cause concern, house sales are dependent upon negative test results for the presence of radon in the house.

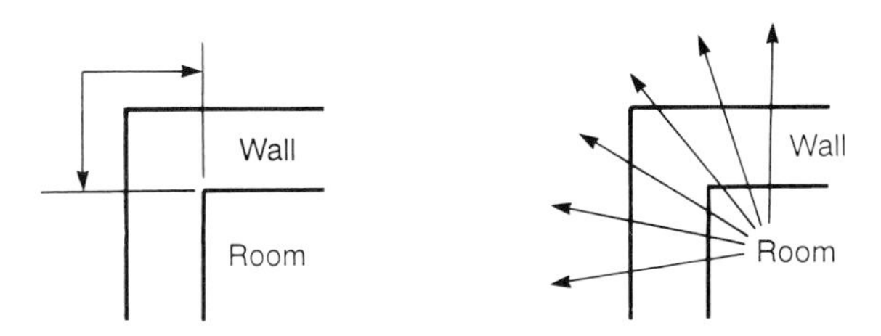

Fig. 10.14 Corner effects: greater surface area results in greater heat flow.

10.4 Thermal bridging

There are numerous instances of thermal bridging in buildings. If a thermal bridge is defined as an area of material with a thermal transmittance higher than that of the material adjacent to it, then a double-glazed window in a well-insulated wall is a thermal bridge. Another way to look at thermal bridging is to consider room surface temperatures; where the temperature drops there is a bridging effect. Generally, wall surface temperatures fall as a window is approached. This phenomenon is pursued later, in sections 10.4.2 and 10.4.3 and is illustrated in chapter 11.

Surface temperatures also fall in the corner of a room, on an outside wall, and at the junction of wall and ceiling at eaves level (Fig. 10.14). In the case of the wall it is because the external surface area is greater than the internal one, so that there is a greater cooling effect at the corner. The solution is always the same – use thicker or more effective insulation, but this is seldom practicable.

Interestingly, if internal surface condensation does occur, it is likely to be found in these two example areas, as well as on any single glazing present.

10.4.1 Mortar joints

Thermal bridging by mortar joints was mentioned in section 10.3.3.1. The 1990 Building Regulations (Approved Document L) permit the designer to ignore the thermal effect of 'thin discrete components'. Mortar bedding is listed as one such component.

A contrary view is expressed in the CIBSE Guide (Section A3–9), which says that 'the effect of mortar joints, which can account for 5% to 20% of the area, should be considered in all cases'. This is a more prudent view. Consider the choice of insulating blocks for the inner leaf of a cavity wall. As the block thickness and density is changed to obtain better insulation, the mortar remains thermally constant. As the U-value of the wall is reduced the thermal effect of the mortar becomes increasingly important. Insulating mortar is available and has only a quarter of the thermal conductivity of the ubiquitous cement mortar (Fig. 10.15).

Type of mortar	λ W/mK
Cement	0.8
Cement/lime	0.8
Insulating	0.2

Fig. 10.15 Thermal conductivity of three mortar types.

10.4.2 Reveals

Window and door reveals are likely to be subject to a variety of cooling effects, so extra consideration is needed by the designer and extra care taken by the builder.

The first consideration is the position of the window or the door frame within the wall thickness. The nearer it is to the outside of the wall the greater is the likelihood of thermal bridging problems. In a masonry cavity wall probably the most practical, while thermally effective, solution is to use a sub-frame to close the cavity and to accept that a wider sill is required on the outside. At present, with standard window frames, the position of the window plane is governed by the width of the sill incorporated into the frame. Any draught sealing between the frame and the wall is an optional extra.

The design of building components such as window frames is optimized for manufacture and cost, whereas the need to link effectively with adjacent components should enter the equation. A window frame put into a prepared opening with a generous tolerance and fitting gap means that designer and builder need to create and accomplish a thermal, draught and weatherproof barrier in the gap. A flange fitting, such as a sub-frame, would be one method of fulfilling this requirement.

BRE Report 143 (1989) sets out the thermal bridging problem at reveals, and also the need for thermal insulation when the frame is set forward. In the Appendix a method is suggested for calculating a limiting U-value of 1.2 W/m^{2}K at the reveals, the same as the limiting value for lintels.

It is now felt that a more thorough approach is needed and that the limiting value should be lower. The calculation method offered does not, for example, include an allowance for ventilation behind dry lining, with its potentially vigorous cooling effect. In summary, a number of features (apart from lintel design) need to be addressed at reveals, as follows:

(1) The return of the inner leaf to close the cavity introduces a thermal bridge. Instead, retracting the window frame to seal the cavity should be considered.
(2) Ideally the wall insulation should be continued, to abut the frame. Added internal surface insulation is an alternative method of reducing the bridging effect.
(3) Effective draught sealing measures should be taken, both around the perimeter of the frame and within the construction adjacent to the reveals.

Material	Density kg\m^3	Thermal conductivity, λ W\mK	$\frac{\lambda_{mild\ steel}}{\lambda_{material}}$
Mineral wool	25	0.04	1250
Softwood	610	0.14	357
Aircrete blocks	480	0.14	357
	650	0.19	263
Cement mortar	1760	0.8	63
Window glass	2500	1.05	48
Dense concrete	2200	1.5	33
Stainless steel	7700–7900	15–25	3–2
Mild steel	7800	50	1

Fig. 10.16 Comparison of heat flow through mild steel with heat flow through other materials.

(4) Inner surface temperatures will fall at the reveals, to an extent dependent upon the magnitude of the bridging effect.

10.4.3 Lintels

Although there are various kinds of lintel, two configurations in steel are widely used in masonry walling.

In timber frame building (used at present in 7% of new houses in England and Wales and 42% in Scotland) thermal insulation thickness is reduced above openings to allow for additional timber sections to span the opening, thus increasing the rate of heat loss at this point and reducing corresponding internal surface temperatures. In masonry construction of houses the hollow steel section is used, whether box or 'T' sections.

There is scant technical data on U-values through the type of lintels currently in use in new housing, although one reference (Chadderton *et al.*, 1981) gives a U-value of about 2.0 W/m^{2}K for a range of types of lintel, other than timber – almost twice the permitted limit in the Building Regulations.

It is instructive to compare the thermal conductivity values for the materials commonly used to build up the wall thickness at the lintel position. They are compared in Fig. 10.16 which shows, for example, that the rate of heat flow through softwood is three times that through an equal thickness of mineral wool, but that the heat flow through mild steel is more than a thousand times as much as through the fibre insulation. These figures suggest that steel lintels will constitute a severe thermal bridge unless insulation is interposed between the steel and the room air, as is shown in the BRE Report 143 (1989). Even filling steel lintel sections with insulation will produce only a limited thermal improvement, chiefly that of draught sealing those box lintels whose otherwise open ends project into the cavity.

The obvious need is to introduce an effective thermal break into the steel, but

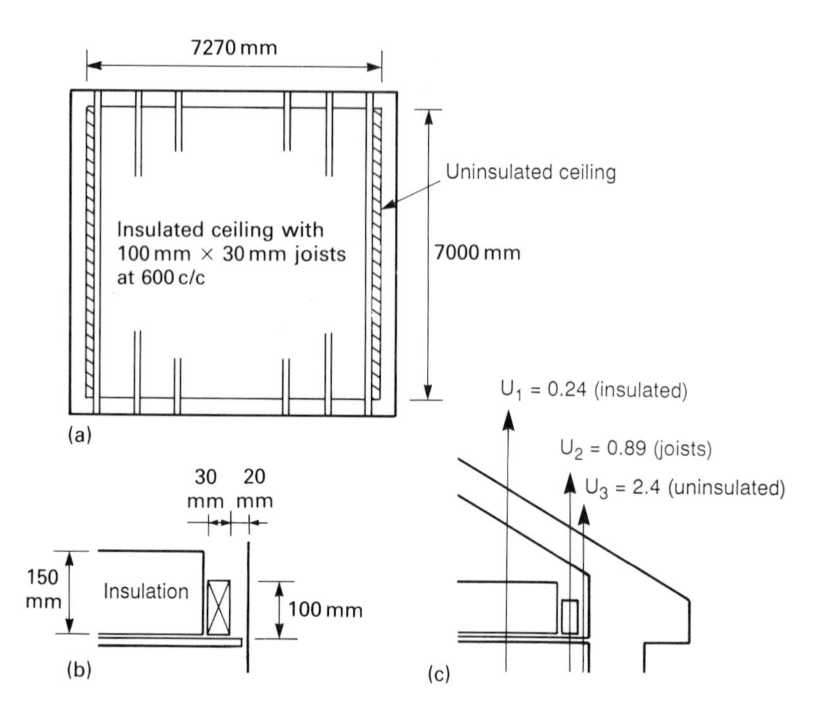

Component	Area m^2	Fraction of whole ceiling area	U-value W/m^2K	Heat loss W/K	Fraction of total actual heat loss
Net insulated area	47.89	94%	$U_1 = 0.24$	11.5	78%
Joists	2.73	5.4%	$U_2 = 0.89$	2.4	17%
Uninsulated area	0.28	0.6%	$U_3 = 2.4$	0.7	5%
Whole ceiling including bridging	50.9	100%	$U = 0.29$ average	14.6	100%
Whole ceiling assumed insulated to $U = 0.24$	50.9	100%	$U = 0.24$ assumed	12.2	84%*

* Actual total heat loss $= \left(\frac{14.6 - 12.2}{12.2}\right) 100\% = 20\%$ greater than assumed total heat loss.

Fig. 10.17 The effects of thermal bridging through a ceiling below a pitched roof.

this means introducing two structural elements instead of one. This in turn means that support is likely to be needed for the cavity gutter. However, it does mean that the recess formed at the base of the cavity, if the lintel is divided into two parts, could be utilized for housing a sub-frame or some other means of sealing the head of the window frame to the soffit of the window opening.

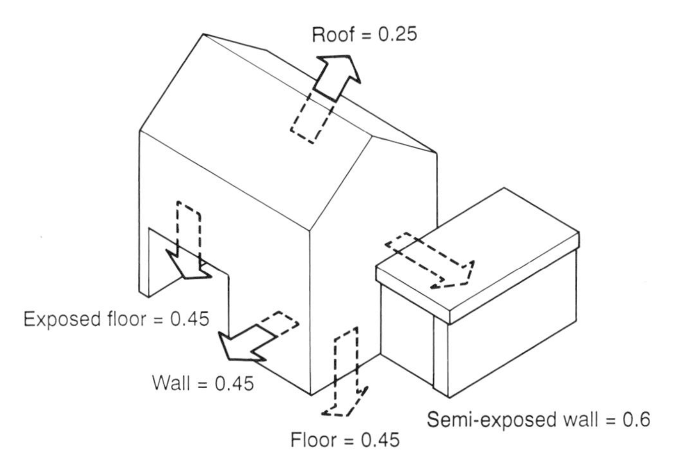

Fig. 10.18 Building Regulation U-value requirements for dwellings (1990).

10.4.4 Timber

A technical note is available which illustrates numerically a variety of examples of thermal bridging by timber and other materials (Siviour & Mould, 1988). In the note simplified examples are given to demonstrate typical overall thermal effects due to bridging.

Figure 10.17 is based upon a domestic pitched roof, some 7 m square. The bottom members only of the trussed rafters are shown (as ceiling joists) 100 mm deep. In this example these joists occupy 5.4% of the total ceiling area. In reality timber occupies 7% or more of the whole area.

The space between the joists is insulated to the present Building Regulations requirements of U = 0.25 W/m^{2}K (Fig. 10.18). However calculation indicates that only 144 mm thickness of glass fibre is needed to achieve this value, whereas the nearest standard product thickness is 150 mm. This greater thickness has therefore been assumed. This gives a calculated U-value of 0.24 W/m^{2}K. In reality two thicknesses of insulation should have been used, 100 mm thickness to fit snugly between the joists and a 50 mm thick mat to be laid on top, across them, so that no cavity is left below the insulation.

Below the end joists and the wall there is always a gap, here shown as 20 mm wide. As is also common practice, this 'insignificant' space is not insulated. Three U-values have been worked out for the three ceiling conditions. The uninsulated, tiled, pitched roof has a U-value of 2.4 W/m^{2}K.

Examination of Fig. 10.17 will show first, the fraction of the total ceiling area occupied by the joists and by the gaps, and second their disproportionate thermal effect on the total actual heat loss. The third, and most important effect, is that by including these thermal bridges in the calculation the actual total heat loss is increased by 20% above the heat loss calculated when ignoring them.

10.5 Confidence in thermal insulation

There are a number of considerations which will influence judgement on the effectiveness of thermal insulation in a building. These fall into three areas; design, building and use.

10.5.1 Design

(1) Is the heat loss calculation (within say a +/− 15% tolerance), a true reflection of the likely actual heat loss of the building through an average heating season?
(2) Is the insulation specification sufficiently robust and detailed? Is the specification really adequate?
(3) Can the design intentions be fulfilled using current building practice? Is the design buildable?
(4) Can all of the insulation work be inspected at its critical stages? Will it be?
(5) Will it be possible to upgrade the insulation when standards are raised in the future?

10.5.2 Building

(1) Are the insulation products specified suitable for their purpose?
(2) Are they sufficiently robust?
(3) Can the insulation be delivered and stored without detriment or are special conditions necessary?
(4) Can the workforce install the insulation within the terms of the manufacturer's and designer's specification?
(5) Will following trades damage the installed insulation?
(6) Is there sufficient knowledge and experience available on site to give effective and authoritative instruction and supervision when it is needed?
(7) Will weather conditions during construction impair the quality and subsequent performance of the insulation?

10.5.3 Use

(1) Will the building need to dry out before the insulation is fully effective?
(2) Will drying out result in drying shrinkage and unwanted ventilation?
(3) Do the occupants of the building need to be instructed in the operation of the building to achieve the designed thermal performance?
(4) Will the insulation last for the life of the building and will it continue to perform as designed for that time?
If not can it be readily replaced?

10.5.4 Further considerations

Looking back, we can see distinct changes which have occurred in house design and construction. We can also see some of the difficulties and high cost of dragging old houses up to be near to modern standards of performance, and the penalties for not doing so. Some of the difficulties arise from the lack of provision for replacing components, such as windows, or for increasing insulation thickness.

The time between acceptable performance and the call for improvement is not long – up to 20 years. An example is that of mandatory U-values for walls. They have dropped to one third of the original level in the last 13 years. Public expectation changes. Satellite dishes have now joined the other two kinds of aerial. Following on after double glazing, whole house ventilation systems might also be peddled by door-to-door salesmen. The relative cost-in-use of the various fuels in well insulated houses might change, and with it householders preferences. A newly-discovered source of pollution in houses might lead to a demand for its excision, perhaps leading to structural disturbance?

Pipes, cables and heating appliances have a limited life and so need to be replaced. Replacement might also be necessary because spare parts are no longer available. Rewiring, two or three times in the nominal 60 year house lifespan, could well lead to damage to the structure, perhaps resulting in damaging interstitial condensation and excessive ventilation. In the UK we do not run house cables in conduit, so enabling them to be rewired simply and without damage to the building. The life expectation of some building components is not great. In commercial buildings seals in some double glazing systems are breaking down in less than seven years.

As has been shown already the historic trend is for thermal insulation levels to be progressively improved. While the accelerating pace of lowering U-values cannot go on, there is still ample scope for increasing levels of insulation and more importantly, of air tightness. There remains also the huge task of improving the existing housing stock. The easiest and cheapest surface to insulate, or to reinsulate, is the ceiling below a pitched roof. This is the area which gets priority in new housing, leaving the difficult and expensive areas to subsequent rehabilitation contracts.

Many of our existing houses, although thermally up-to-date when they were built, are now substandard and unnecessarily costly to run. Should we not now build new houses in a way which facilitates upgrading in the future?

10.6 Inspection, testing and calibration

Throughout this chapter the aim has been to explore the subject of thermal insulation in a way which identifies the impact of critical features of design on installation.

Checking of insulation should begin with the examination of unfixed materials because insulation is generally not robust, so it is prone to damage, even before being built in. Check that the insulation is of the correct type and grade, that it is undamaged and stored in the correct manner, with appropriate provision made for weather protection. Insulating blocks, for example, should not be allowed to become saturated before building-in. If an insulating mortar is specified, consider taking samples for verification by the manufacturer. The perimeter surface of the house needs to be insulated completely. This includes any wall (or floor) which is a boundary between the house and a garage or porch.

Insulation by cavity filling takes several forms. All of them depend ultimately on the carefulness of the workmen for success. The BBA certificate sets out installation procedures. Blown materials should be installed at their specific density and areas of walling should not be left unfilled. Awkward areas of cavity might be left unfilled, such as the wall behind an open porch roof, or above upper windows, or at the tops of gables. Migration of insulation into cavity party walls after the external wall cavities were filled has been known and resulted in the upper parts of the facade being devoid of insulation, resulting ultimately in customer complaint.

The particular problems of achieving effective partial cavity insulation have been described in detail. The insulation material needs to be of the correct type and thickness and be in close contact with the inner leaf. This is the most difficult aspect in which to achieve consistency. The insulation panels must be tightly abutted to each other and, particularly at corners and at reveals, there should be no insulation omitted. Continuity of insulation should be the aim.

The fixing of partial cavity insulation panels is prone to abuse. The joints of both blocks and insulation panels need to course, so that flanged wall ties can be inserted on the top edges of the panels. Bashing ties through the panels or chopping with the trowel are hallmarks of poor quality workmanship and supervision.

Apart from the tests described later, under certain climatic conditions it is possible to see indications of insulation omission in walls. In frosty weather it is sometimes possible to see frost on the walls where there is insulation contrasting with frost-free patches where there is thermal leakage from within. Another phenomenon sometimes occurs with wind-driven snow. The snow adheres to the insulated areas of wall but melts where there is heat leakage.

If specified, check that insulation has been applied correctly to reveals and soffits and that hollow lintels have been fully insulated. A general requirement is that there should be no air cavity on the warm side of the insulation where there is any possibility of leakage ventilation by outside air. An unsealed cavity behind dry lining is an example of this, while another is the void below the ceiling insulation, if the insulation is simply draped over the joists. It should be a snug fit onto the top of the plasterboard.

Check that certain services are insulated and that others are not. The hot water cylinder needs to be thoroughly insulated, preferably not with a petal jacket. Primary hot water circuits should be insulated. Electrical cables, particularly those carrying high currents, should have at most, a minimal length of travel through thermal insulation. Cables in use heat-up. A surround of thermal insulation prevents the normal cooling effect of free air and will lead to a temperature rise in the cable.

Cold water service intakes must be insulated until they emerge into room air. Even then, thermal insulation might be a wise provision, to prevent perhaps severe condensation on the cold pipe.

Any air duct in the loft, or other cold area, carrying exhaust air or room-to-room circulation needs to have a minimum thickness of 50 mm of insulation around the duct continuously tied. The aims are to prevent excessive heat loss in circulating ducts, and the formation of condensation in extract ducts. The loft hatch needs insulation – as much as is practicable.

Unfamiliarity with the new insulation requirements means that ground floor insulation should certainly be inspected before it is covered up. Continuity between any edge insulation upstands and underslab insulation should be watched. Remarks about ceiling insulation in the loft apply equally to insulated timber floor joists. They differ only in being inverted. Concrete floor joists with spanning insulation between them are liable to be damaged by wheelbarrows before the floor surface is laid.

10.6.1 Testing

There are two reasons for seeking to check on the thermal qualities of a house. One is to reassure the designer that his predictions are likely to be met. The other in an occupied house, is to validate or refute a complaint, such as the occurrence of condensation, or that heating costs are excessive. Visual inspection, for example, using an optical fibre instrument to confirm the presence of thermal insulation in a cavity wall, is an initial test.

Tests in response to a complaint aim to discover the cause of any aberration in performance and if so, to allocate blame. It may be that the claimant was trying to shift attention from another cause, such as the inability to pay fuel bills, so consider the background to the complaint before spending money searching for a cause. There are several tests which can be applied to houses to assess their thermal rating. The fairly quick and simple ones are pressure testing and the use of an infra-red camera.

10.6.1.1 Pressure testing

The pressure test equipment consists of a fabric screen sealed into a perimeter framework. The frame is adjustable in size, and fills an open external doorway, where it is clamped to the door frame. A powerful variable speed electric fan is fitted through an aperture in the fabric, together with devices to measure inside and outside air pressure and air flow through the fan. All other external doors and windows are closed. Extract fans, waste plumbing and flues are sealed.

The internal volume of the house is measured and the fan speed varied to give relationship between air change rate and internal pressure. The standard maximum air change rate is 7 ac/h at 50 Pa. At this pressure a higher air change rate means that the house is too leaky. Anything less is a bonus. The test pressure is low and equal to about 5 mm of water gauge (a mere fraction of a child's ability with a drinking straw!), so that even a moderate outside wind speed can exert a considerable effect on the results. Selective sealing of sources of draughts will enable a degree of calibration to be made of individual major leaks.

New, empty houses built in winter with wet internal finishes will be well sealed because the woodwork will be swollen. At the end of the first heating season the house is likely to be in its most leaky state, as it will have dried out and the components shrunk in size.

An experienced person is needed to run the test and to interpret the results. The equipment can be hired, or purchased for between one and two thousand pounds. The higher price includes a small computer to process the readings and to print out the results (Dickson, 1989).

10.6.1.2 Infra-red photography

The infra-red camera shows images of surface temperature, although they may not relate to the pattern of heat flow out through the underlying structure.

Assuming that the house is to be examined externally there needs first to be a significantly higher temperature inside than out, and to have been in operation for at least twelve hours, to establish a sufficient heat flow to generate a discernible picture. The internal air temperature needs to be at least 10 K, above the temperature outside for the majority of existing houses. As insulation thicknesses have increased, so heat flow out through the structure has decreased. Therefore in new, well insulated houses a temperature differential of at least 20 K is needed to generate a clear picture.

In use, the IR camera can quickly give an indication of the general heat flow pattern, together with any local areas of thermal bridging. Any significant lack of insulation, including the omission of partial cavity fill, can usually be seen quite clearly. An IR camera is very expensive, costing some £40 000, and is complex to operate and to maintain. One or two manufacturers of insulating materials and some of the fuel companies, have one at their disposal for helping their customers resolve any serious misgivings about possible short-comings of building insulation. These companies are also likely to have a member of their staff expert in dealing with problems of thermal insulation.

10.6.2 Calibration

Whole house calibration involves skills and techniques which are outside the scope of this chapter. It could be used if it is necessary to obtain accurate information in a particular case or to validate a new house design which will be repeated extensively. Calibration takes time to complete, the length of time depending on the type of information required and on the prevailing weather conditions. A prerequisite is an unoccupied house.

An airtightness test may take as little as one hour to complete, whereas ventilation testing will take a few days, or more if calibration is required over a range of windspeeds, wind directions and temperatures. Whole house thermal calibration will take one week or more and is best done in mid-winter, to maximize the difference between indoor and outdoor temperatures. Greater accuracy, and the separation of solar heat gain would take about six weeks. Quite a range of instruments are required together with electrical power. Some information can be obtained simply from meter readings and temperature recordings over a period of time. This will yield data on total energy consumption and to some extent space heating consumption.

Instrumentation can range from a simple clockwork recording thermometer to check room temperatures, to a remotely controlled and recording thermostat as permanent equipment. (such as is available from Delcomm Limited of Fife (see references for full address)). This type of thermostat can store and off-load data for subsequent computer analysis. The impending 'intelligent house' systems and controls are likely to make energy use scrutiny much easier and more accurate.

10.7 Conclusions

Available knowledge strongly suggests that many new houses, as well as older ones, are to some degree thermally defective. If this is so then their designers are potentially liable for the shortcomings. More precise and independent estimates of thermal performance are now becoming available to the public, against which actual performance can be compared.

Increasing insulation levels lead to increases in the significance of:

(1) Common failings in insulation installation.
(2) Thermal bridging.
(3) The adverse effects of uncontrolled air leakage.
(4) Any mismatch between space heating output and house heat loss.

More education, understanding and on-site inspection are needed to improve product quality.

More research is needed to evaluate the important effects of air movement within the thickness of completed structures and to evolve standard thermal transmittance tests of complete assemblies. These tests should be more representative of actual building practice and conditions than the laboratory tests on small samples used at present. Building component design also needs more development so that components integrate better with each other to produce a more uniform thermal resistance with a minimum of site operations.

References

Building Research Establishment Report BR143 (1989) *Thermal insulation: avoiding risks*, HMSO.

Chadderton, D. V., Mullard, R. J., Brook, S., Green, V. C. U. & Rogers, A. R. (1981). Heat losses through steel and concrete lintels for housing, *Building Services Engineering Research & Technology*, **12**, No. 1.

Delcomm Limited, Battery Quarry House, Forthside Terrace, North Queensferry, Fife, KY11 1JR.

Dickson, J. D. (1989). *House tightness – the fan pressurisation test*, published by E. A. Technology Ltd (formerly the Electricity Research and Development Centre until 2 September 1991) Capenhurst, Chester, CH1 6ES.

Eurisol (1991). *An investigation into actual heat loss through cavity wall constructions*, June.

Siviour, J. B. (1982). Thermal performance of cavity walls insulated with urea-formaldehyde foam, *Building Services Engineering Research & Technology*, **13**, No. 2.

Siviour, J. B. (1989). Plasterboard fixing for thermal integrity, *Building Technical File*, No. 25, April, pp. 35–8.

Siviour, J. B. & Mould, A. E. (1988). Thermal bridging: significance in U-value calculation, *Building Services Engineering Research & Technology*, **9**, No. 3, pp. 133–5.

Chapter 11

Insulating the existing housing stock: mould growth and cold bridging

Tadj Oreszczyn

PhD, MInstE, CEng, MCIBSE

The Bartlett School of Architecture and Planning, University College, London

This chapter argues the need to insulate the current building stock in order that the health and comfort of the occupants is substantially improved and that the energy consumption of the domestic sector is reduced.

Over the next 15 years substantial reductions in energy consumption in the domestic sector has been proposed as one method of reducing the risk of global warming. This will require the insulation of existing buildings as well as new buildings. Currently the Building Regulations, one of the main mechanisms for motivating energy conservation, only apply to new buildings which account for just 1% of the housing stock a year. Thus, they cannot result in a substantial energy reduction over a 15 year time period.

The addition of extra insulation to existing buildings is not always a simple task and can sometimes lead to an increased risk of mould growth in buildings. This chapter examines the effect that the addition of insulation to various constructions has on the severity of cold bridges and develops a methodology for assessing the severity of cold bridges. In particular, the following building details are examined:

- The intersection of a floor slab with an external wall.
- Cantilevered balconies.
- Window recesses.
- The installation of double glazing.

The effectiveness of a novel type of insulation – transparent insulation – as retrofitted to existing buildings, is examined and compared to the installation of conventional external insulation.

11.1 Motivation

11.1.1 Energy and global warming

Over the last decade the argument for energy conservation has shifted from one of conserving limited fossil fuel reserves to reducing emissions of greenhouse gases and other pollutants.

If nothing is done, government projections of energy consumption suggest a growth in carbon dioxide (CO_2) emissions of between 10 and 40% over the next 15 years. Britain has set a target of stabilising CO_2 emissions at 1990 levels by the year 2005 (Department of Environment, 1990). To achieve this target it has been proposed to reduce non-transport energy use by between 20 and 40% over the next 15 years (House of Commons Energy Committee, 1991). Energy efficiency in the domestic sector, which is responsible for 30% of UK CO_2 emissions, is seen as one of the main mechanisms for reducing CO_2 emissions.

The main mechanism for controlling energy consumption in the domestic sector is Part L of the Building Regulations (Department of the Environment and the Welsh Office, 1990) which covers the conservation of fuel and power. However, the current Building Regulations only apply to new buildings and some extensions, therefore they only have a marginal effect in reducing the energy consumption over the short term. This is because the UK builds less than 200 000 new buildings a year, equivalent to less than 1% of its existing housing stock, and at the same time the number of households has been increasing by 1% a year. Thus energy efficient new dwellings only serve to limit the rise in energy consumption and do not reduce the current energy consumption.

Approximately 60% of the current UK housing stock was built before there were any regulations controlling the energy consumption of houses, and it has been estimated that by the year 2000 some 70% of the housing stock will still be pre-1974 vintage (House of Commons Energy Committee, 1991). These older houses, if left uninsulated, can consume five times the energy for space heating that a new building can use. Demolishing these buildings and replacing them with new ones is not an option in the short term because of the energy required to construct the building and its materials (energy use within a building only overtakes the energy embodied in the materials in two to five years – the energy payback period – Howard, 1991).

Even adding insulation to an existing building consumes energy. The return, in terms of energy saved, obtained by adding insulation is greater in uninsulated dwellings than in dwellings which already have some insulation. Howard (1991) argues that if the UK Regulations for new homes were improved to the insulation levels of the Swedish Building Regulations of 1986, with our climate it would take 15 years to recover the manufacturing energy for a polymeric insulation product, and about six years for a mineral fibre insulant. However, adding this additional insulation to uninsulated houses would pay for itself much quicker.

Fuel pricing as a mechanisam for promoting energy efficiency is ineffective because of the relatively low cost of energy. This is particularly true for the rich who spend less than 3% of their disposable income on fuel. On the other hand, the poor, who spend a greater percentage of their income on fuel, do not have access to the necessary capital required to invest in energy efficiency.

Clearly there is a need for additional mechanisms to control energy consumption in existing dwellings if the target values of 20 to 40% reductions in energy consumption are to be met. Such mechanisms should be primarily targeted at refurbishing our existing stock. Possible options include:

- Legislation – changing the Building Regulations to include refurbishment, compulsory labelling etc.
- Financial inducements (e.g. grants) and/or penalties (e.g. a CO_2 tax).

The potential for energy efficiency in existing buildings is highlighted by the following conclusions taken from the 1986 English House Condition Survey:

- Some 86% of our existing stock has no wall insulation, and less than 20% of cavity wall dwellings have cavity insulation.
- Similarly, 86% of dwellings do not have full double glazing.
- As many as 11% are still without roof insulation and 40% of dwellings with loft insulation have less than the currently recommended thickness.

Perhaps even more disturbing is the fact that 9% of houses still have no adequate heating system. If they did, it is likely that their energy consumption would increase and not decrease! In fact, although energy improvements have been undertaken in the UK housing stock over the past 20 years, there has been no net reduction in energy consumption because of the increased heating standards largely resulting from the installation of central heating and the increase in the number of households (Henderson & Shorrock, 1989).

Henderson & Shorrock (1990) have estimated that implementing all the technically possible energy saving measures in the domestic sector would reduce the CO_2 emissions in the domestic sector by 35% resulting in an overall reduction in UK CO_2 emissions of 10%, other emissions remaining constant. Of all the technically possible and cost effective measures, wall insulation has the greatest potential, and is the main subject of this chapter. Wall insulation accounts for over half the possible energy and CO_2 savings due to insulation and draught proofing and results in a greater saving than from the installation of energy efficient boilers and appliances.

11.1.2 Comfort and health

Energy efficiency is not the only motivation for insulating our existing building stock. A study in 1972 of houses occupied by people over the age of 65 showed that 74% of the occupants woke up to temperatures below 18°C; 54% woke to temperatures below 16°C; and 10% woke to temperatures below 12°C (Mant, 1986).

The 1986 English House Condition Survey covering all types of owner occupied houses showed that 22% of households had a living room temperature below 16°C and that there was a relationship between the level of insulation and the incidence of cold homes.

The low temperatures recorded in UK housing may go someway to explaining in excess of 40 000 deaths which occur in the UK in the winter. Such statistics do not occur in countries such as the USA and Sweden, where there are better levels of insulation and higher internal temperatures, despite colder weather.

11.1.3 Mould growth

11.1.3.1 The problem

Reduced temperatures is not the only problem resulting from the lack of insulation. Condensation mould growth affects an alarmingly high percentage of

UK housing. The English House Condition Survey in 1986 estimated that a fifth of all households (nearly three and a half million homes) experience some mould or damage to decorations due to damp.

Mould growth caused by high humidities and surface condensation, can result in dampness, a deterioration in the building fabric and a range of psychological conditions and physiological illnesses in the occupants. Stephen Platt (1989) of the Medical Research Council's Epidemiology unit at the Royal Edinburgh Hospital and colleagues have found that children in damp and mouldy homes had significantly more wheezing, sore throats, persistent headaches, fevers and runny noses than children in dry homes.

In damp homes adults were significantly more likely to have symptoms of nausea, vomiting, constipation, blocked nose, breathlessness, backache, bad nerves, and other symptoms than those in dry homes. The worse the damp the greater the number of symptoms. Such studies are always controversial because damp houses are more likely to be overcrowded and their occupants on low income or unemployed and with a higher incidence of smoking. However, Platt's study used statistical techniques to eliminate the effects of these variables.

The Department of Health (Environmental Committee, 1991) now believes that damp surfaces which promote the growth of, and long term exposure to, mould, bacteria and mites, can lead to hypersensitivity with severe reactions, including breathing difficulties. Generally, the persons affected in these ways by mould spores are those in the population who are particularly liable to be sensitized to allergy producing substances; that is some 10% of the population. The most at risk are children, the elderly and those receiving immunosuppressive drugs. It is now believed that mould growth may make an important contribution to the ill-health of many people with asthma.

Because of the health risks associated with mould growth, refurbishment is often recommended for mouldy buildings. Many of the benefits of such refurbishment can be undone if mould growth reappears. This has happened in many rehabilitated buildings and is partly a result of lack of understanding of mould growth and its prevention.

11.1.3.2 The cause

The main requirements for mould to grow are:

- Mould spores which are always present in the air – there are several hundred per cubic metre in the outside air, even in winter.
- Oxygen – not a problem in most houses!
- A temperature between 0°C to 40°C – individual species of moulds have an optimum temperature but in practice all moulds will tolerate wall temperatures encountered in buildings, even freezing.
- Some nutrients – most moulds can thrive on the nutrients present in dust and other deposits, even in well cleaned and maintained houses.
- Moisture.

Results from tests carried out in environmental chambers suggest that for surfaces such as wallpaper and plaster, mould growth may start at humidities above 80% (the main exception to this is non-porous materials such as tiles and glass where mould growth only occurs at relative humidities close to 100%). Some species of mould have been reported to grow at lower relative humidities, but these are very rare.

11.1.3.3 The solution

To avoid mould growth in houses the relative humidity at the surfaces of materials needs to be kept below 80%. It is not essential to have condensation on the surface, so one could argue that it is a misnomer to call it condensation mould growth. Mould will grow in a room even if the relative humidity is below 80%. This is because the surface of the wall is colder than the room air, thus an outside wall has a higher relative humidity at its internal surface than the relative humidity of the bulk of the air in the room. The reduction in temperature of the air close to external walls results in a higher relative humidity because cold air can hold less moisture than warm air. Therefore, the general design guidance, which you find in most texts, is that the room relative humidity should be kept below 70%.

However, for an uninsulated wall the temperature difference between the air and the wall is greater and hence the surface relative humidity is higher and consequently the risk of mould growth is greater. In reality, the insulated house is more likely to have a higher internal temperature because it costs less to heat. Therefore, there are two mechanisms by which insulation can reduce the risk of mould growth. Firstly, insulated houses are cheaper to heat and so likely to maintain a higher temperature and hence lower relative humidity, and secondly, the temperature difference between the wall and air temperature is less thus the difference in relative humidity between the room and the surface is less.

The way that insulation reduces mould risk is reasonably well understood. Theoretical calculations by Boyd *et al.* (1988) have shown that the addition of cavity wall insulation and an extra 100 mm of roof insulation to a mid-terraced house built to the Building Regulations in force in 1971, can halve the space heating required to prevent mould growth in the building. How this transfers to results in practice can be demonstrated by the 1986 House Condition Survey which investigated the occurrence of mould growth. The survey found that the occurrence of mould growth was substantially less (40% reduction) in those dwellings with the walls insulated as compared with uninsulated walls (Sanders, 1989).

Insulation is not the only means of reducing the risk of mould growth. Reducing moisture production, increasing the heat input, and increasing the ventilation rate of the building can also be used and should all be considered in combination with insulation. For example, insulation will not significantly improve the temperature if there is no adequate heating.

There is considerable guidance to assist the designer with the most appropriate measures for any particular situation (British Standard Institution, 1989) and computer packages which predict the effect of different measures (Boyd *et al.*, 1988). Insulation is, however, one of the more robust methods as it is not dependent on occupants behaviour, like for example, increasing ventilation by the installation of extract fans. In addition, insulation also reduces energy consumption and improves comfort which the other three methods do not.

11.1.3.4 Inappropriate solutions

Adding insulation can actually encourage mould growth: for example if it creates localized cold spots (see section 11.2) or affects any of the other parameters necessary for mould growth. An example of this is the installation of sealed double glazing which is often installed as a refurbishment measure when the existing window frames need replacing. Double glazing marginally reduces heat loss and results in higher temperatures and greater comfort. However, if the double glazing is improperly installed it can sometimes result in mould growth. The most common cause is the fact that old leaky windows have been replaced by modern tight fitting windows which can reduce the ventilation rate substantially.

Other factors which may also encourage mould are occupants controlling ventilation less in double glazed buildings than in single glazed ones. Condensation appearing on the glazing is often the stimuli for opening windows. Condensation occurs at a higher relative humility for double glazed windows than for single glazed windows, so occupant-controlled ventilation is more likely to occur at lower relative humidities and more frequently in single glazed dwellings. Also, the installation of double glazing in uninsulated buildings may shift the location of the coldest surface from the window to the corner of a wall or some other cold bridge. When this happens, condensation on the glazing will no longer be a sufficient warning of high relative humidities in the building.

It is sometimes thought that mould growth can occur after installing double glazing because double glazing is a less efficient dehumidifier. According to this reasoning single glazing encourages condensation which results in the removal of moisture from the air, and this in turn reduces the relative humidity.

However, the quantity of water that condenses on the surface of glazing is in fact small compared to the quantity of water produced in the building by the occupants (which is measured in litres). In addition, any moisture removal that single glazing causes is offset by the slightly lower temperature in a single glazed room compared to that of a double glazed room. The net result of the combined effect of lower temperature and less moisture is that the relative humidity remains little changed.

Care should also be taken when adding insulation to an existing structure, that interstitial condensation does not occur. This can be a particular problem with dry lining. Calculation methods for predicting interstitial condensation are well documented in several texts (e.g. Chartered Institute of Building Services Engineers, 1988). Some doubt has been cast on the accuracy of some of these techniques and work is currently being undertaken at Strathclyde University to refine these

(McLean & Galbraith, 1988). The remaining part of this chapter concentrates on the addition of insulation to existing buildings. The reader is referred to *Thermal insulation: avoiding risks* (Building Research Establishment, 1989) for information regarding problems with new buildings.

11.2 Cold Bridging

Mould growth usually starts on the coldest surface of a building. This may be a point of weak thermal resistance, such as a concrete post bridging the insulation in a wall, or in the corner of an external wall where heat escapes in two directions reducing the temperature at the corner below that of the rest of the wall. When adding insulation to a building it is important to increase the temperature of these colder surfaces.

11.2.1 Characterizing cold bridges

One method of characterizing the severity of a cold bridge is the temperature difference ratio (TDR), a coefficient which specifies how cold a surface is relative to the inside and outside temperature (Oreszczyn & Littler, 1989). The TDR is a dimensionless coefficient between 0 and 1.

$$\text{TDR} = \frac{\text{Internal air temperature} - \text{Cold bridge temperature}}{\text{Internal air temperature} - \text{External air temperature}}$$

The more severe the cold bridge the greater the TDR. Depending on the TDR cold bridges can be roughly divided into four categories – negligible, moderate, severe and unacceptable (see Fig. 11.1). Cold bridges with a TDR greater than 0.3 are unacceptable because the cold bridge will be colder than the surface temperature of double glazing, hence, condensation will occur on the surface of the cold bridge before it occurs on the surface of double glazing. Thus conden-

Cold bridge category	TDR	Examples
Negligible	<0.15	Plain walls U-value less than 1.2 W/m^2 K. External corners U-value less than 0.6 W/m^2 K. Insulated lintels.
Moderate	0.15–0.2	Plain walls U-value greater than 1.2 W/m^2 K. 3D corner U-value greater than 0.6 W/m^2 K.
Severe	0.2–0.3	External corners U-value 0.9 to 1.5 W/m^2 K. Uninsulated lintels. Concrete party wall or floor.
Unacceptable	>0.3	2D corners U-value >1.5 W/m^2 K. 3D corners U-value >1.0 W/m^2 K. Party floor and wall of drylined wall. Window reveal of drylined wall.

Fig. 11.1 Severity categories for cold bridges.

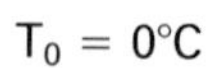

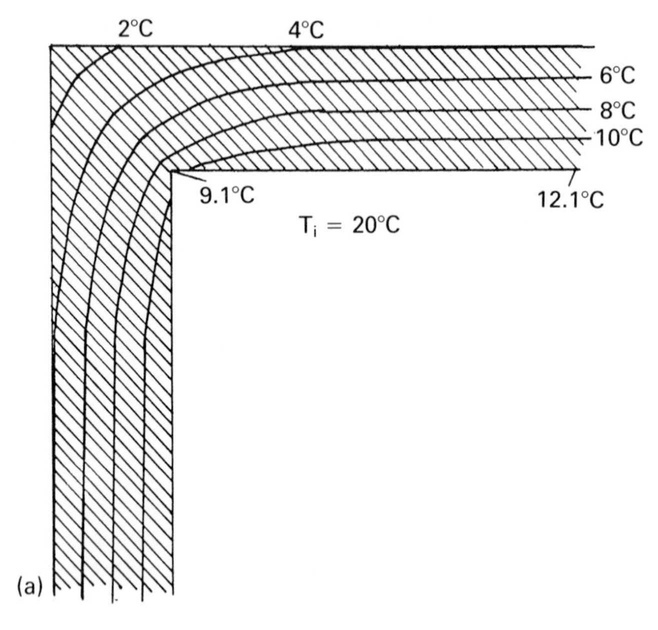

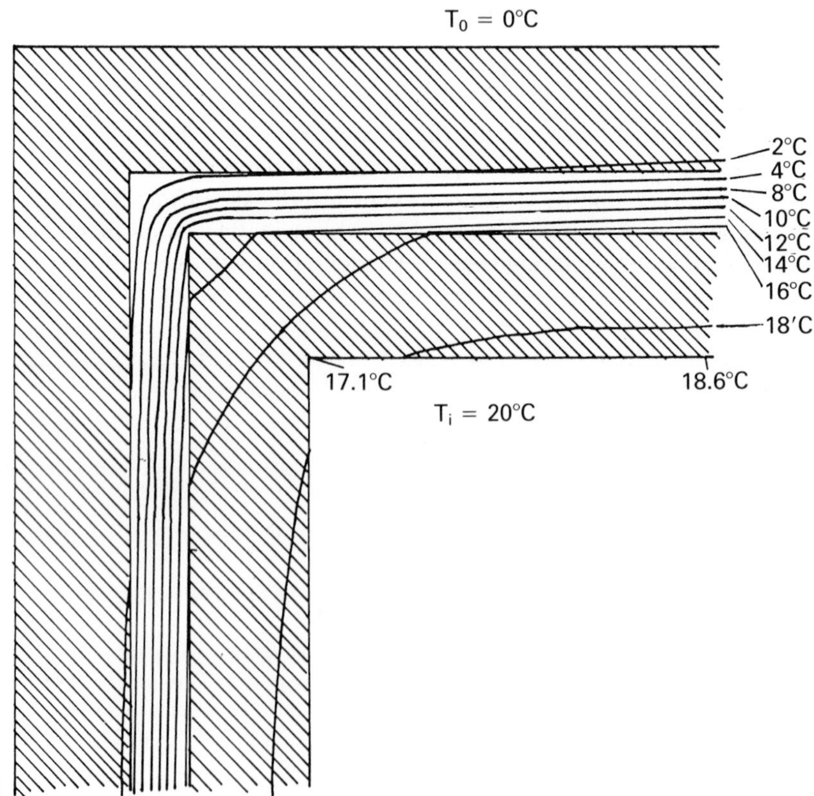

Fig. 11.2 Isotherms and temperature difference ratio (TDR) at the corner of an insulated and uninsulated wall (a) 100 mm brick, U = 3.3 W/m^{2}K, TDR = 0.54 (b) 100 mm brick – 50 mm insulation – 100 mm brick, U = 0.6 W/m^{2}K, TDR = 0.14.

sation on the glazing would no longer act as a warning of dangerously high humidities.

Figure 11.2 compares the TDR for two corner constructions. Also shown are the temperature contours corresponding to an internal temperature of 20°C and an external temperature of 0°C. The coldest surface temperature for the single skin brick wall is 9.1°C at the corner and this corresponds to a TDR of 0.54, i.e. an unacceptable cold bridge. By comparison the corner of the cavity wall is 8°C warmer and has a negligible cold bridge TDR of 0.14.

The TDR for a particular construction can be calculated with the aid of computer programs designed specifically for such a task (Standaert, 1986). These programs are suitable for use by consulting engineers and can run on personal computers, the analysis of a particular construction only taking a couple of hours. Such programs have been used to construct catalogues containing the most common constructional details in countries such as Belgium, Norway and Germany and work commissioned by the Energy Efficiency Office is currently underway to construct a catalogue of UK constructional details.

11.2.2 Adding insulation

There are many situations where it is impossible to completely insulate an existing building, for example, where external walls are punctured by floor slabs, party walls, cantilevered balconies, or windows. In general designers have an intuitive feeling for the best location for insulation. Most serious cold bridges can be avoided provided care is taken to avoid breaks in the insulation so that the path of least resistance is always kept above a minimum, as proposed in various design guides (Building Research Establishment, 1989).

However, in areas where two-dimensional heat flow occurs, particularly with novel constructions, detailed analysis is recommended. Also, during refurbishment the ideal solution is often too expensive and a compromise solution is required. Computer analysis can help in determining the optimum solution, which often proves not to be the same as the intuitive solution. For example, the addition of insulation can sometimes make cold bridges more severe, see Fig. 11.3. Here the solution is to carry insulation in along the floor and ceiling. The question is how far does this insulation need to be carried?

Figure 11.4 shows the effect of adding insulation to an external brick wall which has a concrete floor slab penetrating through it. The temperature difference ratio of the coldest location, normally where the floor slab penetrates the wall, is also plotted on the table. For the uninsulated solid wall the TDR is 0.265. Drylining the wall raises this to 0.325, i.e. an unacceptable cold bridge. This is because the insulation is penetrated by the floor slab.

In order to prevent this cold bridge, insulation must also be carried along the internal surface of the floor and ceiling. Only 200 mm is sufficient to avoid a cold bridge. Additional insulation results in little or no extra benefit. Simply insulating only the ceiling, although easy to do, does not improve the cold bridge. Figure 11.4 also shows that the largest reduction in cold bridging (TDR 0.115) is achieved by the addition of external insulation.

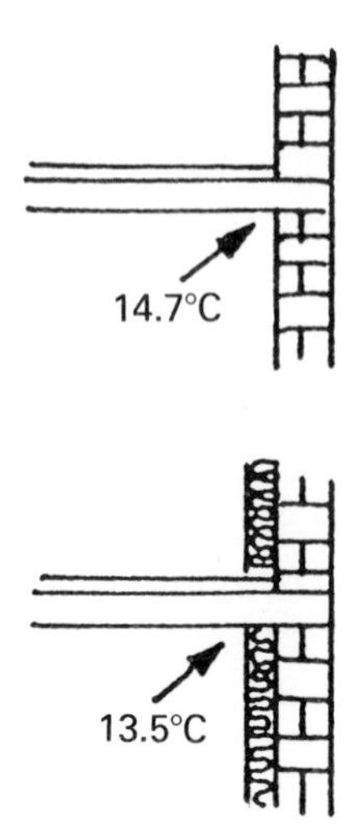

Fig. 11.3 The cold bridge temperature for a wall both with and without insulation.

In the 1950s and 1960s cantilevered balconies were a popular construction. These have often been thought of as 'cooling fins' on the outside of the building which would increase the heat loss and cold bridging. However, P. Cooper (1987) has shown that for most building materials this is unlikely to be the case. If the fin were made from a high thermal conductivity material such as metal, then you would expect an increase in heat loss. This is why you find metal fins on car radiators to increase the heat loss. However, if the fin were made from a low thermal conductivity material, such as polystyrene, then you would expect a reduction in heat loss.

In practice most balconies are made from materials whose thermal conductivity lies midway between these two extreme cases. Thus a balcony fin plays an insignificant role in the heat loss. Figure 11.5 shows the effect of adding insulation to a wall penetrated by a cantilevered balcony. The 'no fin' column shows the temperature difference ratio if the floor slab was terminated as soon as it penetrated the external wall. The 'plus fin' column shows the effect the added fin has, i.e. the dashed lines. In all cases, the addition of a fin actually reduced the TDR of the construction.

Window reveals, particularly on north facing windows, can often be the site of cold bridges resulting in condensation and hence staining and mould growth. They are also one of the few cold bridges that can get substantially worse when insulation is added to a wall, see Fig. 11.6, unless it is added to the exterior of the wall. Thus, every effort should be made to provide some insulation along the reveals when wall insulation is carried out. Narrow window frames, as found in metal framed windows, can make this a difficult detail. Even the installation of cavity insulation may cause window reveals to become colder if the window frame is set back towards the outside face of the wall.

Description	TDR_{cb}	
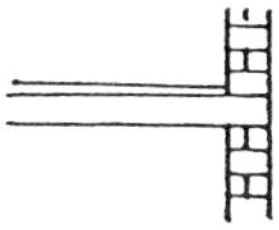 Uninsulated solid wall	0.265	
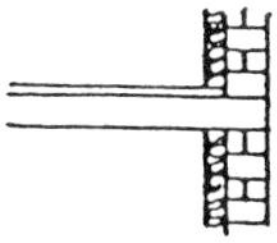 25 mm drylined solid wall	0.325	×
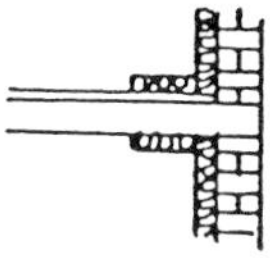 Solid wall plus 187 cm floor and ceiling drylining	0.145	√
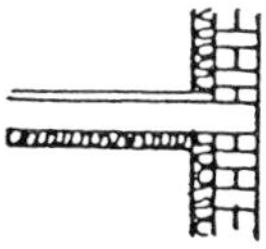 Solid wall only ceiling and wall drylined	0.315	×
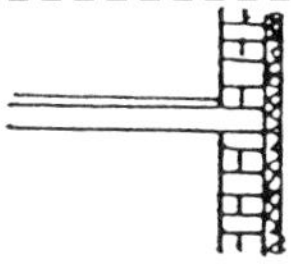 Solid wall external insulation	0.115	√
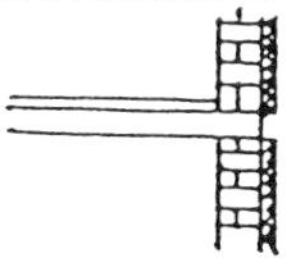 Solid wall partial external insulation	0.175	√
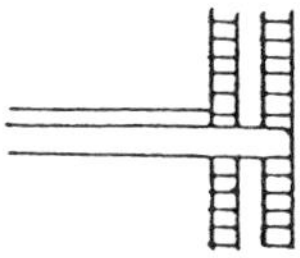 Uninsulated cavity wall	0.240	
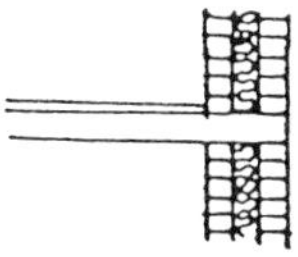 Insulated cavity wall	0.210	√
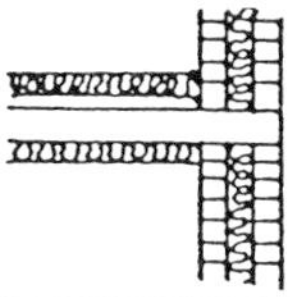 Insulated cavity wall with floor and ceiling insulated	0.250	×

Fig. 11.4 The temperature difference ratio for various insulated and uninsulated constructions consisting of a brick wall penetrated by a concrete floor slab. Description of wall; *Cavity* 105 mm brick – 50 mm cavity 105 mm brick, *Solid* 260 mm brick, *Floor* – 150 mm dense cast concrete, *Insulation* 25 mm drylining or external.

	Description	TDR_{cb} No fin	TDR_{cb} Plus fin
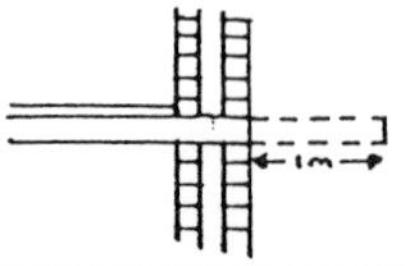	Uninsulated cavity wall	0.240	0.235
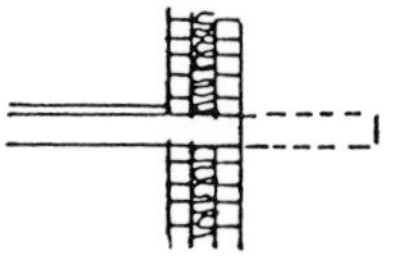	Insulated cavity wall	0.210	0.205
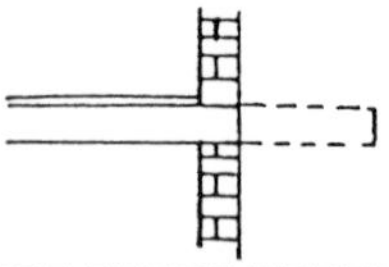	Solid wall	0.265	0.260
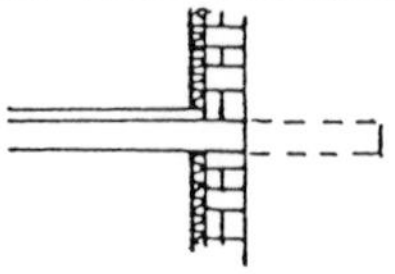	Drylined solid wall	0.325	0.320
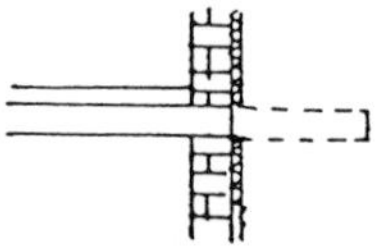	External cladding solid wall	0.175	0.170
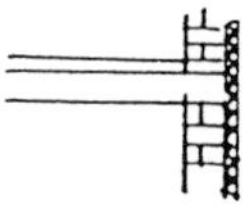	(insulation over edge of floor)	0.115	

Fig. 11.5 The temperature difference ratio for a concrete floor slab both with and without fins/balconies. Description of wall; *Cavity* 105 mm brick – 50 mm cavity 105 mm brick, *Solid* 260 mm brick, *Floor* – 150 mm dense cast concrete, *Insulation* 25 mm drylining or external.

11.3 Transparent insulation

A new insulation product, transparent insulation (see chapter 8), has recently appeared on the market. This appears to be particularly attractive for the refurbishment of existing buildings with mould growth problems.

Transparent insulation has similar insulating properties to that of conventional insulation but has the added advantage of allowing the transmission and absorption of solar energy by walls, see Fig. 11.7. Consequently the wall warms up

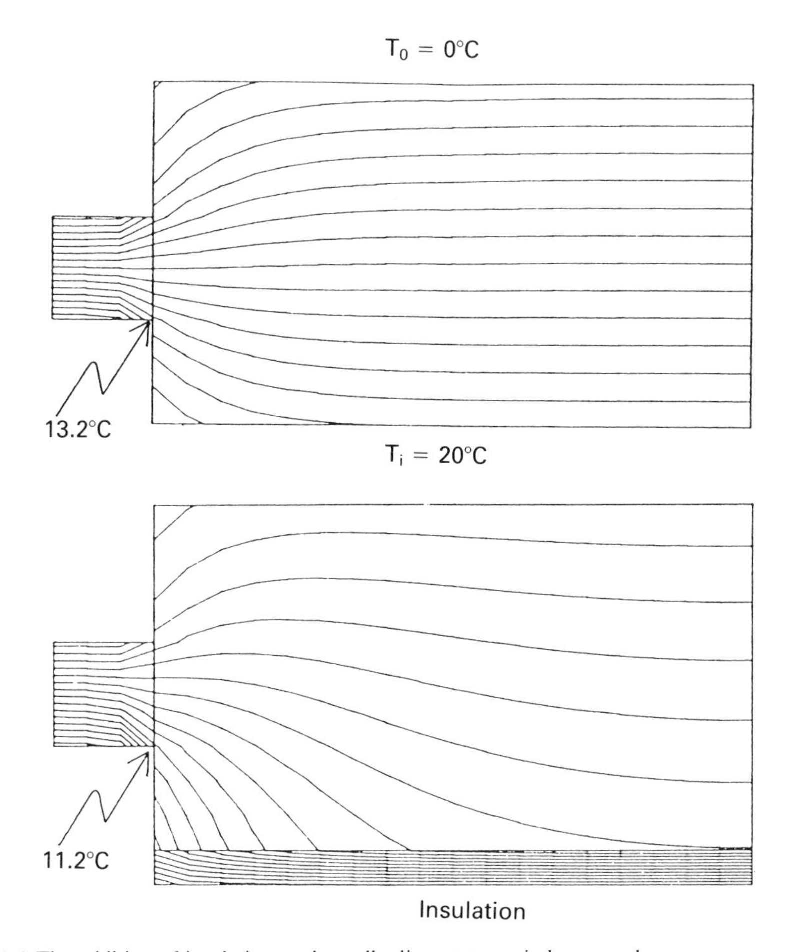

Fig. 11.6 The addition of insulation to the wall adjacent to a window reveal.

and heats the building. Even in the cloudy UK a wall covered in transparent insulation can act as a heat source for the majority of the year. Because the wall is a heat source it will be warmer than the air in the house and so condensation cannot occur on the wall, even if the building is not heated. Thus transparent insulation offers the possibility of providing heating to buildings from a renewable energy source in a way that avoids the risks of mould growth.

If transparent insulation has all these advantages why has it not found its way into the refurbishment market? The main reason is the cost. The material costs alone are in excess of £40 per square metre, then there is the cost of the glass cladding system essential for protecting the insulation from the elements. Also, in the summer there is a real risk of the building overheating, consequently expensive blinds must be installed. These need to be automatically controlled, and this

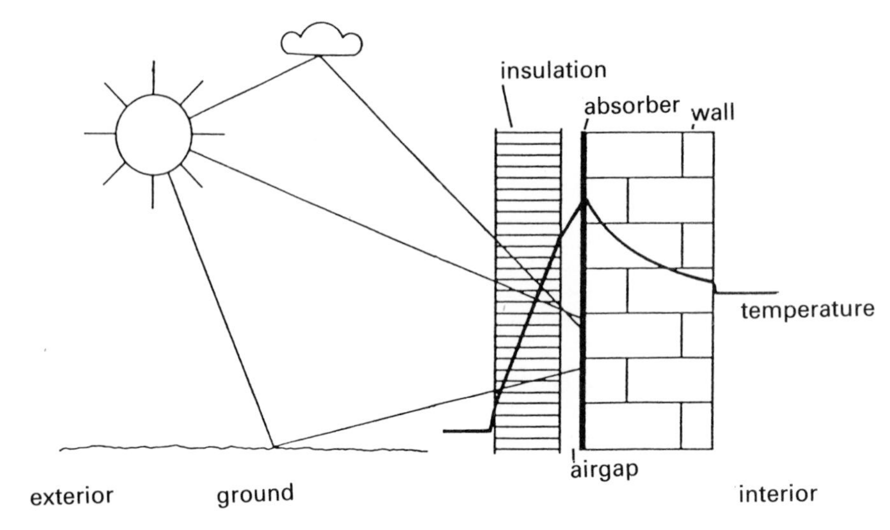

Fig. 11.7 A wall clad with transparent insulation.

adds to the expense, both in capital and maintenance. Finally, the installation of a dark glazed surface on many buildings is simply not acceptable for aesthetic reasons. Until these problems can be overcome conventional external insulation will be used in preference.

Acknowledgements

Thanks go to SERC for funding the research into cold bridging; The Bartlett School of Architeture for financial support and the comments of colleagues; Sue Oreszczyn and Mike Flood for typing and amendments.

References

Boyd, D., Cooper, P. & Oreszczyn, T. (1988) Condensation risk prediction: Addition of a condensation model BREDEM. *Building Serv. Eng. Res. Technol.* **9**(3), pp. 117–25.

British Standards Institution (1989) Code of practice: the control of condensation in buildings. British Standard BS 5250:1989, London BSI.

Building Research Establishment (1989) *Thermal Insulation: Avoiding risks. A guide to good practice building construction*. Building Research Establishment Report BR 143, HMSO.

Chartered Institution of Building Services Engineers (1988) CIBSE Section A10: Moisture Transfer and Condensation. In CIBSE Guide, Vol. **A**. The Chartered Institution of Building Services Engineers, London.

Cooper, P. (1987) Fin-type cold bridges: Heat loss and surface temperature. *Building Serv. Eng. Res. Technol.* **8**, pp. 221–7.

Department of the Environment (1990) *This Common Inheritance: Britain's environmental strategy*. Presented to Parliament by the Secretary of State for the Environment HMSO, London.
Department of the Environment and the Welsh Office (1990) *The Building Regulations 1985*, Part L, Conservation of fuel and power, 1990 Edition, HMSO.
Environmental Committee (1991) *Sixth Report: Indoor Pollution*, **1**, House of Commons Session 1990–1, HMSO.
Henderson, G. & Shorrock, L. D. (1989) *Domestic Energy Fact File*. Report BR151, Building Research Establishment, Garston, Watford.
Henderson, G. & Shorrock, L. D. (1990) *Greenhouse-gas emissions and buildings in the United Kingdom*. BRE Information Paper IP 2/90, Building Research Establishment, Garston, Watford.
House of Commons Energy Committee (1991) *Energy Efficiency*, Third Report, **1**, HMSO.
Howard, N. (1991) Energy in Balance, *Building Services*, May 1991 pp. 36–8.
Mant, D. C. & Muir, J. A. (1986) *Building Regulations and Health: a report to the Building Research Establishment concerning the influence on health of building fabric and services controllable by the Building Regulations*, Building Research Establishment, Garston, Watford.
McLean, R. C. & Galbraith, G. H. Interstitial condensation: Applicability of conventional permeability values *Building Serv. Eng. Res. Technol.* **9**(1), 29–34.
Oreszczyn, T. & Littler, J. (1989) *Cold Bridging and Mould Growth*, Final Report to the Science and Engineering Research Council, Research in Building Polytechnic of Central London, 35 Marylebone Road, London.
Platt, S. C., Martin, J., Hunt, S. M. & Lewis, C. W. (1989) Damp Housing, Mould Growth, and symptomatic health state. *British Medical Journal*, **298**, pp. 1673–8.
Sanders, C. H. (1989) Dampness data in the 1986 English House Condition Survey, *International Energy Agency – Annex XIV: Report UK–T7–35/1989*, Building Research Establishment Scottish Laboratory, East Kilbride, Glasgow.
Standaert, P. (1986) *KOBRU86 Manual*, Heirwig 21, B–9990 Maldegem, Belgium.

Housing

Chapter 12

A design guide to energy efficient housing

Geoffrey Pitts
BArch, MPhil
Deputy Chief Architect, Timber Research & Development Association
AND
John Willoughby
BSc, MPhil, CEng, MCIBSE
Energy & Environmental Design Consultant

In order to design a house which has low energy running and maintenance costs, energy efficient considerations have to be taken into account at every stage of the design process. This chapter outlines these considerations, including site, building form, construction details, ventilation strategies, insulation, passive solar design and the selection of heating and lighting systems and appliances.

12.1 Introduction

Until recently houses were poorly insulated. Better standards of insulation have only existed for the past 10–15 years. Figure 12.1 shows the effect of improvements in UK building regulations on the energy costs for a $70\,m^2$ semi-detached house; it demonstrates that heating costs have fallen from around £200 a year in 1976 to less than £100 in 1990. The figure also shows that with further improvements in insulation levels heating costs can be reduced to insignificant levels.

But the designer would be quite wrong to think that low energy house design is simply a matter of designing houses and adding insulation. This fact was dramatically illustrated in a 1980s field trial which compared energy costs for houses in two adjacent neighbourhoods of Milton Keynes (Lowe & Chapman, 1985). Both sets of houses were insulated to the same standards with U-values in the range $0.4–0.6\,W/m^2\,K$. Monitoring fuel consumptions over a two year period revealed that, although electricity costs were similar on the two estates, the cost of gas varied by around £100 a year (Fig. 12.2).

The difference between the two sets of houses was that those at Pennyland were designed with energy savings in mind while those at Neath Hill were not. The results demonstrate the value of taking care with:

- Orientation and site layout.
- Building form and internal planning.
- Construction details to reduce infiltration.
- The control of ventilation.
- Insulation details.

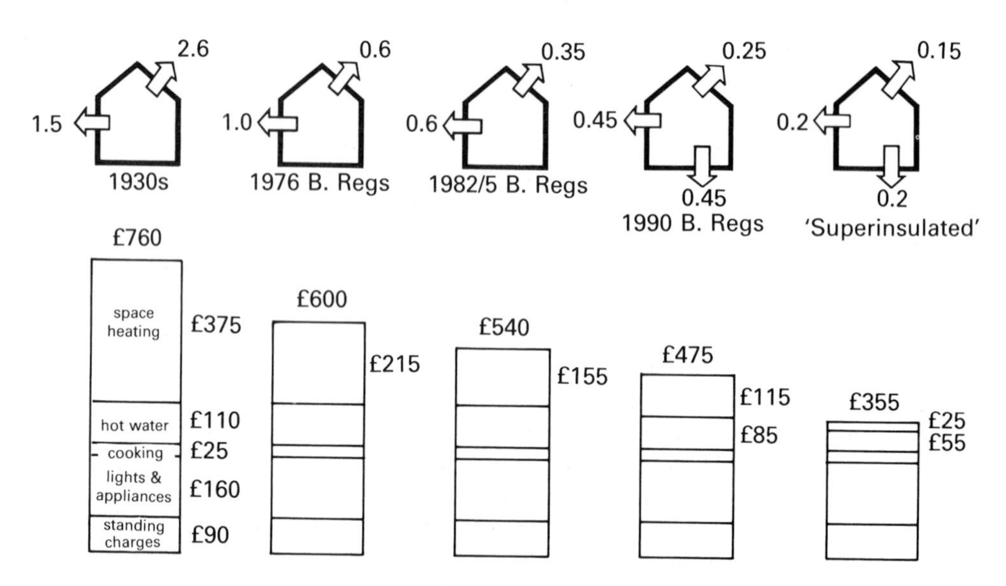

Fig. 12.1 The effect of different insulation levels on energy costs for a 70 m² semi-detached house.

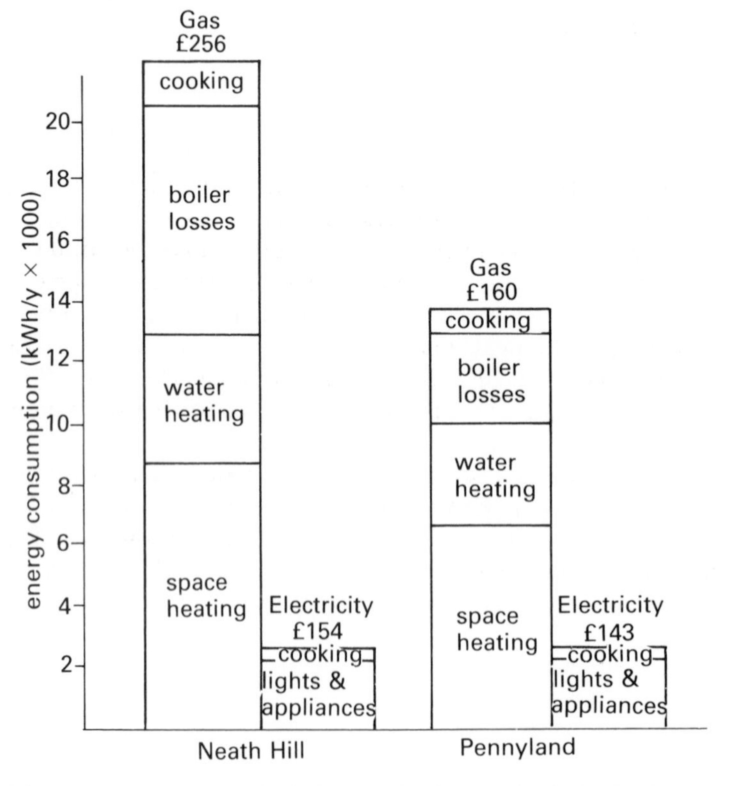

Fig. 12.2 Measured energy consumption in houses with the same insulation level.

- Passive solar design.
- The selection of heating, hot water systems, lights and appliances.

Practically every design decision affects energy consumption. Energy efficiency needs to be kept in mind throughout the design process; from the broad issues, such as estate layout and building form, through to the details such as the positioning of the light switches or thermostats.

The 1990s have started with an increased awareness of environmental issues, particularly the greenhouse effect and global warming. However, the link between these environmental issues and energy use in housing is seldom made (Hedges, 1991). But burning fossil fuels, either directly in the home or at the power station results in the production of CO_2 – the major greenhouse gas. Nearly 30% of UK CO_2 production results from energy use in housing (Shorrock & Henderson, 1990). Thus energy efficiency results not only in cost savings but also in reduced damage to the environment.

This chapter looks briefly at the issues involved in the design and construction of low energy houses. Most of the effort over the past few years has been directed towards reducing energy costs for heating and hot water supply. Sections 12.2 to 12.7 of this chapter look at the issues involved in reducing these energy costs. In modern houses electricity used for lights and appliances often involves greater costs than providing heating and hot water. Section 12.8, therefore, shows how energy costs for lights and appliances could be dramatically reduced.

12.2 Orientation and site layout

Orientation and site layout decisions are taken early in the design process and will have a significant effect on overall energy efficiency. In cool temperate climates such as the UK these decisions normally involve allowing sunlight into the dwelling in winter, but providing shading in the summer and, conversely, providing wind shelter in winter, but allowing wind into the dwelling for summer cooling.

In addition to the general climatic factors, local microclimate must also be considered. For example, windspeed is increased on hill sites; surrounding buildings or vegetation may affect windspeed and direction or overshade parts of a site and so on.

A particular attraction of energy efficient site layout is that there is generally little or no cost penalty associated with the decisions while the potential energy benefits can be as high as going to the next level of insulation (see Fig. 12.14). Thus designers should consider:

- Orientation.
- Overshading.
- Wind shelter.

Fig. 12.3 Willow Park, Nr Chorley, Lancs, UK.

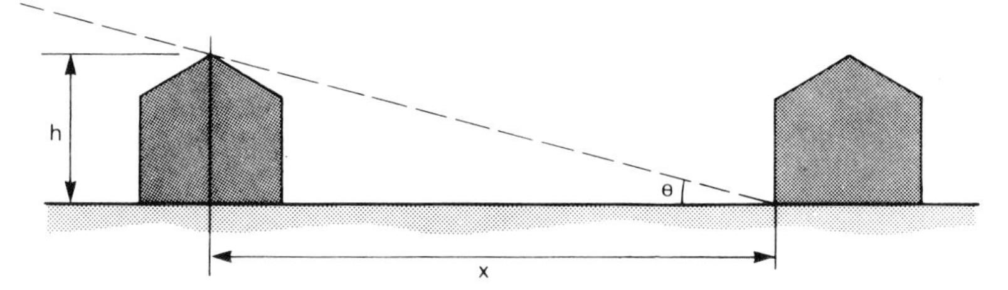

h = ridge height
x = ridge to face spacing to prevent overshading for a given sun angle
Θ = solar altitude

Example Assume h = 7.5 m (typical two storey house); Use midday solar altitude for latitude 50°N on December 22 which equals 17°

$$\text{thus } x = \frac{7.5}{\tan 17°} = 24.5\,\text{m}$$

No winter overshading will occur with ridge to face distance of 24.5 m.

Fig. 12.4 Calculations for minimum overshading.

12.2.1 Orientation

Orientation simply refers to the aspect of windows in habitable rooms. Orientation in response to local climate is an ancient design principle used extensively in vernacular architecture (Butti & Perlin 1981). More recently this principle has been ignored due to the combination of volume housing and relatively cheap fuel for heating. Good orientation not only reduces heating costs but sunny living spaces provide high amenity value. High density housing need not compromise good orientation (NBA Tectonics, 1988).

To make the best use of winter solar gains houses should be orientated towards the south. However, where this is not possible, deviations up to 30° east or west of south reduce the theorectical energy savings by only 2–5%. Deviations up to 45° can reduce savings up to 10% (Turrent *et al.*, 1980).

Figure 12.3 illustrates the Willow Park Estate, built in 1983 in north–west England where all 31 houses are orientated due south while still maintaining an interesting site layout.

12.2.2 Overshading

Good southerly orientation is of little benefit if the south windows are overshaded by adjacent buildings for most of the winter. The minimum distance required between buildings to prevent overshading can be calculated using local solar data as shown in Fig. 12.4. The restrictions imposed on layout by distance between buildings for minimum overshading will normally be more than adequate to satisfy the requirements for privacy.

The Willow Park scheme (Fig. 12.3) illustrates the principle of minimum overshading where a solar altitude angle of 17° resulted in a distance of approximately 25 m between the ridge of a house to the south and the south face of a house to the north of this.

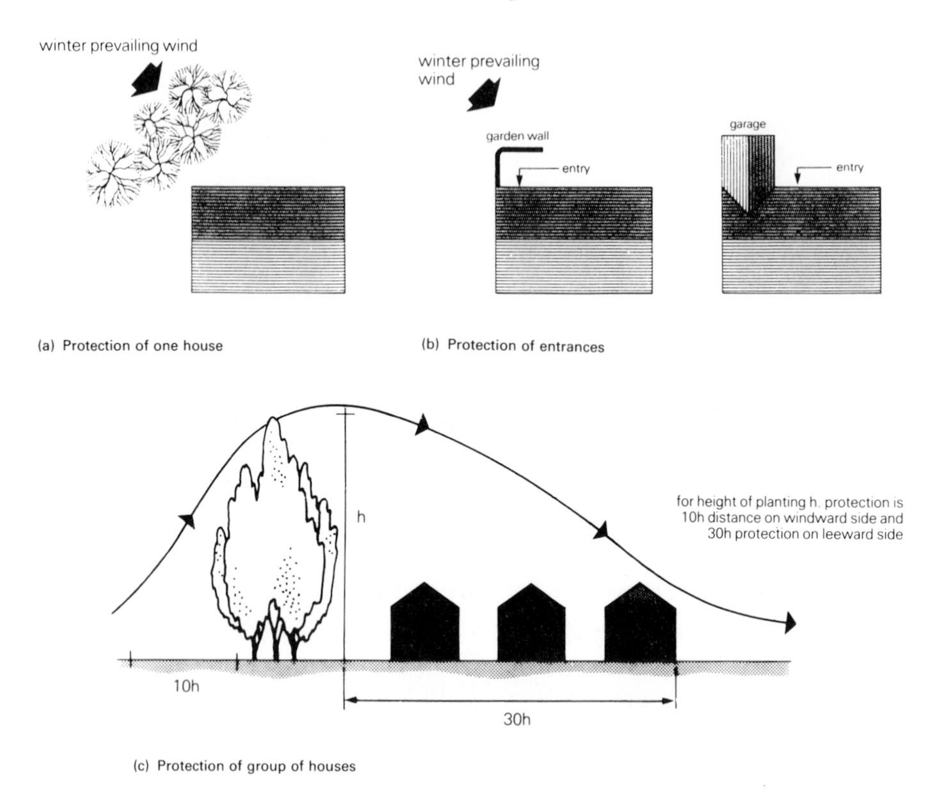

Fig. 12.5 Wind protection with landscaping.

12.2.3 Wind shelter

Winter wind can affect building energy consumption by increasing convective heat loss and, of much greater importance, increasing air infiltration heat losses.

These effects can be reduced by providing shelter from winter prevailing winds to the whole site, to individual buildings or simply, and perhaps most importantly, to building entrances. Figure 12.5 shows a number of wind shelter options.

Again the Willow Park scheme (Fig. 12.3) illustrates some wind shelter planning. The tree planting to the south–west shelters the site from prevailing winds and most of the house entrances are on the leeward side.

12.3 Building form and internal planning

Building form is influenced by planning regulations, client brief and site constraints. The major energy considerations are the overall surface area, which

For each m^2 increase in surface area the increase in heat loss for a component with a U-value of $0.3 W/m^2K$ is $0.3 W/°C$
This represents a heat loss rate of 2.9W over a heating season (September to April with average temperature difference 9.7°C) which is equal to an increase in the seasonal heat loss of only $16.8 kWh/m^2$

(a) **Plan aspect ratio 1 : 1** Wall area : 140 m^2; Take heat loss from Table 1.3 as = 9902 kWh.
(b) **Plan aspect ratio 1 : 2** Wall area = 148.5 m^2; Heat loss increase is 143 kWh/season = 10045 kWh, ie 1.4% increase.

Note:
The high level of insulation ensures that the % increase due to configuration variation is small.

Fig. 12.6 Plan: aspect ratio and heat loss.

affects fabric heat loss; the volume, which affects ventilation heat loss and the plan: aspect ratio, which affects the potential to receive solar gain.

Internal planning also affects the potential to receive solar gain and more importantly, determines the usefulness of this gain. Again variations in built form and internal planning can enhance energy performance at little or no extra cost.

12.3.1 Building form

A compact building form such as a cuboid will result in the minimum practical surface area and thus the minimum fabric heat loss. However, this form does not necessarily give optimum solar access to a building.

With good insulation levels (e.g. $0.3 W/m^2 K$ or lower) variations away from the compact cuboid form only result in small increases in fabric heat loss. Figure 12.6 shows that moving to a plan: aspect ratio of 1:2 will only increase fabric heat loss by about 1.5%. Thus, with such a form the designer, has plenty of scope to introduce solar gain without a significant increase in heat loss.

Another technique which can be used to reduce heat loss is to build the top storey into the roof space, i.e. build attic rooms or 'rooms in the roof'. Thus, for a given floor area, the overall surface area and volume are decreased resulting in reductions of both fabric and ventilation heat loss.

12.3.2 Internal planning

The general aim for energy efficient internal planning is to group the living spaces on the sunny south side of the dwelling and to use service and circulation spaces as a buffer to the north. Thus the areas which require higher temperatures and therefore can make best use of solar gains have a southerly aspect. Heat gen-

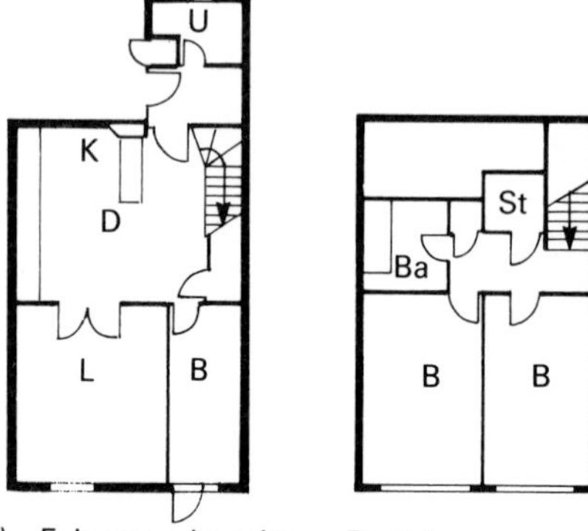

(a) Felmore housing, Basildon c. 1978 by Ahrands Burton & Kovalek.

Narrow frontage, north entry via lobby. Services, stairs to north and living and beds to south. Good interzone heat transfer from living to dining.

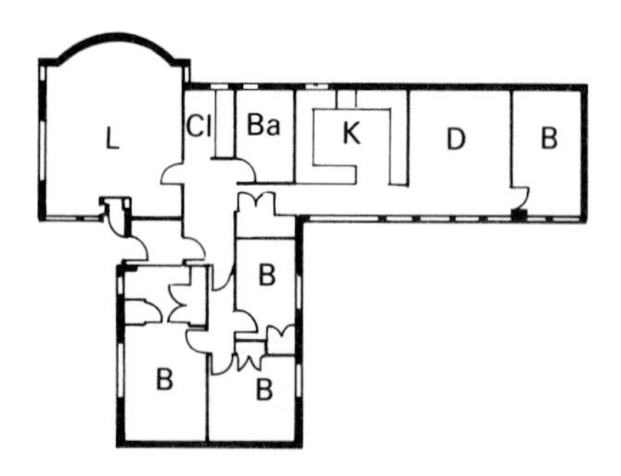

(b) Shenley Lodge, Milton Keynes c. 1985 by Feilden Clegg Design.

Wide frontage bungalow to give good solar access to all habitable rooms. Side entry via lobby. Additional sunlight to Living via high level lights.

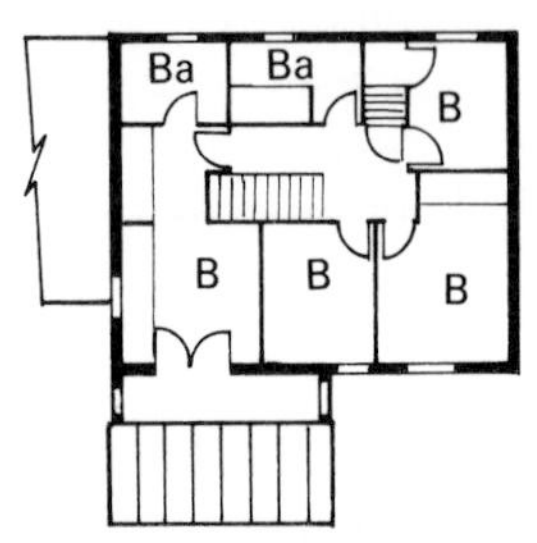

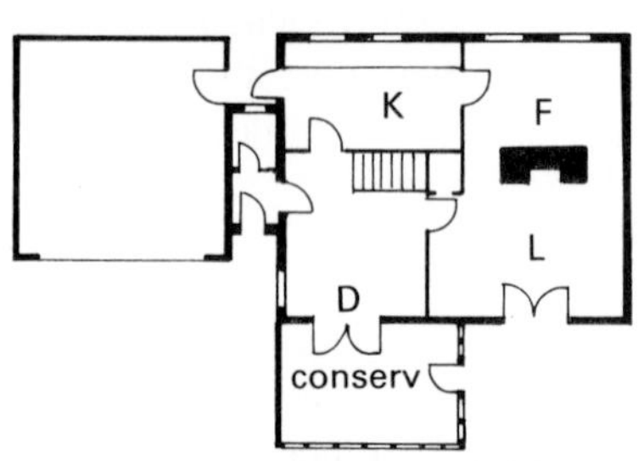

(c) Willow Park housing near Chorley c. 1983 by TRADA Architects.

Wide frontage, south entry via lobby. Services to north and most habitable rooms to south. Attached conservatory as buffer and 'solar collector'. Good interzone heat transfer from Living to Family.

(d) Shenley Lodge, Milton Keynes c. 1985 by David Tuckley Associates.

Compact plan with south entry through conservatory as lobby. All habitable rooms face south.

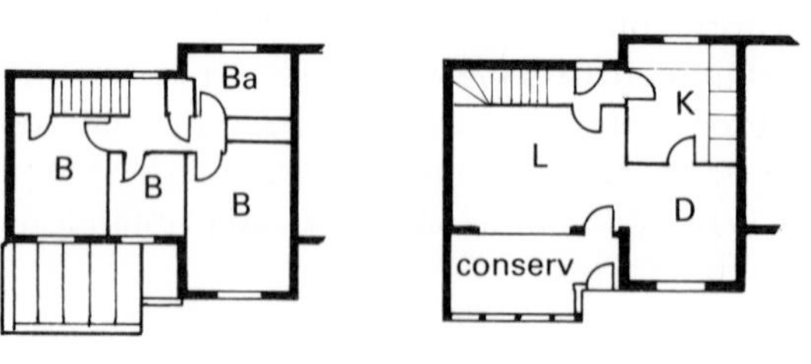

Fig. 12.7 Energy efficient internal planning options.

erating areas such as the kitchen are best located to the north where they are less affected by solar gain.

Unheated spaces such as garages may also be employed as a buffer against north heat loss. Internal planning should always include lobbies to external doors to reduce ventilation heat loss.

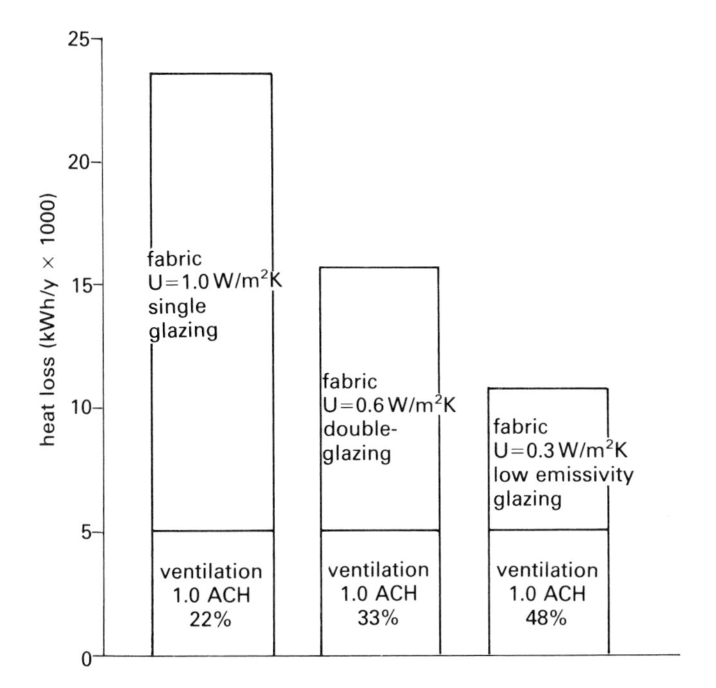

Fig. 12.8 Percentage increase in ventilation heat loss as fabric insulation improves.

Another energy efficient planning technique which should be considered is the distribution of solar gain throughout the house. This inter-zone heat transfer can be achieved with open plan and/or air movement between rooms by careful location of internal walls and doors. Figure 12.7 illustrates many of the options for energy efficient internal planning techniques. Air movement is important to encourage cross ventilation for summer cooling.

Another important aspect of internal layout is the strategy for services distribution. Heating and hot water systems, including appliances and pipework, should be contained within the insulated building shell. A central location for such appliances not only minimizes service distribution runs and pipe standing losses, but also ensures that heat gains are usefully contained within the dwelling. This strategy removes services from the cold loft space which has significant implications for electric cable distribution and cold water storage.

12.4 Infiltration and ventilation

In well insulated houses (e.g. U-values of $0.3\,W/m^2K$ or lower) heat loss through uncontrolled infiltration or excessive ventilation can undermine the energy efficient strategy. Unless steps are taken to control infiltration and ventilation the resultant heat loss will be an excessively high proportion of the total (Fig. 12.8).

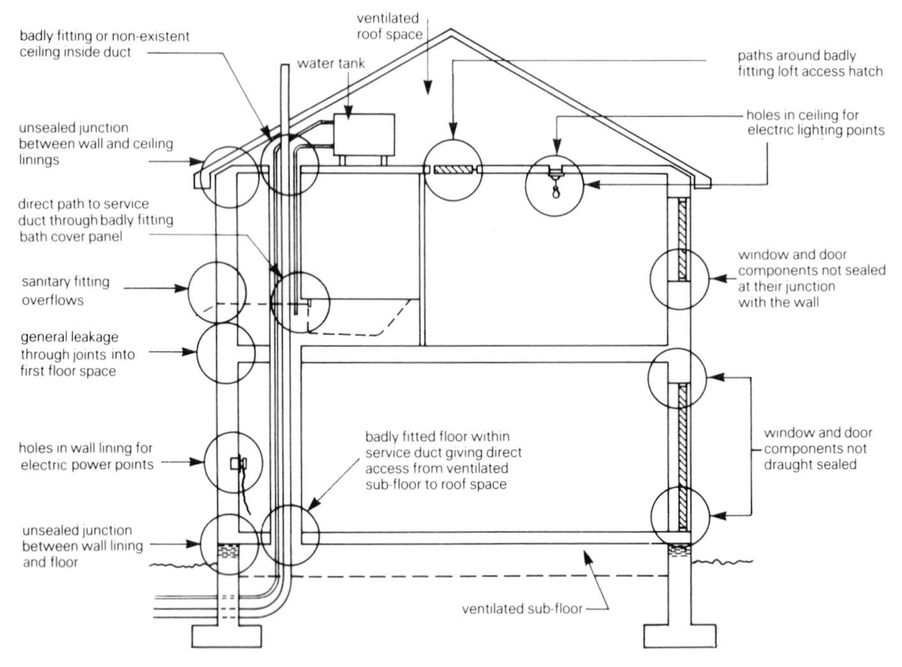

Fig. 12.9 Typical air leakage points.

There are two processes controlling infiltration and ventilation; the wind effect and the stack effect. Wind pressures on the building shell cause cold air to leak into the house and warm air to leak out. The stack effect on the other hand is caused by the buoyancy of the warm internal air which tends to rise and leak out of cracks or openings at high level. The wind effect is dependent on windspeed, direction and exposure, while the stack effect depends on inside/outside temperature difference and the number of storeys.

It is important to understand the difference between infiltration and ventilation. Infiltration is the uncontrolled air leakage through cracks and other openings in the building fabric (see Fig. 12.9). Ventilation is the controlled introduction of fresh air to combat odours and condensation, for general healthy living and, in summer, for cooling.

12.4.1 Infiltration

Traditionally, infiltration has been relied on to provide house ventilation and the resultant air leakage rates have been very high (see Fig. 12.10). Infiltration is uncontrollable – it varies with wind speed and temperature and occurs all the time, day and night whether the house is occupied or not. The key is to reduce infiltration to a minimum and to provide a controllable source of fresh air for ventilation.

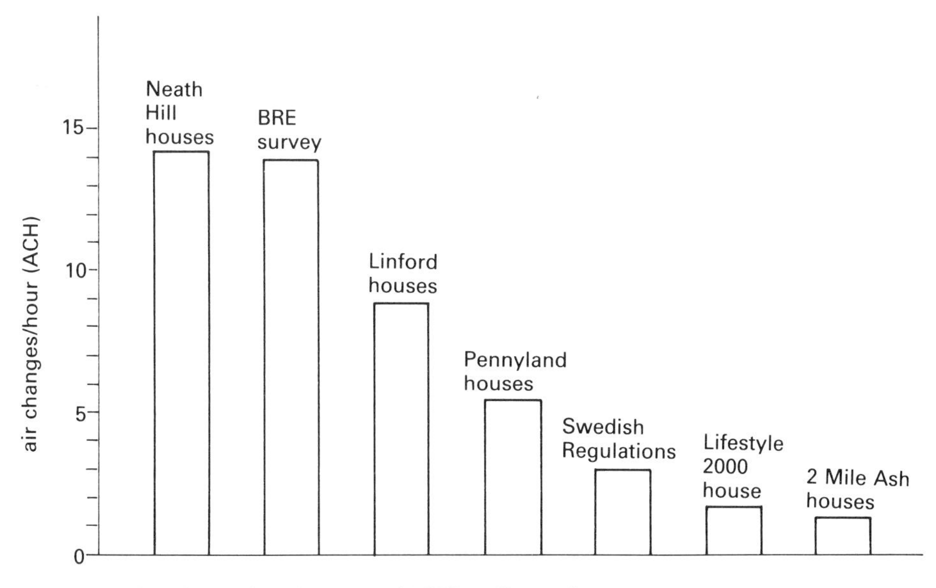

Fig. 12.10 Whole house air leakage rates (ACH) at 50 pascals.

Reducing infiltration is not an easy task as there are a great number of potential air leakage routes. An experiment to trace leaks in 15 masonry houses found that 67% of leakage occurred via unidentifiable routes!

Reducing air infiltration in housing requires a strategic approach. The first part of the strategy is to recognise that the inner skin of the insulated envelope should be considered as an air barrier. To ensure the integrity of this air barrier the second part of the strategy is to minimize the number of penetrations through the insulated envelope and, where penetrations are unavoidable, seal these against air leakage. The strategy will involve:

- Where possible avoiding services in loft spaces, in wall cavities and in ground floor cavities e.g. use alternative water systems which require no loft tanks; specify internal services cavities to walls and run services in sealed ducts in floors.
- Avoiding building structure through the insulated envelope e.g. avoid building floor joists into and through masonry walls.
- Where service penetrations are unavoidable ensuring that the gaps around the service are well sealed e.g. seal around a balanced flue or overflow pipe.
- Ensuring that component junctions are well sealed e.g. provide a silicone or mastic seal to the window to wall junction; ensure that the loft access hatch is draughtsealed and latched to compress the seal.
- Specifying high performance windows with good quality draughtseals to opening lights and ensure that draughtseals are provided to door frames (both the external door and the inner lobby door).

12.4.2 Ventilation

A 'tight', well sealed house has energy advantages in terms of infiltration control but for the comfort and safety of occupants and for condensation control, planned controllable ventilation must be provided (Uglow, 1989 & Kadulski, 1988). In practical terms, designers cannot plan for a specific ventilation rate and should ensure, first that the house is well sealed to minimize infiltration heat loss, and second that the planned ventilation has sufficiently fine control to enable occupants to find their own level of ventilation for economy, comfort, and condensation control. Window opening lights do not normally provide sufficiently fine control, and even at a minimum opening, heat loss can be excessive and draughts may be caused. Slot ventilators can provide better control and because of their position at the head of windows do not normally cause draughts.

Planned, fine controlled ventilation can be provided by:

- Natural means under the control of occupants.
- A combination of natural and mechanical means still largely under occupant control.
  or
- A full mechanical system which requires minimal occupant control.

In each of these cases, the fresh ventilation air may be pre-heated by solar heat gains (see section 12.6).

12.4.2.1 Natural ventilation

For all rooms, except kitchens, bathrooms and utilities, natural ventilation can be provided by a combination of fine control slot ventilators in window heads and window opening lights. The ventilators should allow opening without rain or snow penetration and should provide security against access when open. The ventilation can be manually operated or incorporate passive devices which throttle the air at a given inlet flow rate or internal relative humidity. The opening lights are used for occasional high ventilation rates to 'flush out' rooms (for example when high odour levels occur) and for summer ventilation. Neither slot ventilators nor periodic window openings are adequate for kitchens, bathrooms or utility rooms where frequent high moisture output can occur.

Figure 12.11 illustrates a passive ventilation system which uses a combination of stack effect and wind effect to provide higher levels of ventilation at the moisture source, i.e. where cooking, bathing or washing occurs. The normal provision of slot ventilators and opening lights is still required with the passive system.

12.4.2.2 Combined natural and mechanical ventilation

Combined systems normally require the provision of slot ventilators and opening lights as for natural ventilation, in addition to the provision of mechanical extract

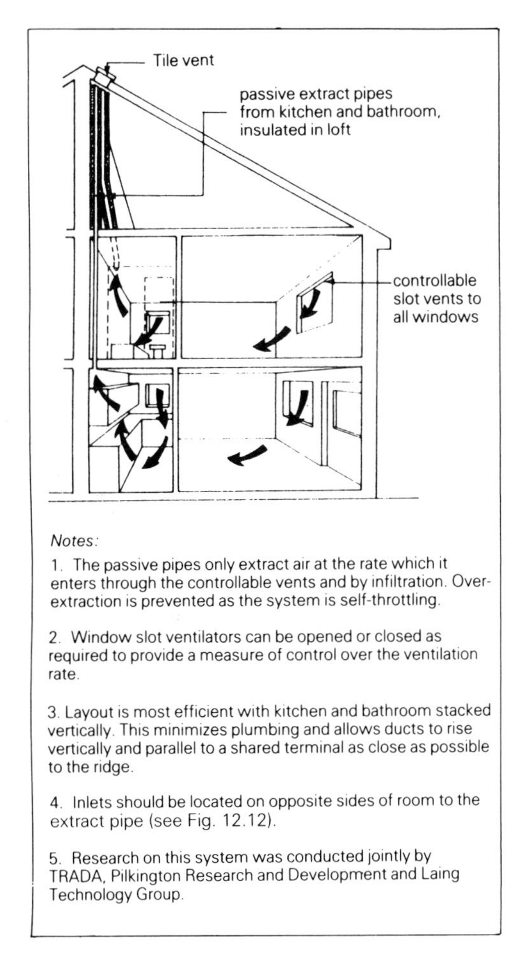

Notes:

1. The passive pipes only extract air at the rate which it enters through the controllable vents and by infiltration. Over-extraction is prevented as the system is self-throttling.

2. Window slot ventilators can be opened or closed as required to provide a measure of control over the ventilation rate.

3. Layout is most efficient with kitchen and bathroom stacked vertically. This minimizes plumbing and allows ducts to rise vertically and parallel to a shared terminal as close as possible to the ridge.

4. Inlets should be located on opposite sides of room to the extract pipe (see Fig. 12.12).

5. Research on this system was conducted jointly by TRADA, Pilkington Research and Development and Laing Technology Group.

Fig. 12.11 Passive ventilation system.

fans in kitchens, bathrooms and utility rooms. Kitchen extract fans should ideally be located above the cooker and on an external wall at high level. Where this is not possible, high level ducts from a cooker hood to the fan in an external wall should be used.

Sufficient inlet area is required to avoid de-pressurising the room and the inlet should be positioned to avoid short-circuiting the ventilation path (Fig. 12.12). Kitchen extracts should be automatically controlled with a humidity sensor, preferably with an inbuilt degreasing facility. It is important that the high extract rate of mechanical fans does not adversely affect air supply to any heating appliance in the kitchen (the relevant fuel authority should be consulted).

The moisture in bathrooms and utility rooms is more evenly distributed than in kitchens and thus the fan positioning is less critical. However, as for kitchens, the inlet and outlet should be as far apart as possible to avoid short-circuiting. Again the extract fans should be controlled with a relative humidity sensor.

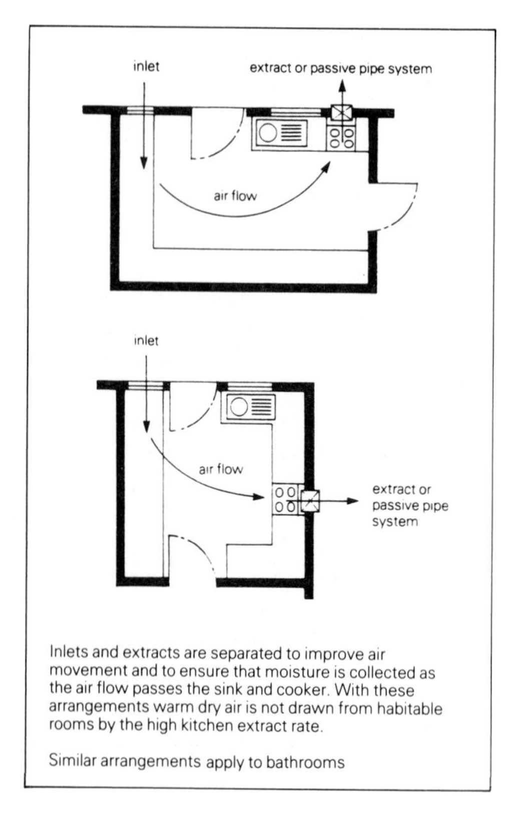

Fig. 12.12 Inlet/extract arrangements.

12.4.2.3 Mechanical ventilation

Full mechanical ventilation can provide a precisely controlled ventilation rate. If the system incorporates heat recovery this can result in energy savings, but the benefits must be carefully weighed against the cost of installation, the running costs of the fans and the maintenance costs (Ruyssevelt, 1987a).

The main advantage of mechanical ventilation is in terms of improved air quality. Condensation control is also improved, as fresh air is distributed evenly through all rooms and moist air in bathrooms and kitchens is extracted at source. Opening lights to windows are still required for summer ventilation and it may be advantageous to specify slot ventilators both to give occupants the option of occasional natural ventilation or for use in the event of mechanical breakdown.

12.5 Insulation

The 1990/1991 UK Building Regulations call for fabric U-values of $0.45\,W/m^2\,K$ for walls and floors and 0.25 for roofs. Despite these relatively high insulation levels, the Regulations still allow single glazing to be used. There are now numerous examples of housing schemes which show that it is well worthwhile to improve on these insulation standards.

For instance a group of superinsulated houses at Two Mile Ash, Milton Keynes have U-values of 0.24 for the walls, 0.2 for the floor and 0.12 for the roof. The windows are triple glazed, argon filled and incorporate a low emissivity coating, giving a U-value of 1.2 $W/m^2 K$. Monitoring has shown that the heating costs for these 70 m^2 semi-detached houses average around £10 a year (Ruyssevelt, 1987a). The additional cost of the insulation added only £2200 to the building cost compared with similar houses built to the then current Building Regulations. It was predicted that the savings on the heating costs would exceed the extra mortgage repayments in only four years.

It is clearly worth thinking carefully about specifying insulation levels better than the current standards; particularly if the 60–100 year life of the building is taken into account. Another reason to consider high insulation levels is that the opportunity to install the insulation may never arise again. For instance, if a solid ground floor slab is not insulated during construction, it will be difficult to improve its performance at a later date.

To incorporate high levels of insulation the designer probably needs to reassess standard familiar insulation details. It may be worth starting this process with a review of the range of insulation materials which are available. Figure 12.13 shows that there is a huge variety of materials available.

When incorporating the appropriate insulation material into the building fabric, it is important to think about the following principles:

- Distribute the insulation around the building shell. It is far better, in terms of both economy and comfort, to have a good overall level of insulation rather than, say, a highly insulated roof with no wall or floor insulation.
- Achieve continuity of insulation and avoid cold bridges. Cold bridges in well insulated constructions cannot only represent a significant increase in heat loss but can also cause condensation problems. Avoiding cold bridges will require careful attention to the detailing around openings and at junctions between walls and floors.
- Integrate insulation details with the strategy for the services distribution. It is difficult to attain a well insulated, air-tight construction if it is riddled with holes from services entries. Services entries should be controlled and carefully detailed and the services distribution should be contained within the insulated shell. The most important ramification of this approach is to remove all services from the cold draughty loft space.
- Consider condensation. Surface condensation in well insulated houses should be a thing of the past, if the heating and ventilation are adequate. But the designer needs to eliminate the risk of interstitial condensation.

The range of details for well insulated constructions is too numerous to be covered here. Pitts (1989) deals with timber frame energy efficient house construction and the BRE (1989) covers masonry details. With a strategic approach

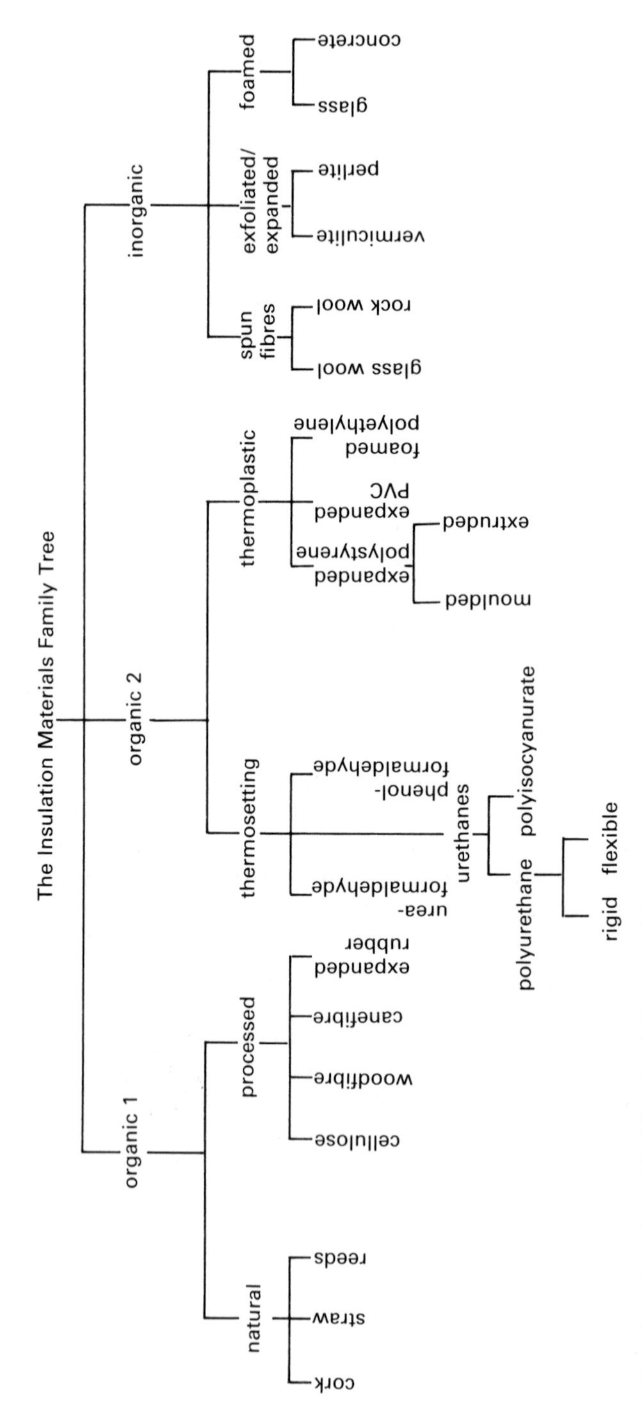

Fig. 12.13 A review of building insulation materials.

to house design there will be less reliance on traditional standard details. A strategic approach which recognizes the need for a piece of construction to be loadbearing, insulating, breathing and water-shedding should result in a logical solution.

12.6 Passive solar

Trombe walls, roofspace collectors and atria have all been used to improve passive solar gains in houses. While, in general, these techniques have worked well, their capital costs have often been excessive compared to the energy saved. Direct gain and conservatory systems, on the other hand, have proved to be a very cost effective way of reducing running costs in dwellings.

12.6.1 Direct gain

Four simple techniques are employed in direct gain houses:

- Face the house south and avoid overshading.
- Concentrate the glazing on the south side.
- Ensure that the window is not obscured with blinds or net curtains.
- Improve the performance of the glazing so that the solar gains are maximized and the heat losses are minimized.

These simple techniques can make a substantial impact on energy consumption. This was illustrated in some passive solar direct gain houses in Milton Keynes which were closely monitored over three heating seasons (Everett *et al.*, 1985). Some 24% of the energy needs of the houses were met by solar gains. This amounted to 1500 kWh over the year which was attributed equally between the first three techniques above.

It should be noted that these three techniques can be used to improve performance without increasing capital costs. Figure 12.14 shows computer predictions of the effect of moving glazing from the north side of a house to the south (Riddle, 1991). The results are given for three different levels of insulation and it can be seen that moving the glazing from north to south has a similar effect to moving to the next level of insulation.

Improving the performance of glazing systems has been the subject of much interest over the past decade. Movable insulation which reduces heat losses at night but allows solar gains during sunny days has been the subject of some research. But, apart from a few propriety blind systems, movable insulation has been designed to suit specific projects.

Low emissivity coatings have been developed to enhance glazing performance by reducing radiant heat losses across the air gap in double glazing. One experiment used quadruple glazing with two low emissivity argon filled gaps and an integral low emissivity blind to produce a window with a U-value of less than

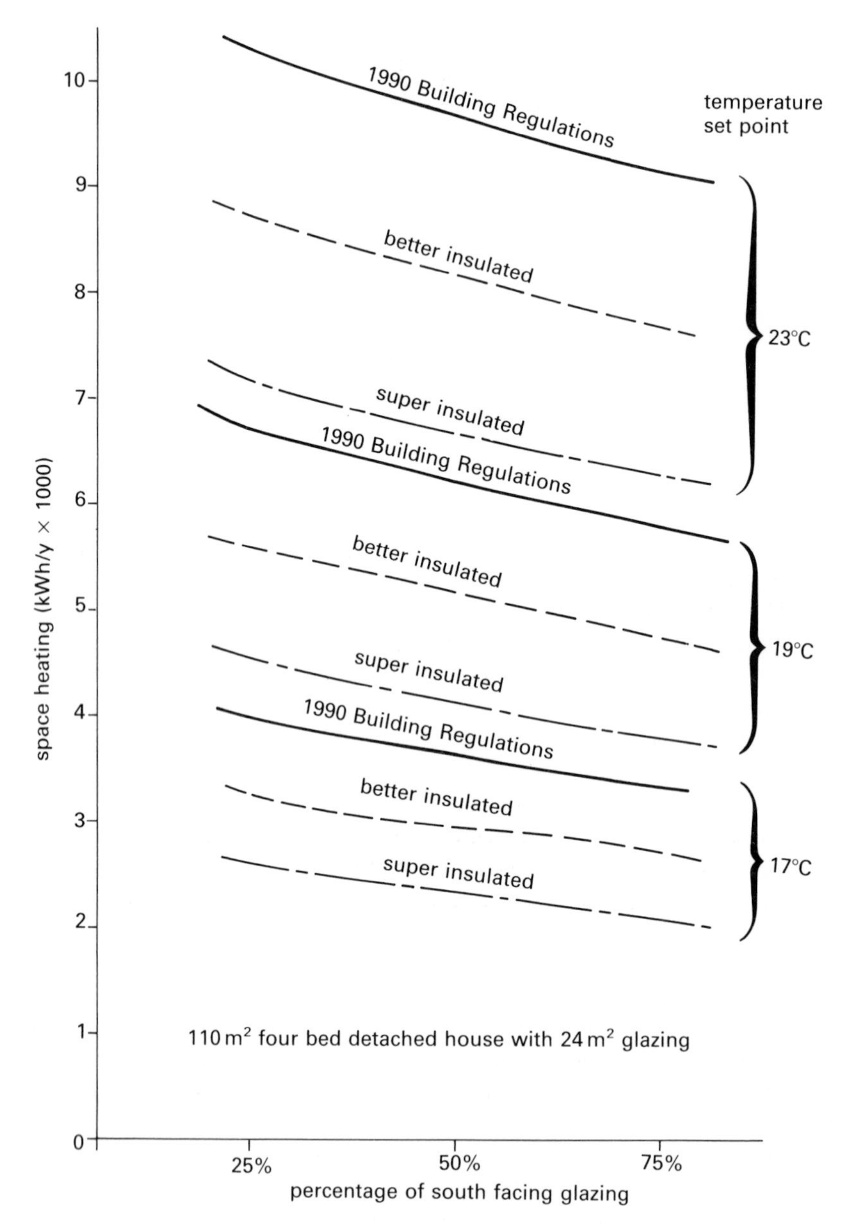

Fig. 12.14 Space heating as a function of south facing glazing percentage.

1.0 W/m^2 K (Robinson & Littler, 1991). It was hoped that with such low U-values this system would out-perform other glazing systems. However, it was found that the solar transmission was reduced to such an extent that it outweighed the improvement in insulation. Thus current research is concentrating on improving solar transmission while maintaining good insulation.

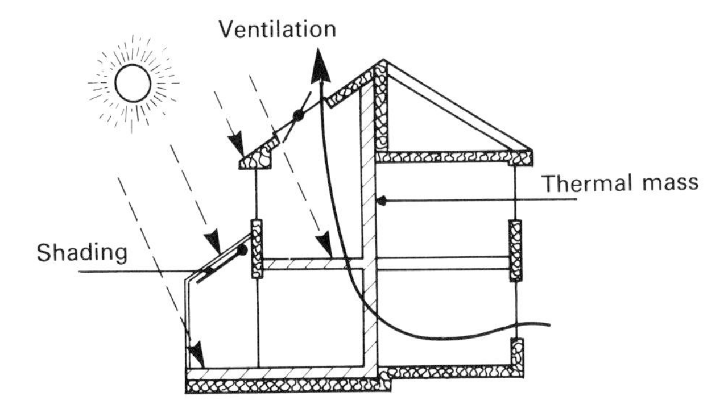

Fig. 12.15 Solar control by a combination of shading, ventilation and thermal mass.

Summer overheating is a danger when using large areas of south facing glazing. This danger can be overcome by shading, thermal mass and ventilation (see Fig. 12.15). These techniques need to be kept in mind when planning the elevations, the construction and the ventilation strategy.

12.6.2 Conservatories

Carefully designed conservatories can not only improve the thermal performance of a dwelling but can also add much in terms of amenity and extra space. However, unless care is taken in the design, the conservatory can increase heating costs. It is important to understand how conservatories are meant to function (see Fig. 12.16).

- The buffer effect. Temperatures in the conservatory will be higher than outside. This will reduce heat losses from the wall covered by the conservatory. In addition air leakage from the house will be reduced by the conservatory.
- Heat gains. During sunny weather the temperature in the conservatory will rise above the house temperature with the potential for heat gains to the house.
- Ventilation preheat. If the ventilation air for the house is introduced through the conservatory it will be preheated thereby reducing the need to heat ventilation air for the house.
- Summer cooling. Shading the conservatory is important to reduce heat gains in summer. Venting the hot air from the top of the conservatory can be used to assist house ventilation.

The conservatory should be thought of as a passive element which can improve the performance of the house but will only be usable for part of the year. If the conservatory is heated to extend its use then it will always increase energy costs since heat loss will outweigh the heat gain and preheat benefits.

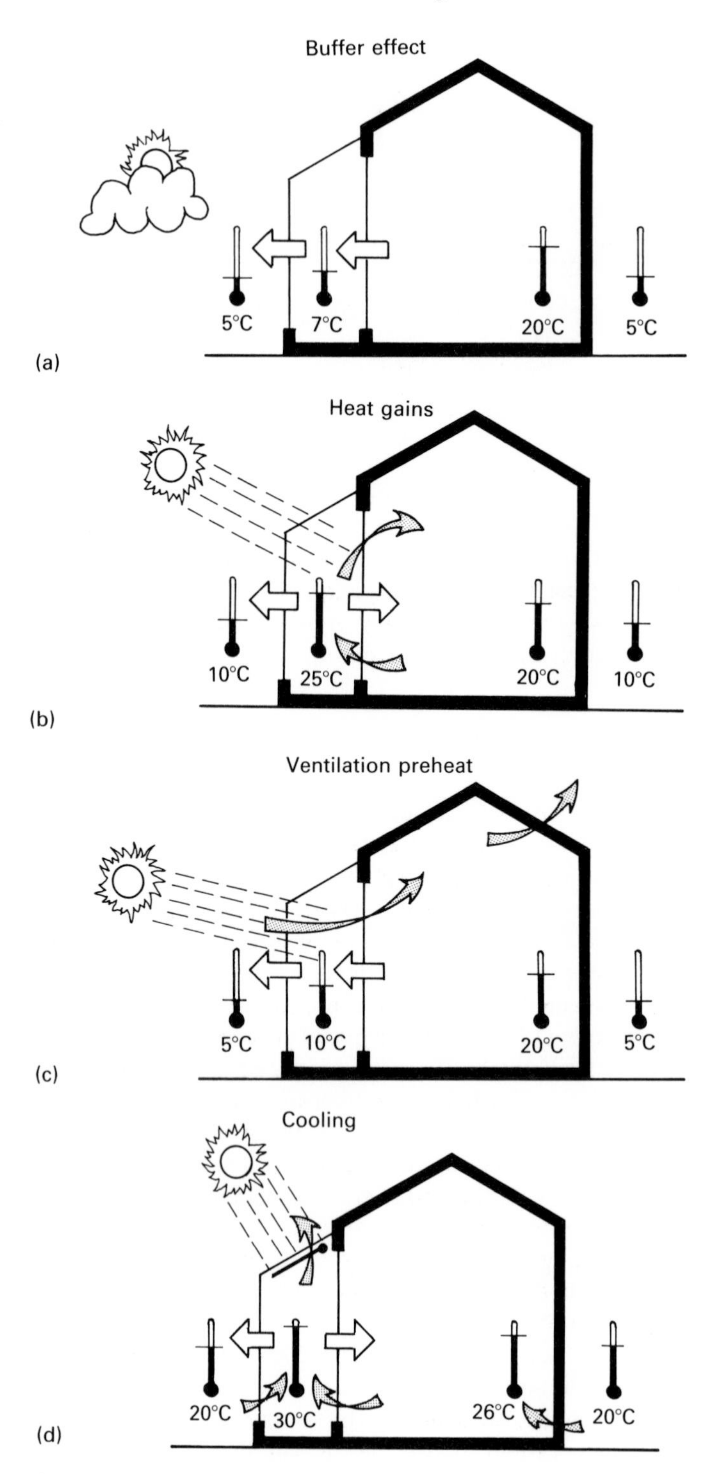

Fig. 12.16 How conservatories function.

12.7 Heating and hot water systems

Many well insulated carefully designed houses have been let down by the choice of heating system. A poor choice of heating system and the type of control can easily double the running cost of a house. The design of the heating system needs to be an integral part of the overall house design.

In small, well insulated houses, heating loads will be only a few kilowatts. For instance, the peak heating load of the 70 m^2 superinsulated house illustrated in Fig. 12.1, is only 1.2 kW. With such small loads it is difficult to justify using full central heating.

Small gas or electric room heaters have a role to play in heating small houses. Using these appliances can substantially reduce the cost of the heating system which can help to pay for additional insulation. A 1990 study (Connaughton & Musannif, 1990) has shown that, by using the trade-off between the cost of the heating system and insulation costs, low energy housing can be built for little or no extra cost. Even in larger houses with peak heating loads of around 5 kW it is difficult to justify using full central heating, but market demands often necessitate the use of this form of heating.

12.7.1 Central heating

The following checklist may help in the design of the central heating system:

- Use the most efficient heat source consistent with the fuel available.
- Keep the heat source (e.g. boiler) within the insulated envelope where the casing losses can add to the useful heat gains.
- Use a time controller which suits the expected occupancy. A simple-to-read, 24 hour time switch might be the most appropriate solution for the elderly; while more 'mobile' occupants might do better with a seven day programmer.
- Think carefully about temperature control. Figure 12.14 shows how small increases in temperature can dramatically increase energy consumption. It is essential to maintain close control over temperature. It is also important to make sure the system responds to incidental gains – for instance if the thermostat is in an unheated north facing lobby it will not take advantage of solar gains to the south facing rooms. Thermostatic radiator valves will give temperature control on a room-by-room basis but it is important that the boiler is switched off when all the rooms are up to temperature.
- Domestic boiler energy managers can be used to give overall control of boiler flow temperatures. They are used to vary the flow temperature in accordance with external temperature.
- In larger houses think about zone control with independent time and temperature control.
- Think carefully about the domestic hot water system.

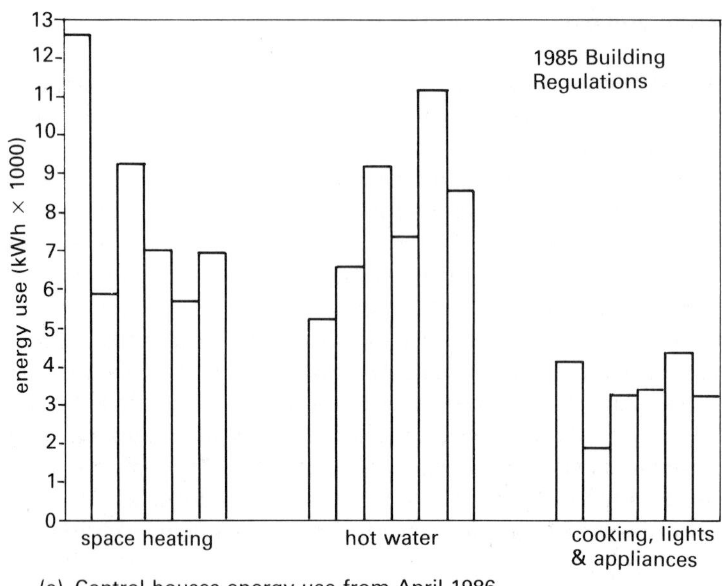

(a) Control houses energy use from April 1986 to March 1987.

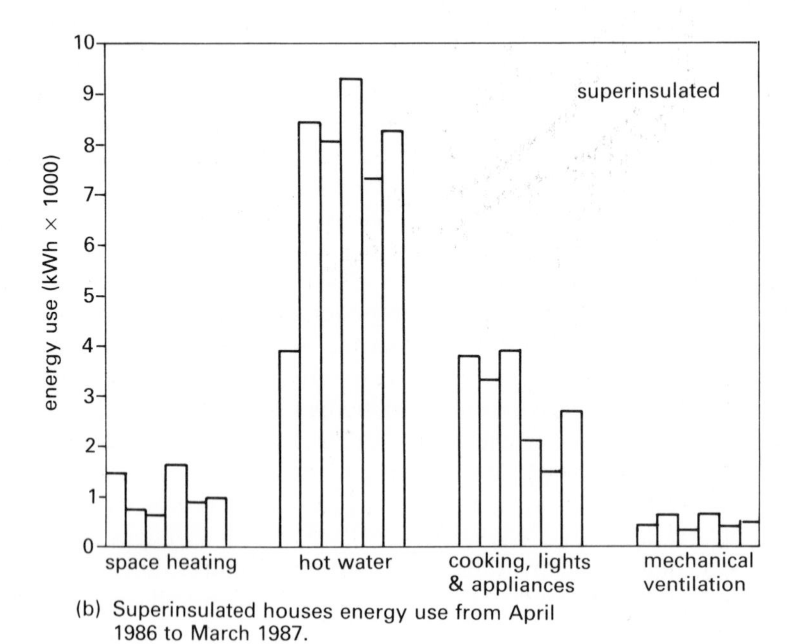

(b) Superinsulated houses energy use from April 1986 to March 1987.

Fig. 12.17 Energy used for hot water is of the same order as for space heating in moderately insulated houses (a) and can be substantially more than space heating use in well insulated houses (b).

12.7.2 Hot water supply

In well insulated houses energy used for hot water heating is usually of the same order as that used for space heating. Often, as shown in Fig. 12.17, it is substantially greater. Thus, in this case, the major use of the central heating system is not to heat the house but to heat hot water. This fact, together with the desire to remove the cold water storage from the cold loft space, combine to make the provision of hot water in low energy dwellings much more significant than ever before.

The problem is two-fold; first to generate hot water in such a way that the efficiency of the heating and hot water system is not adversely affected and second, to get sufficient head to run a shower.

Combining the heating and hot water supply in a conventional heating system using a hot water cylinder can lead to inefficiencies because the boiler will need to be sized to cope with both loads. For most of the time the boiler will be running on part-load at reduced efficiencies. This is less of a problem with condensing boilers but could well reduce the seasonal efficiency of a conventional boiler by anything up to 5%.

Another inefficiency in conventional systems is the heat transfer into the cylinder. This can be improved by using a cylinder with an extended heat exchange surface such as the IMI 'Supercal' range of cylinders. Conventional cylinder-based systems also have other problems. They lose heat, particularly via the primary pipework which is seldom well insulated, and it is difficult to provide sufficient head for the shower. Mains pressure cylinders are available but these leave the householder with a maintenance problem. Another solution is to use a

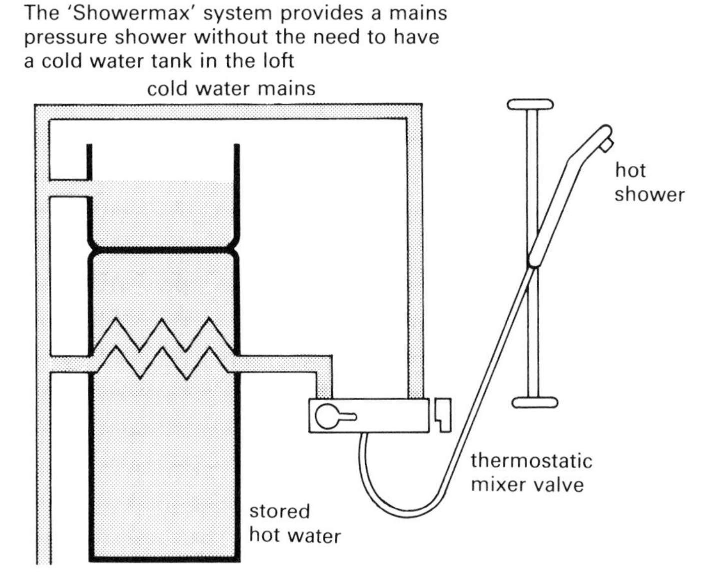

Fig. 12.18 The 'Showermax' system.

Appliance	Electricity consumption (kWh/yr)	
	Typical	Best available technology*
Fridge/freezer	1727	62–184
Freezer	789	91
Clothes dryer	786	60–264
Electric cooker	600	210
Colour TV	339	113
Dishwasher	249	0
Washing machine	70	35
Total	4260	593–897 kWh/yr
Cost	£325	£45–£68 per year[+]

* These figures are based on a survey of American appliances carried out in 1984 (Olivier, 1986).
[+] Electricity price 7.64 p/kWh.

Fig. 12.19 Typical energy use for electrical appliances.

Item	100 W tungsten	20 W plce lamp
Tungsten bulbs 8 @ 50 p	£4.00	
Electricity costs		
100 W × 8000 h × 7.64	£61.12	
Total	£65.12	
Plce lamp 1 @ £13.50		£13.50
Electricity costs		
20 W × 8000 h × 7.64 p		£12.22
Total		£25.72
Saving		£39.40

Fig. 12.20 A comparison of costs of tungsten and fluorescent lamps.

'Showermax' cylinder which has a separate coil to heat mains pressure cold water for the shower (Fig. 12.18).

But perhaps it might be better to do away with the cylinder altogether. Other options are to use instantaneous gas or electric water heaters or a combination boiler. The latter is an attractive option in smaller dwellings particularly that now condensing 'combi's' are available. Further savings can be made by specifying spray taps and low flow showers.

12.8 Lights and appliances

Designers tend to concentrate on reducing energy costs for heating and hot water production. But, in well insulated houses, electricity costs for lights and appliances are of the same order, if not more, than those for heating, cooking and hot water (see Fig. 12.1).

It may be thought that there is not a great deal that the designer can do to reduce energy used in appliances. However, as Figure 12.19 shows, the potential for reducing costs is enormous.

The designer can do much to cut energy costs of electric lighting by specifying low energy fluorescent lamps instead of the usual tungsten lamps. Figure 12.20 shows the total costs of compact fluorescent lamps compared to tungsten lamps over the life of the fluorescent lamps.

During the life of the fluorescent lamps the initial cost is saved three times over by savings in electricity costs. With these economics it is a wonder that compact fluorescent lamps are not used in all new homes.

References

BRE (1989) Thermal Insulation: *Avoiding Risks*, Report BRE 143 Building Research Establishment, Watford, HMSO, London.

Butti, K. & Perlin, J. (1981) *A Golden Thread*. Marion Boyars, London.

Connaughton, J. N. & Musannif, A. A. B. (1990) *Low Energy Housing at little or no extra cost*. Energy Efficiency Demonstration Scheme Report No. RD 53/42. Building Research Energy Conservation Support Unit, Watford.

Everett, R., Horton, A., Doggart, J. & Willoughby, J. (1985) *Linford Low Energy Houses*. Report ERG050, Open University Energy Research Group, Milton Keynes.

Hedges, A. (1991) *Attitudes to Energy Conservation in the Home*. HMSO, London.

Kadulski, R. (1988) *Residential Ventilation: Achieving Indoor Air Quality*, Solplan Review, Vancouver, Canada.

Lowe, R. & Chapman, J. (1985) *The Pennyland Project*. Report ERG053, Open University Energy Research Group. Milton Keynes.

NBA Tectonics (1988) *A Study of Passive Solar Housing Estate Layout*. Report No. ETSU S 1126, Energy Technology Support Unit, Harwell.

Olivier, D. (1986) *Energy Efficiency and Renewables: Recent North American Experience*. Energy Advisory Associates, Hereford.

Pitts, G. (1989) *Energy Efficient Housing: A Timber Frame Approach*. Timber Research and Development Association, High Wycombe.

Riddle, R. A. J. (1991) Private Communication referring to work by YARD for the Department of Energy R & D Programme.

Robinson, P. & Littler, J. (1991) *Courtyard Passive Solar Houses*, CEC Report No. SE 432/85. CEC DG XVII.

Ruyssevelt, P. (1987) Ventilation and Heat Recovery in Superinsulated Houses. In *Superinsulation* (Ed. P. Ruyssevelt), pp. 54–67 UK International Solar Energy Society Conference (C47).

Ruyssevelt, P. (1987a) Experience of a Year Monitoring Four Superinsulated Houses. In *Superinsulation* (Ed. P. Ruyssevelt), pp. 76–89 UK International Solar Energy Society Conference (C47).

Shorrock, L. D. & Henderson, G. (1990) *Energy Use in buildings and Carbon Dioxide Emissions*. Building Research Establishment, Watford.

Turrent, D., Doggart, J. & Ferraro, R. (1980) *Passive Solar Housing in the UK*. ECD Partnership, London.

Uglow, C. E. (1989) *Background Ventilation of Dwellings: A review*. Report BR162, Building Research Establishment, Watford.

Chapter 13

Energy design of dwellings using the BRE domestic energy model

George Henderson
BSc, MSc, CEng, MIEE
Head of Energy Economics and Statistics, Building Research Establishment

In the last 20–30 years the standard of heating anticipated in dwellings has risen very considerably; architects now need to give their attention to how these increased standards can be achieved in an economic and effective way. Tools for quantifying the energy design have become essential to ensure that the most effective measures are used: the Building Research Establishment has developed a model which aims to be both accurate and easy to use. This chapter describes this model and its application.

13.1 Introduction

A warm home has always been valued in the cool British climate. In recent decades, however, our expectations have risen and we have come to expect all rooms to be heated to a high standard. Cold bedrooms are no longer considered acceptable and the traditional system of heating by one or two open fireplaces is generally incapable of giving the performance we expect. The extent and speed of this change is evident if we consider that in 1960 well under 10% of the housing stock had central heating: that had risen to 75% by 1990.

The expectation of better heating standards has coincided with greater awareness of the need to conserve energy. The energy 'crisis' in the 1970s awakened considerable interest in means of saving energy, including energy used in housing and prompted a series of revisions to thermal insulation requirements set by the Building Regulations.

The problem of poor standards of heating in low income households also received considerable attention, particularly in relation to the elderly. This has given rise to the idea that dwellings should be constructed, or modified, so that they can be heated at a cost that less affluent households can afford.

More recently, the problem of global warming has led to renewed interest in reducing energy demand in general. Since housing accounts for well over a quarter of all energy use in Britain, further pressure to raise energy efficiency standards is inevitable.

The demand for better standards of heating and affordable running costs means that architects have to give greater attention to how that can be realised. The idea that the energy performance of dwellings is something requiring *design* is fairly recent: for most architects it has been simply a matter of meeting the thermal insulation requirements set by the Building Regulations.

Three decades ago, the influential Parker Morris report (Ministry of Housing and Local Government, 1961) identified the importance of improved heating standards and described heating systems capable of achieving them. It did not, however, go on to advocate levels of insulation that would ensure that households could afford to run those systems. The experience of the last two decades tells us that affordable running costs are indeed an important consideration.

Energy design for dwellings has grown in significance to become an important part of the overall design process. To carry it out competently requires not only a good understanding of the general principles of heat flow but also tools for *quantifying* those flows of energy. Quantitative analysis is desirable to ensure that the measures applied are the most effective in reducing energy demand and give the best value for money.

13.2 BREDEM

BREDEM (the Building Research Establishment Domestic Energy Model) was developed to make estimates of energy requirements in dwellings, aiming to be both easy to use and to give reliable results (Anderson *et al.*, 1985). It is based on practical experience of how energy is used in dwellings, gleaned from measurements made in many houses and backed up by laboratory measurements where appropriate. It represents a synthesis of the information available and is updated as new information becomes available.

The approach adopted by BREDEM is to identify the various uses of energy in dwellings and to make an estimate of requirements for each use. Figure 13.1 shows the main flows of energy in a typical dwelling. Space heating requirements are calculated from the physical details of the dwelling, the performance of the heating system, and internal and external temperatures. Particular emphasis is given to obtaining realistic estimates of internal temperatures throughout the dwelling, taking account of the living patterns of the occupants. Other uses of energy are estimated from average consumption derived from surveys, taking account of the composition and activities of the household.

Internal energy gains resulting from other uses of energy are taken into account in the calculation of space heating energy needs, along with solar and metabolic gains. This has a very significant effect on the amount of energy required for space heating, even in a dwelling with modest levels of insulation, and becomes increasingly important as insulation levels improve. BREDEM puts particular emphasis on obtaining realistic estimates of those energy gains and on the heating patterns adopted by households in practice. It is this emphasis on how the dwelling is used that distinguishes BREDEM from earlier methods that were suitable for sizing heating systems but unsuitable for estimating annual fuel requirements.

There are a number of different versions of BREDEM, each tailored to a particular application and identified by a version number. For example, a version designed to estimate energy requirements for a specific house and household requires information of the heating standards and appliance ownership of that

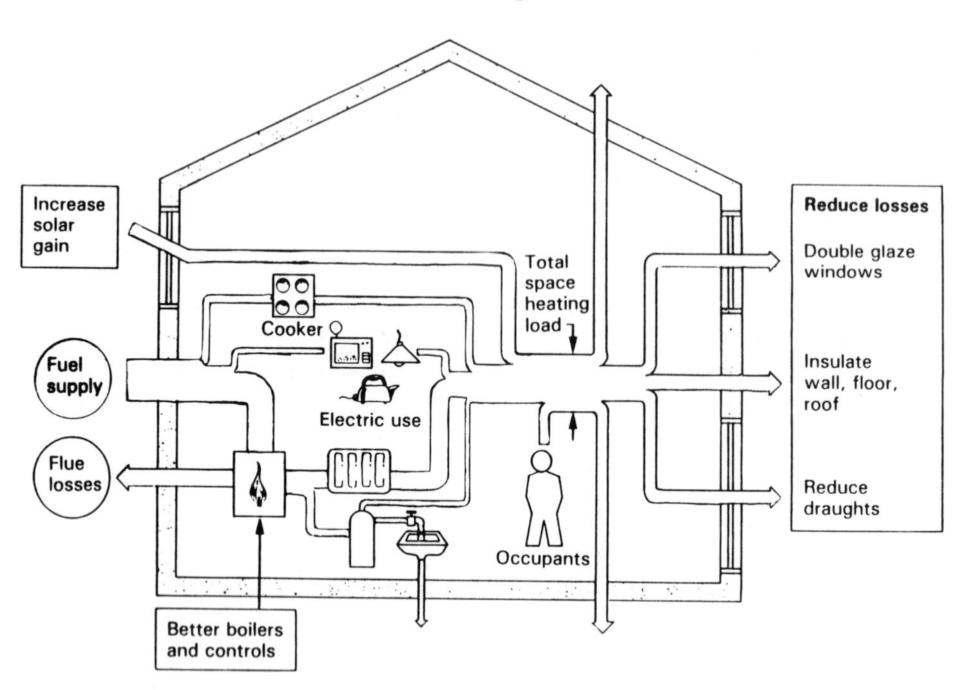

Fig. 13.1 Main flows of energy in a typical dwelling.

particular household. However, a version aimed at evaluating design options in new houses would require standardized estimates relating to typical households. Versions also need to suit the level of computational support available, ranging from worksheets and hand-held calculators to desk-top personal computers.

BREDEM is subject to continuing development to expand its range of application and to take account of new research. Work is currently in progress to develop BREDEM as a design tool for passive solar houses. This is being undertaken in conjunction with the Energy Technology Support Unit, making use of the findings of its extensive programme of design studies, simulation and monitoring of passive solar houses.

BREDEM performance has been assessed by making comparisons with actual energy consumption in a large number and variety of dwellings. Recent comparisons using data from low energy houses in Milton Keynes have shown it to be applicable to houses with levels of insulation that are well beyond the current requirements of the Building Regulations (Shorrock *et al.*, 1991).

13.3 Important applications of BREDEM

BREDEM was originally developed to give the Departments of Environment and Energy a consistent basis for making estimates of annual energy use in dwellings and the effects of applying energy efficiency measures. It has also been widely used outside Government and is incorporated in a number of commercial computer programs.

The Building Regulations for England and Wales, and their counterparts for Scotland and Northern Ireland, were upgraded in 1990. The revision introduced requirements for better levels of insulation but also greater flexibility in how the Regulations may be met. As previously, it is possible to meet the Regulations by specifying prescribed U-values for individual elements such as roofs and walls. However, it is also possible to show compliance by using calculations to show that the annual energy requirements of the building are no greater than they would have been had the prescribed U-values been applied.

For dwellings, a special version of BREDEM has been approved for this purpose. This is described in detail in a BRE booklet, which also gives instructions on how to use accompanying worksheets (Anderson, 1989). It is also available as the computer program *Regulations Calculator*, available from Energy Advisory Services.

BREDEM has also been included in the British Standard Code of Practice on the *Energy Efficient Refurbishment of Housing* (BS 8211, Part 1) and its associated *Designer's Manual* (BSI, 1988). This version of BREDEM is available in worksheet form and also as programs for hand-held, lap-top and desk-top computers. BRE is also contributing to the development of European standards under the auspices of CEN, building on the experience gained from BREDEM.

BRE has encouraged commercial exploitation of BREDEM through licensing agreements with suppliers of software. Apart from *Energy Calculator* and *Regulations Calculator* referred to above, a number of other programs have been published based on BREDEM. *Energy Designer* is a program for architects and others involved in the design of new houses. *Energy Auditor* and *Energy Assessor* are suitable for assessing existing houses on an individual basis. *Energy Targeter* is principally intended for Local Authorities and Housing Associations who wish to assess and improve the energy performance of their stocks of existing houses.

This use of BREDEM has arisen as Local Authorities and Housing Associations have come to attach increasing importance to energy performance in their stock in recent years. Hard-to-heat properties have had numerous problems with condensation and mould growth, making them unattractive to tenants and costly to maintain. Many authorities now have an energy policy for their housing stocks, aimed at ensuring that they can be heated at an affordable cost.

13.4 Energy design

The energy performance of a dwelling is only one of many aspects that the architect has to take into account and only rarely will it be given a high priority by the client. It is nevertheless an aspect of design where significant improvements can often be made at little or no additional cost. What is needed, therefore, is a means of quantifying energy performance that can be used on a routine basis by people who are not energy specialists. It must be neither unduly time consuming nor demanding of specialist knowledge or equipment. However, it must be able to tell the user how features of the design affect energy performance, whether or not those features are specifically aimed at improving energy efficiency.

The aspects of energy performance of a house that matter most are the ability of the heating system to maintain comfortable temperatures throughout the house under all weather conditions and the amount of energy required for heating in an average year. The former is the traditional province of the heating engineer who will usually be able to find a system to cope with whatever level of insulation is installed.

Annual energy requirements, however, are largely determined by the design of the house and, particularly, the level to which it is insulated. The layout of the house, and the orientation and size of its windows are also aspects of design that affect energy requirements. The options for influencing energy efficiency economically are far greater at the time of construction than when the building is completed, so it is important that they are fully considered during the design process.

A good design tool needs to be easy to use and to be able to give estimates of how much annual energy requirements change as aspects of the design are applied or modified. The level of detail adopted by BREDEM has been found to be appropriate for a general purpose design tool that can be used on a routine basis by architects. Much more detailed energy simulation procedures are available for studies of particular aspects of design but, for the moment at least, they are too time-consuming and need too much specialist knowledge to be suitable for general use.

At the other end of the scale, 'rules of thumb' may be useful for general guidance on what measures are likely to be effective but are incapable of estimating the size of the benefits that can be obtained from applying different measures in different dwelling types. For example, improved wall insulation will have a much larger benefit in a detached house than in a flat and a more efficient boiler will produce greater reduction in energy demand in a poorly insulated house than in a well insulated house. Those particular examples may be self evident but others are not, especially those involving comparisons between increased insulation and improved heating system efficiency.

Good design tools are needed if energy calculations are to become a routine part of the design of dwellings. The next chapter describes an energy design tool based on BREDEM, for use with a personal computer. It also includes a method for specifying the energy performance of dwellings in terms of annual running cost.

References

Anderson, B. R., Clark, A. J., Baldwin, R. & Milbank, N. (1985) *BREDEM – The BRE Domestic Energy Model: background, philosophy and description*, BRE Report BR 66, BRE.

Anderson, B. R. (1985) *Building Regulations: conservation of fuel and power – the 'energy target' methods of compliance for dwellings*, BRE Report BR 150, BRE, 1989.

British Standards Institution (1988) *A designer's manual for the energy efficient refurbishment of housing*, (published with BS 8211, Part 1), BSI.

Ministry of Housing and Local Government (1961) *Homes for today and tomorrow*, HMSO, London.

Shorrock, L. O., Macmillan, S., Clark, J. & Moore, G. (1991) BREDEM 8, A monthly calculation method for energy use in dwellings: testing and development, In *Proceedings of Building Environmental Performance Conference*, Canterbury 1991, BEPAC.

Chapter 14

The Milton Keynes Energy Cost Index: a low cost energy design tool

Peter F. Chapman
BA, MA, PhD, FRSA
Professor of Energy Systems, Open University

The Milton Keynes Energy Cost Index is an energy performance standard used by the new town Development Corporation in Milton Keynes to ensure dwellings have an energy performance better than that now required by the Building Regulations. Several thousand dwellings have now been processed using this standard. The standard is applied using a microcomputer program based on the Building Research Establishment's domestic energy model. The success of the scheme in Milton Keynes has led to the widespread use of BREDEM-based programs and the recent introduction of national energy labelling.

14.1 Introduction

The new town of Milton Keynes in the United Kingdom has been planned and supervised by a new town Development Corporation. The Milton Keynes Development Corporation (MKDC) was designated in 1967 and work on the green field site began in 1969. The city now has a population in excess of 160 000 and has a target completion population of 200 000 by 1999. The MKDC is financed by purchases of agricultural land which it then sells for development at commercial rates. The Corporation exercises a tight control on all development and imposes requirements over and above the normal building and planning regulations.

A series of energy projects carried out in the city between 1972 and 1980 demonstrated the viability and cost effectiveness of energy conservation and passive solar techniques (MKDC, 1982, Chapman & Lowe, 1985, Everett *et al.*, 1985, Everett, 1986). Following on this the MKDC designated one development area as an 'energy park' in which all developments would need to demonstrate a high level of energy efficiency.

The Energy Park, when complete, will consist of 1200 dwellings and 1 000 000 ft^2 of commercial development on a site of some 300 acres. To meet the energy objectives of the Energy Park an energy performance standard was devised, based upon the total annual running costs of a dwelling under standard occupancy assumptions. In order to be able to assess the energy performance of the dwellings the Corporation, in collaboration with the Energy Efficiency Office (Dept. of Energy), commissioned the development and production of a microcomputer program. The specification for the program included the following:

(1) Architects with no previous experience of using computers should find the system easy to use.
(2) It should take no longer than 30 minutes to provide an assessment of a dwelling and should require only data normally available to architects.
(3) The program would be endorsed by the Building Research Establishment, the official government research laboratory concerned with all aspects of buildings.
(4) The basis of the program should be such that it was acceptable to the major fuel industries.

In addition to the function of imposing a performance standard it was recognised that the program would also have an educational value in that it should provide users with a learning environment in which they can quickly assess the energy effects of alternative measures.

The resulting computer program is known as *Energy Designer* and incorporates the Milton Keynes Energy Cost Index (MKECI). It was developed in 1984 and was first used in 1985 to assess the 50 dwellings on the 'Energy World' exhibition site used to publicize and open the Energy Park. It was then used for the 550 houses of the first phase of the Energy Park. By late 1987 it was clear that the system was so well received, by developers and householders alike, that it was extended to all dwellings constructed within the city of Milton Keynes. This means that every dwelling proposed for construction within the designated area of Milton Keynes has to meet an energy standard defined by the MKECI program. To date the program has been used in the assessment and improvement of more than 5000 dwellings.

14.2 The definition of the Energy Cost Index

There are three stages involved in the definition of a performance standard. The first is the definition of an appropriate measure of performance; in the case of an energy standard this is not straightforward. The second is the definition of the standard to be achieved (in terms of the measure of performance). Finally a procedure has to be defined whereby the proposed dwelling is assessed.

14.2.1 The measure of performance

During the design of the MKECI program considerable attention was paid to what was meant by an 'energy efficient' dwelling. The various measures of performance, or options, that were considered, or that had been used by other schemes, were:
(1) A checklist of fabric and heating system features. In order to 'pass' the dwelling would have to equal or better each item on the list.
(2) An extension of the then current Building Regulations that would include

lower U-values and specifications for items not then included in the Regulations (such as the floor U-value and air infiltration).
(3) A standard expressed in terms of delivered energy per m^2. The energy used to be calculated by an agreed procedure.
(4) A standard expressed in terms of primary energy per m^2, calculated by an agreed procedure.
(5) A standard based on the calculated energy running costs for the dwelling.

Options (1) and (2) were rejected because they are too prescriptive. They do not give the architect/developer scope to make trade-offs between different aspects of the dwelling. It was also felt that any scheme that was too prescriptive would be opposed by developers.

Options (3) and (4) were rejected because they give misleading results for electrically heated houses. There are two reasons for this. Using the delivered energy effectively masks the power station losses. Electric systems appear to have the lowest rating but are not actually more efficient in either cost or energy terms. To overcome this a primary energy standard has been used. However this fails to reflect any difference between on and off-peak electricity.

Option (5), basing the assessment on estimated energy costs, overcame all these objections. It also had the very significant advantage that it related directly to the interests of the final consumer, the householder. As such, this made the system more acceptable and open to use as a marketing device.

Once the decision to use cost as a measure of performance had been taken then a host of subsidiary questions needed to be resolved. The most important were as follows:

(1) What costs should be included? The final system included an estimate of all the energy-related running costs for the dwelling. These were identified as the space heating fuel, water heating fuel, cooking fuel, electricity used in lights

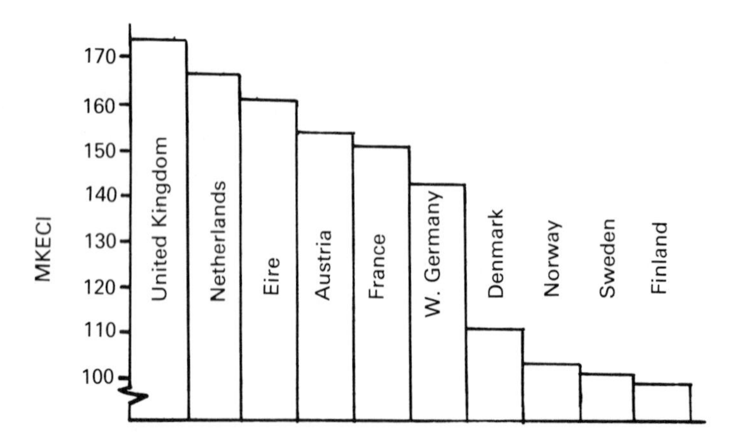

Fig. 14.1 The MKECI for dwellings built to the Building Regulations applying in various European countries in 1984.

and appliances, all standing charges (imposed as part of the fuel tariff) and essential maintenance costs.

(2) How to cope with varying fuel prices. Since the system was to be used to impose a standard it was necessary that the standard did not change significantly with changes in fuel prices. The solution adopted was to make use of a standard set of fuel prices which were retained for at least a year. The standard was also indexed such that one house always had a fixed rating.

(3) What occupancy to assume? The aim of the assessment was that it should be the dwelling, in its particular location and with its stock of insulation and equipment, that was being assessed – not any characteristics of the occupants. Thus a standard occupancy was defined in much the way that a standard driving cycle is defined for assessing the fuel performance of cars.

14.2.2 The Milton Keynes Energy Cost Index

The obvious basis for the index is the total running costs per square metre of floor area of the dwelling. However this does not quite work because the fixed charges make a larger contribution in small dwellings than in dwellings with a large floor area. In order to avoid this 'size effect' the annual costs are adjusted by subtracting a fixed term. The resulting adjusted costs per square metre are then indexed and multiplied by a scaling factor so that the MKECI for a typical house is in the range 100–200. The technical basis of the MKECI calculations has been described elsewhere (Chapman, 1990).

The resulting scale means that a dwelling conforming to the Building Regulations in 1985 achieved an MKECI in the range 200–150; a dwelling meeting the 1990 Regulations would have an MKECI in the range 170–130. A Scandinavian standard dwelling achieves an MKECI in the range 80–90; super-insulated dwellings can get down to 60–70 if they have efficient heating systems (Fig. 14.1).

14.2.3 The standard

Once the basis of the index had been established it was then necessary to determine the standard to adopt. The standard was the performance level that developers had to prove before they could build in the Energy Park. Three approaches to this issue were adopted:

(1) The first was to look at the results of the many field trials carried out in Milton Keynes and decide what level of insulation and improved heating appeared to be cost effective and easy to incorporate into current building practice.

(2) The second approach was to examine the absolute level of the additional costs implied by different levels of MKECI standard. For reasons of acceptability it was decided that the standard should not require extra costs in excess of 2% of building costs.

(3) The third approach was to compare the MKECI with the Building Regulations and standard practice both in the UK and a number of other countries. For the international comparison a standard semi-detached house had its MKECI evaluated using the prescribed U-values for the different countries.

The three approaches all pointed to an MKECI standard at the level of the Pennyland better insulated case. The Pennyland field trial showed significant energy and cash savings for an extra £400 per dwelling. The overall payback time was about two years. The MKECI scale was fixed to make this Pennyland house have an MKECI of 120. The international comparison shows that the 120 standard is better than European practice, but not as good as Scandinavian standards.

14.3 The energy calculation model

Since the proposed energy performance standard was to be incorporated into the planning permission cycle it was clear that the assessment would have to be based upon plans and drawings and not upon physical measurements. The assessment therefore required a calculation procedure which could reliably estimate the energy consumption of a dwelling. In the context of this standard it was more important that the calculation procedure accurately reflected the differences between dwellings rather than predict the absolute level of energy use.

The major constraints on the calculation procedure were that it had to be completed within half an hour and accessible to architects and developers. This placed constraints on the level of data input that was feasible. To be accessible the procedure should be implemented on desktop computers, which in 1984 meant machines with 64k RAM and floppy disk drives. This effectively eliminated simulation models.

The procedure used was one based upon the two-zone, variable base annual degree day model developed by the Building Research Establishment. This model, known as the Building Research Establishment's Domestic Energy Model, BREDEM, has been developed over the last decade (Anderson *et al.*, 1985). The development has been consistently based upon field trial information on the performance of dwellings. One of the early computer versions of BREDEM was checked against data on more than 50 dwellings and found to perform very well (Henderson & Shorrock, 1986).

For application to the MKECI a new version of the model was developed. This involved extensions to the heating system and controls, water heating, ventilation, solar gains and auxiliary fuel use algorithms. At the request of developers algorithms were also introduced for handling conservatories, mechanical ventilation systems and novel heating systems. The technical basis of the calculations has been published by the software developers (Energy Advisory Services, 1987).

14.3.1 Heating systems and controls

The fundamental difference between an energy standard and the thermal part of the UK Building Regulations is that the energy standard includes the heating and other appliances in the dwelling. This feature of the MKECI has caused some resistance since it has called into question some of the traditional practices for specifying heating systems. However, it is clear that in dwellings designed to have a low energy demand the role of the heating system and its controls is critical. For example in a dwelling designed to exploit solar gains it is essential that the heating system is responsive enough, and well enough controlled, to reduce its output when the solar gains are high. This is not the case for a number of heating systems.

It is also obvious that in terms of running costs the role of the heating system is critical since it determines the fuel used and the conversion efficiency. One of the strongest lessons learned by users of the program is that greater savings can be made by improving the heating system and controls than by any level of solar design; indeed the heating system is usually more significant than insulation in terms of savings and pay-back time.

14.3.2 Auxiliary fuel use and internal gains

The use of auxiliary fuel is significant in the MKECI model for two reasons. First the fuel used in water heating, cooking and lights and appliances usually accounts for more than half the total fuel cost. Second the internal gains arising from these fuel uses make a significant contribution to meeting the space heating demands. For example, in the Pennyland house about 30% of the total space heating requirement was satisfied by internal gains and a further 25% by solar gains. Prior to the development of the MKECI, which was concerned solely with low energy dwellings, the auxiliary fuel algorithms had not been developed to any great extent. Thus new algorithms were developed for water heating systems, lights and appliances use and for cooking. These are described in outline below and explained in detail in the Technical Manual of the MKECI.

14.4 The assessment system

The first use of the MKECI system was at the Energy World exhibition in 1986. Both the exhibition and the application of the MKECI were very successful. There was an unofficial competition between the developers to achieve the lowest MKECI figure. This encouraged all the architects and developers to become aware of the factors that affected the energy performance of dwellings. It is likely that the most significant impact of the MKECI has been this educational effect. Most of the houses on the Energy World site achieved MKECI values in the range 80–100; the lowest MKECI, of 45, was achieved by a well insulated dwelling heated by an off-peak heat pump.

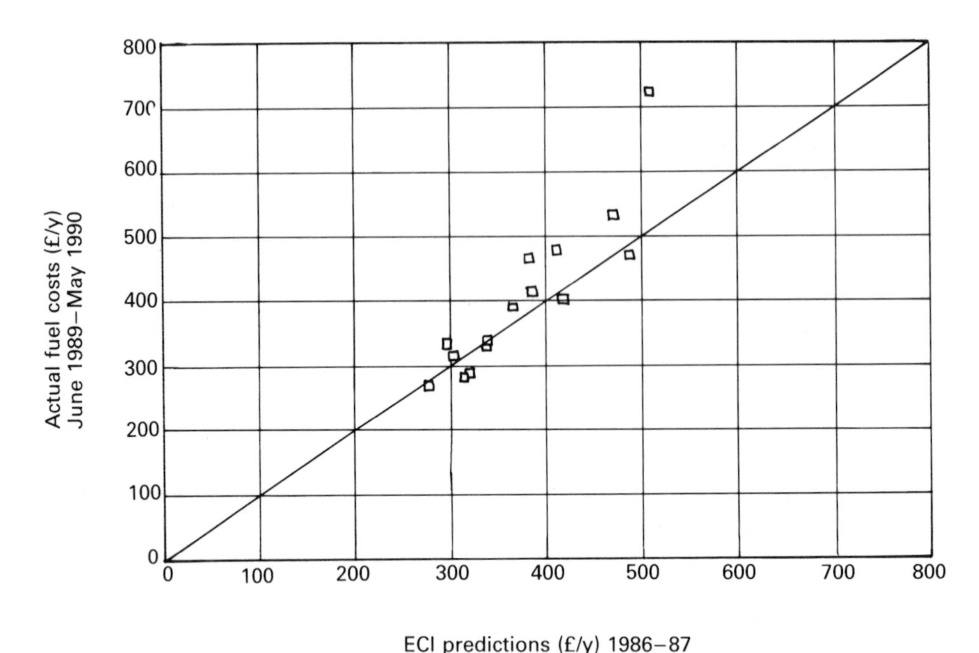

Fig. 14.2 Comparison between the predicted running costs for dwellings on the Energy Park with the values measured 1989–90. Each point represents a group of up to 20 dwellings. The data has been corrected for differences in fuel prices between the predictions and measurements but not for actual occupancy.

The MKECI was formally introduced into the Corporation approval system for the first phase of the Energy Park project in 1987. After preliminary discussions with Corporation staff the developer had to prepare a detailed brief for the scheme and that brief had to include an MKECI evaluation for each house type in the scheme. The MKECI evaluation had to be carried out by one of the Corporation's approved consultants. These consultants included specialist architectural practices as well as energy consultancy firms. Some architects and developers purchased their own copy of the MKECI program.

For a typical detached house the 120 standard could be achieved with a roof U-value of 0.3, a wall U-value of 0.45, 50 mm of insulation in the floor, double glazing and weather stripping throughout and an efficient and well controlled heating system. Some designers opted to increase the solar gain by paying attention to orientation; others used a conservatory since this also provided a sales feature for the house. Some retained higher wall U-values and achieved the standard by using condensing boilers.

Almost two hundred of the houses on the Energy Park have been monitored by a central monitoring facility. Results from this exercise show that the predicted running costs were within 10% of the observed running costs for all but one of the groups of dwellings (this group was 17% underestimated and is the source of complaints by residents) (Fig. 14.2).

The MKECI computer program was originally written in BASIC and made available on CPM, MS-DOS and Apple II computers. The program has evolved significantly since then and is now restricted to MS-DOS computers. The program is easy to use and the complete evaluation of a dwelling can be completed in about an hour; 20 minutes extracting data from plans; 20 minutes entering the data into the computer and 5–15 minutes evaluating combinations of options. The commercial version of the MKECI program is known as *Energy Designer* and is marketed by Energy Advisory Services Ltd. The educational aspect of the program is apparent from the fact that it is currently in use in many schools of architecture and building science departments in the UK.

14.5 Reactions to the use of the MKECI

There have been three main channels of feedback from developers and architects on the use of the MKECI, and an additional channel of feedback from the approved consultants:

(1) At the end of 1985 and early 1986 feedback was formally collected, by questionnaire and interview, from the developers and architects associated with the Energy World site and the initial schemes on the Energy Park.
(2) The second formal channel of feedback was organized by the Milton Keynes Development Corporation which arranged a one day meeting in the autumn of 1987 to receive feedback and comments on the use of the MKECI to date.
(3) The final channel of information is the day to day contact between the officers of the Development Corporation and the developers and their representatives. With a few exceptions the response has been consistently positive and encouraging to go further with the scheme.

Indeed it was the extremely positive response of the developers and builders in 1987 that encouraged the MKDC to extend the application of the MKECI to *all domestic development* within the city of Milton Keynes.

The successful and user-friendly implementation of BREDEM on microcomputers has also encouraged the use of BREDEM in the Building Regulations (which now includes an energy target calculation option). Perhaps most significantly the system gave rise to initiatives to develop a national energy labelling scheme. There was widespread agreement that the MKECI worked, but was too complex to understand for the general public. What was needed was a simpler label, something that graded houses on a ten point scale and was directly related to running costs. The National Home Energy Rating scheme was launched in 1990 and is expected to have a significant impact on both the design of new homes and the refurbishment of existing dwellings (Chapman, 1991).

14.6 Conclusions

The strongest conclusion from the use of the MKECI over the last six years is that developers and builders are willing and able to build to better energy standards provided that there is a system for ensuring that everyone meets the same standard. There is no doubt that for the system to work it was essential that MKDC had both a strong commitment to improved energy performance and the power to enforce compliance with an energy standard.

Another key component in the success of the system was the availability of micro-computers and an appropriate model of house energy performance. Without the access to micro-computers the calculations involved would have either been very time consuming or would have had to have been handled in a relatively inaccessible computing facility. With the machine on the desk architects, developers and consultants have been able to explore and learn very quickly.

The BREDEM model was essential because it combines the right level of accuracy with moderate data requirements. It also has a low computational overhead. The more traditional building simulation models require far more data, are far more computationally intensive and could not have been made available on the desk-top machines of 1984–5. It was also essential that the program was endorsed by the BRE and accepted by the major fuel industries. The monitoring of the Energy Park houses has also given confidence in the BREDEM approach as well as providing further field data to support future development.

At a more practical level there are strong grounds for believing that over the coming years 5000 households in Milton Keynes will be spending £250–£150 less than they otherwise would have done on fuel bills. It is also significant that many of the local developers and architects now have a far greater appreciation and understanding of the factors that affect the energy performance of dwellings.

References

Anderson, B. R., Clark, A. J., Baldwin, R. & Milbank, N. O. (1985) *BREDEM – BRE Domestic Energy Model: background, philosophy and description*, Building Research Establishment, Watford.

Chapman, P. F. & Lowe, R. J. (1985) *The Pennyland Project: Report on Studies carried out 1977–84*, (2 vols), Energy Research Group, Open University, Milton Keynes.

Chapman, P. F. (1990) The Milton Keynes Energy Cost Index, *Energy and Buildings*, **14**, pp. 83–101.

Chapman, P. F. (1991) *The development of the National Home Energy rating*, Building Environmental Performance 1991; BEPAC Conference, Canterbury 1991 (available from BRE, Watford).

Energy Advisory Services Ltd (1987) *The technical specification of the Milton Keynes Energy Cost Index*, EAS Ltd, Milton Keynes.

Everett, R. C., Horton, A., Doggart, J. V. & Willoughby, J. (1985) *Performance of Passive Solar Houses at Great Linford*, Energy Research Group, Open University, Milton Keynes.

Everett, R. C. (1986) The Pennyland and Linford low energy housing projects, *Heating & Ventilating Engineer*, **59**(680), pp. 21–4.

Henderson, G. & Shorrock, L. D. (1986) BREDEM – The BRE domestic energy model: Testing the predictions of a two zone version, *Building Services Engineering Research & Technology*, **7**(2), pp. 87–91.

MKDC (1982) *Energy Projects in Milton Keynes: Energy Consultative Unit Report 1976–81*, Milton Keynes Development Corporation.

Non-Domestic Buildings

Chapter 15
Solar design in non-domestic buildings

Ian McCubbin
BSc, PhD
Energy Technology Support Unit, Harwell Laboratories

Over the last ten years the Department of Energy has sponsored a programme of research directed towards investigating and promoting the application of solar design principles in non-domestic buildings. This programme has complemented work sponsored by the Energy Efficiency Office, and managed by the Building Research Establishment, which has promoted many other aspects of energy conscious design in building, refurbishment and operation. The results from these two programmes of work, which are very closely related, will be disseminated in the Energy Efficiency Office's Best Practice Programme.

The field of design is vast, in terms of building type, desired performance and energy use, and a careful review of both the building brief and energy use profile has been the starting point of the studies. The advantages of incorporating solar design principles are obtained both through energy savings and improved amenity.

This chapter describes the work that has been carried out and seeks to develop a strategic approach to solar design from the information obtained so far.

15.1 The Department of Energy's passive solar programme

The early stages of the passive solar programme explored the range of solutions that might be suitable for use in this climate. Design teams and acting clients, experienced in specialist fields of building, were set up. Theoretical energy inputs for designs using solar features were compared with designs that did not have these features. Building designs were assessed using simulation studies to identify those that might contribute the most significant energy benefits.

The first series of design studies considered eight types of building: low and medium rise offices; light industrial buildings; sports halls; nurses hostels; hotels; schools (retrofit) and DIY superstores.

Each of the design teams included within their design appropriate solar features from the following range:

- Building orientation.
- Built form.
- Window location, size and shape.
- Thermal mass.
- Shading.
- Movable insulation.
- Reflective surfaces.
- Atria.

Building type	Reference building annual energy consumption (kWh/m² GFA)[1]	Capital cost	Design building annual energy consumption (kWh/m² GFA)[1]	Capital cost	% reduction (increase) in energy usage	% reduction (increase) in energy cost[2]	% increase in building cost
Light industrial building	Heat – 49.8 Light – 65.6 115.4	£871 220	Heat – 40.9 Light – 27.5 68.4	£903 580	40	49	3.7
Low-rise office	Heat – 41.0 Light – 53.2 94.2	£2 574 052	Heat – 54.5 Light – 20.4 74.9	£2 633 324	20	43	2.3
Medium-rise office	Heat – 66.0 Light – 55.4 121.4	£7 622 130	Heat – 59.3 Light – 20.4 79.7	£8 051 240	37	51	5.6
Hotel bedroom block	Heat – 88.6 Light – 21.2 109.8	£384 860	Heat – 59.1 Light – 19.0 78.1	£417 670	24	27	8.5
Nurses hostel[3]	Heat – 187.1 Light – 23.1 210.2	£249 410	Heat – 70.7 Light – 26.4 97.1	£264 560	53	31	6.1
Secondary school (retrofit)	Heat – 138.0 Light – 17.2 155.2	–	Heat – 123.8 Light – 23.7 147.5	–	(9)	(18)	–
Sports hall	Heat – 97.8 Light – 52.7 150.5	£540 030	Heat – 80.2 Light – 34.9 115.1	£593 920	23	27	10
DIY superstore	Heat – 20.6 Light – 126.6 147.2	£847 680	Heat – 56.8 Light – 78.2 135.0	£958 430	8	28	13.1

(1) Delivered energy.
(2) Calculation based on rates of 1.26 p/kWh (gas) and 5.52 p/kWh (electricity) (efficiencies of 70% and 100% respectively).
(3) Design located in Scotland and analysed using appropriate weather and cost data.

Fig. 15.1 Designs produced in phase 1 non-domestic design studies.

- Conservatories.
- Trombe walls.

The designs were studied using energy simulation, cost analysis and client commentary and were judged from the viewpoints of energy, cost and amenity. Figure 15.1 summarizes the results of these studies. In those building types where lighting is a major consumer of electrical energy, such as the office designs and the industrial building design, good daylighting coupled with electrical lighting control systems resulted in significant energy savings. Space heating saving from passive solar design appeared somewhat less promising, especially in buildings where internal heat gains from equipment or machinery were likely to be significant.

Subsequent design studies have built on the information collected in the first phase. This subsequent generation of studies is looking at both direct gain and atrium solutions, with priority being given to office, industrial and school buildings.

Conclusions from these studies are that solar space heating is unlikely to produce a major contribution to energy saving in many non-domestic buildings but that solar driven natural ventilation can be a useful way of reducing summer overheating and the need for expensive air-conditioning. Aspects of this particular issue are under scrutiny within atrium design studies.

Detailed studies are now underway to complement this information by monitoring a variety of non-domestic buildings. By the end of 1991, 16 different passive solar non-domestic buildings, ranging from offices with atria through to schools and sports halls, will have been monitored for their energy performance. Their capital costs will be analysed and occupants interviewed regarding their view of the buildings amenity. Figure 15.2 lists the buildings under study.

Results emerging from these monitoring projects support the design studies conclusions that daylighting and natural ventilation are worthwhile areas of development. This is due not only to the energy issues involved but also because building occupants broadly express a preference for the environmental conditions achieved by well-designed natural lighting and ventilation. It has become clear that the design of the building controls for human comfort are particularly important. Monitoring studies have often demonstrated lack of user satisfaction with both manual and automated forms of control in relation to both thermal and visual environments. It is clear that designers must pay particular attention to the design of the controls to achieve a system that will be reliable and functional.

15.2 Lessons for non-domestic passive solar design

A number of the important points emerging from the Department's solar programme have been touched on above. The detail of these results will be spread to architects and designers by means of the Energy Efficiency Office's Best Practice Programme. This work brings together passive solar considerations and other efficiency measures, to indicate ways of reducing energy demand.

Title	Type	Market sector	Passive solar measures	Location
South Staffordshire Water Co. HQ	Office	Private – commissioned and occupied by client	Direct gain; daylighting (light – shelves); natural ventilation; lighting EMS; insulating glazing	Walsall, B'ham
Looe school	Junior & infants school	Public – Cornwall LEA	Direct gain; mini 'Trombe' bench massive; high insulation	Looe, Cornwall
JEL HQ	Light industrial (factory and office)	Private – commissioned and occupied by client	Direct gain; daylighting EMS; 'atrium' with warm air distribution	Hazel Grove, Stockport
Heaps	Light industrial (distribution and office)	Private – refurbishment commissioned and occupied by client	Direct gain; daylighting	Hoylake, Birkenhead
Gateway II	Office	Private – commissioned and occupied by client	Atrium; daylighting; natural ventilation	Basingstoke, Hants
Ystradgynlais community hospital	Hospital	Public – Powys Area Hlth. Auth.	Daylighting	Ystradgynlais, Swansea
Boyatt's hostel	Sheltered accom. for physically handicapped	Public – Hampshire CC	Modification of microclimate	Eastleigh, Hants
Nabbott's school	Junior school	Public – Essex LEA	Atrium	Chelmsford, Essex
Netley Police HQ	Office/ educational	Public – Hampshire CC	Daylighting; natural ventilation	Netley, Hants
Romsey sports hall (1)	Sports hall	Public – Hampshire CC	Daylighting; top lit	Romsey, Hants
Romsey sports hall (2)	Sports hall	Public – Hampshire CC	Daylighting; top lit	Romsey, Hants
Warsash college	Engineering college	Public – Hampshire CC	Daylighting; natural ventilation	Warsash, Hants
Christopher Taylor court	Sheltered housing for elderly	Private – speculative development by BVT	Direct gain; night insulation shutters	Bourneville Village, B'ham
Netley school	Junior school	Public – Hampshire CC	Conservatory	Netley, Hants
Hasbro	Office	Private – speculative builder	Daylighting	Stockley Park, Heathrow

Fig. 15.2 Monitoring studies on non-domestic buildings.

A broad overview of the studies can be used to develop a design strategy to help designers in their approach to solar design. There are three important elements to this strategy:

- Context.
- Target.
- Measure.

15.2.1 Context

Any design takes place in a situation defined by a wide range of impinging factors. The context of energy considerations may be both global and specific.

The global context in this situation might include information on differing fuel types and their significance in buildings. For example, the reduction of electricity consumption, because it is the most CO_2 intensive fuel, can lead to environmental benefits as well as energy savings.

It is also important to understand from precedent studies the energy characteristics of the building under consideration, so that the magnitude of possible savings through design can be appreciated. This information is increasingly available in the form of the 'best practice' guides and case studies. The reduction of space heating demand may be a significant benefit in a building where space heating is the major consumer of energy, but in a highly serviced building a more effective reduction in energy demand may be achieved by reducing the lighting and air conditioning load.

The specific context is provided by the individual demands of the building, the client and the site location:

(1) The client context includes items which are usually part of the brief, such as occupancy patterns, flexibility in use, equipment use, cost constraints, internal comfort conditions and any building zoning required by the client.
(2) The locational context is developed from a detailed site appraisal, and includes consideration of microclimate, access to site, service constraints etc.

Development of a detailed list of contextual issues can lead not to a narrowing of the design opportunities but to a flowering of the design from the increased awareness of the possibilities. This is well illustrated by a recent study published in the *Architect's Journal* (Technical Studies, 1990), where the environmental zoning of the building for exploitation of solar energy was considered in parallel with the functional zoning of the building for the client's needs (Figs. 15.3 and 15.4).

15.2.2 Target

The 'target' is the performance objective that the designer sets out to achieve in that 'context'. In this situation targets can be considered as being either energy or environmentally related.

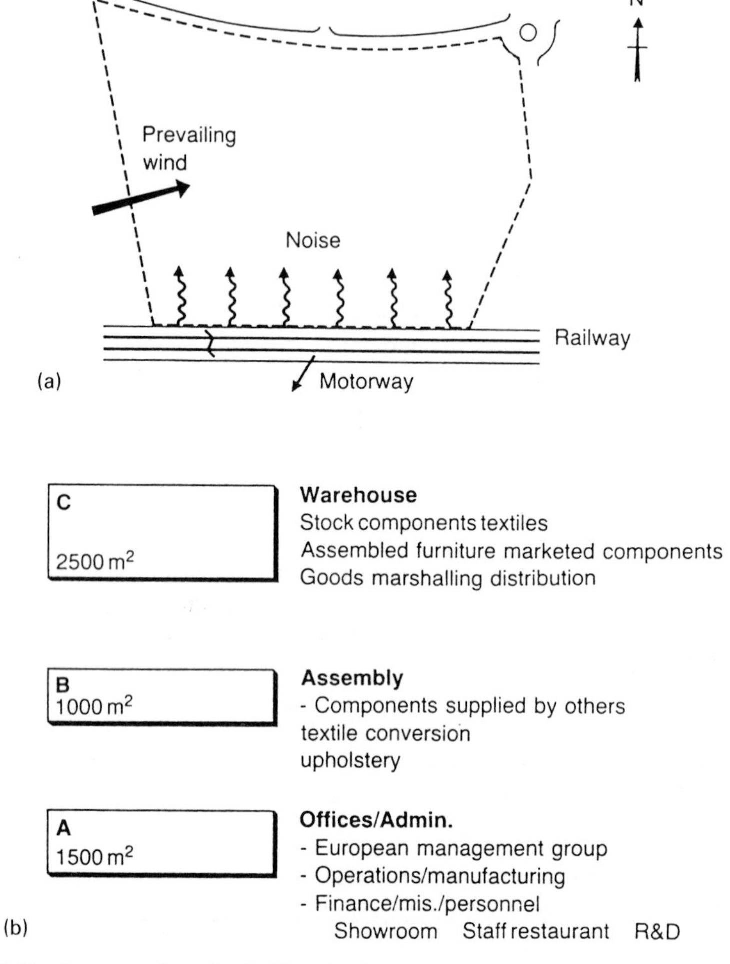

Fig. 15.3 Furniture manufacturing facility: development of scheme (a) site plan (b) space requirements.

Passive solar design targets might be set to reduce energy consumption by:

- Reducing space heating demand.
- Maximizing natural lighting and thus minimizing electrical lighting demand.
- Reducing or eliminating the need for energy use in cooling the building by the use of solar driven natural ventilation.

Environmental targets might include:

- Provision of enhanced comfort or amenity by better distribution of natural lighting levels.
- Avoidance or amelioration of discomfort from overheating, glare etc.

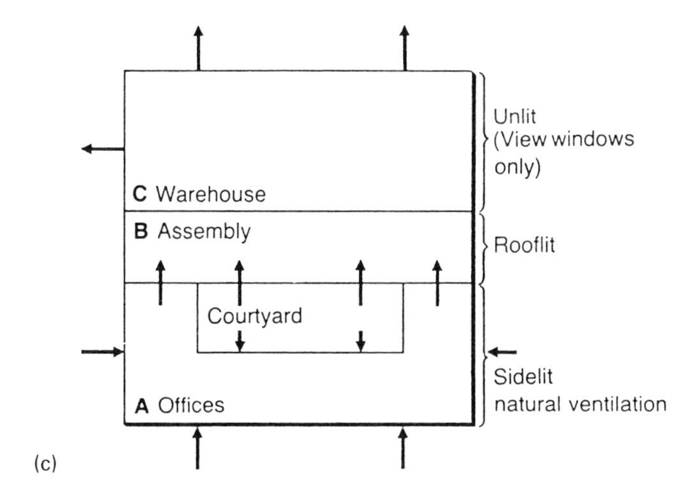

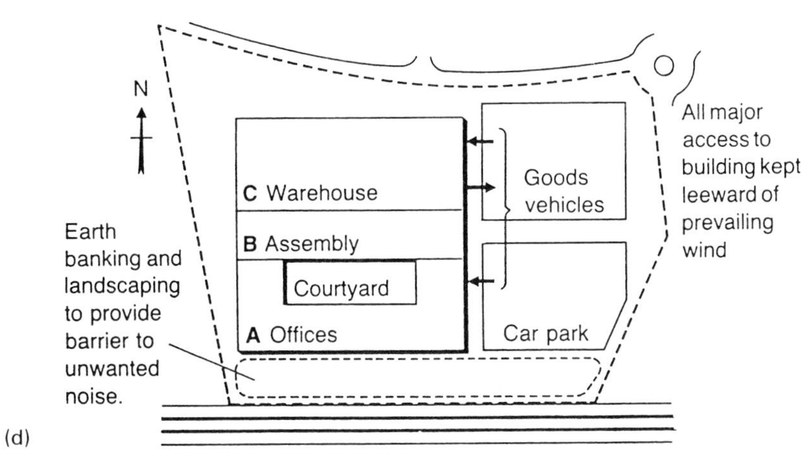

Fig. 15.3 continued. Furniture manufacturing facility: development of scheme (c) plan matching solar and spatial requirements (d) final factory location.

15.2.3 Measure

The 'measure' is the mechanical means by which the designer might achieve the target, for example by manipulation of the window area in a facade. However, as any change in window area will have repercussions on solar gain, daylight quality and quantity, risk of glare, and the view of outside, it can be seen that 'target' and 'measure' are interrelated.

15.3 Concluding points

These are intended to complement the energy strategy developed above.

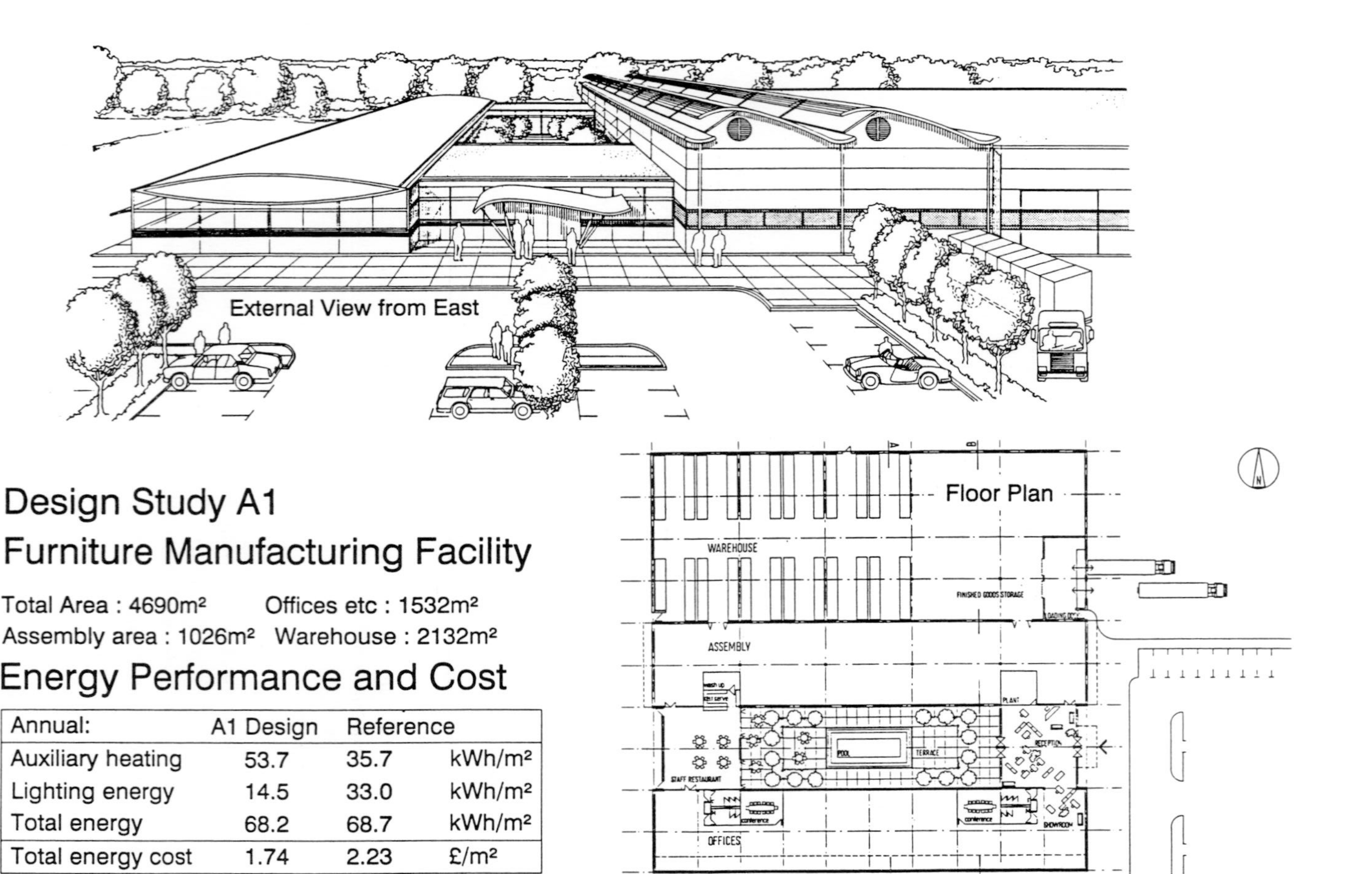

Annual:	A1 Design	Reference	
Auxiliary heating	53.7	35.7	kWh/m²
Lighting energy	14.5	33.0	kWh/m²
Total energy	68.2	68.7	kWh/m²
Total energy cost	1.74	2.23	£/m²

Fig. 15.4 External view and data on Jestico and While's final design for a furniture manufacturing facility.

(1) Relate the cost of the measures under consideration to the likely scale of energy cost savings. If the use of energy for lighting is £2/m^2 gross floor area, the target is to save 50% of lighting energy, and the client desires that energy savings are achieved within a five year payback, an appropriate capital cost for this measure might be of the order of £5/m^2 gross floor area.
(2) Try to ensure that the design concepts and measures are simple and robust; they are more likely to be successful. This applies not only to their operation and understanding by the occupants but also to issues such as maintenance.
(3) Be aware of the effect of design decisions at the earliest possible stage by using available design tools and techniques, for example the CIBSE window design manual (*CIBSE Applications Manual*, 1987) or the use of an artificial sky for daylight analysis (Littlefair, 1989). The early stages of design, involving the form and massing of the building will determine the potential of a design to exploit passive solar options.
(4) Integrate passive solar measures within the logic of the design as a whole. 'Add on' features are often less effective and may compromise some other aspect of design.

Inherent in the decision to utilize solar radiation and daylight as free sources of energy has to be the acceptance of the variability of external weather conditions. The internal conditions may consequently be subject to greater variability than that of a highly regulated environment. This could be perceived to be a key benefit of solar design, but it is essential that the design accommodates this variability as well as any changes imposed by modifications to the building use. The designer needs to be aware of any limitations of the design, through the development of design scenarios using means such as dynamic simulation models, so that the best options can be developed from an early stage.

References

Architect's Journal (1990) Passive Solar Factories, *Architects' Journal*, **191**(5) (31 Jan).

CIBSE (1987) Applications Manual – Window Design.

Littlefair, P. J. (1989) Measuring Daylight – The Effective Use of Scale Models, In *Conference Proceedings*, *UK-ISES conference on Daylighting Buildings*.

Chapter 16

Low energy strategies for non-domestic buildings

Nicholas Baker
BSc, MA, PhD
Director of The Martin Centre for Architectural and Urban Studies, University of Cambridge

The energy performance of non-domestic buildings varies widely, and is the result of a range of interacting factors. Some of these factors relate to decisions made early in the development of the design. Here we examine the need for strategic guidance at this stage. The role of prescriptive advice and quantitative energy design tools is discussed.

16.1 Introduction

Energy consumption in non-domestic buildings, of similar use type, varies over a range of about tenfold without any obvious correlation between user satisfaction and productivity, and this energy use. We would be surprised to find two equivalent family saloon cars with a fuel consumption ranging from 6 to 60 litres per 100 km! How much of this range of energy performance can be accounted for by the design, and can the key design parameters be identified?

We can propose that the performance of a building is due to three factors:

(1) That which is inherent in the building design.
(2) That which relates to services systems design and efficiency.
(3) That which is due to the effect of occupant behaviour.

We refer to these as *building*, *systems* and *occupant* factors.

16.2 Independent factors

First we can consider the building, systems and occupants as *independent* factors. The final performance of the building would be a product of these three factors. For example the decision to adopt a deep plan will result in a certain building performance. This creates a demand for heating, cooling and lighting in order to satisfy the conditions for an idealized occupant.

Next, the systems will require energy to provide these environmental conditions. The delivered energy demand will depend upon the system efficiency. This will vary with the choice of system design, e.g. centralized air handling with heat recovery compared with perimeter heating with tempered air and comfort cooling. These efficiencies are relatively predictable, but may carry an uncertainty,

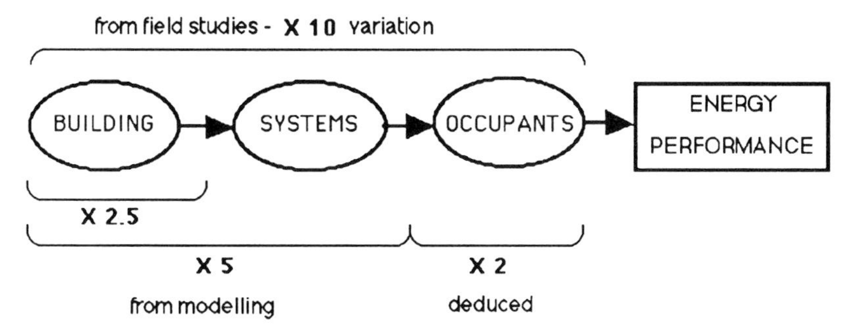

Fig. 16.1 The relative contribution of building, system and occupant factors on energy use.

e.g. quality control, commissioning etc., the main causes of a system to operate less efficiently than anticipated.

Finally, the occupants (and management) will operate the systems in a non-optimal fashion – e.g. having set temperatures too high, heating unoccupied rooms, and leaving lights on. These factors are the least well known, and modellers of building performance have only just begun to recognize their importance.

We could imagine a building with inherently good environmental performance but being equipped with poor services systems and with poor management. This may use more energy than a building of poor design but with highly efficient services and management. Similarly we could imagine other combinations, and only when all three factors are optimized will we get the best performance.

What are the relative magnitudes of the building, systems and occupant factors? In studies (Steemers, Baker & Hawkes, 1991) on energy consumption in relation to urban built form, using an integrated energy model we concluded that if systems and occupant factors are fixed, the range of energy use is unlikely to be greater than about 2.5 fold. This is comparing a fully air-conditioned artificially lit tinted glass 'horror story', with a naturally ventilated and lit building with an atrium. If the system parameters such as lighting loads, ventilation rates, etc., are allowed to range over plausible values the energy consumption range increases to fivefold. This implies that occupant factors typically account for a twofold variation, in a total variation of tenfold (Fig. 16.1).

Thus the building factors influence the building performance little more than half as much as the joint influence of system and occupant factors. This could be seen as rather disappointing for those endeavouring to develop strategic design tools, responding to basic building parameters. The designer may question the need to optimize glazing ratio if the energy performance is going to be swamped by the luminaire efficiency or environmental management policy. He may be tempted to resort to stylistic, or short term economic considerations to generate the building concept, believing that the final energy performance will rest in the hands of the engineer and the management.

However, one possible argument for special consideration for the building factors is that these factors are the least easy to change. The building factors could only be changed by major remodelling of the building whereas system efficiency could be improved by retrofitting new systems or modifying or simply improving tuning of existing systems, and occupants could be persuaded and trained to operate the building more efficiently.

16.3 Interacting factors

It can be argued that these factors may not be entirely independent. For example, where occupants have access to openable windows they are likely to be much more tolerant to ranges in thermal conditions than in sealed air-conditioned spaces. This has been observed as long ago as 1971 by Humphreys. It is also readily observed that as daylight levels fall at the end of the day, people will happily work at much lower lighting levels than would be acceptable in an artificially lit space. It appears then that people are much more ready to accept what mother nature has to offer than what the engineer thinks is good for them! These are examples of where building factors interact positively with occupant factors.

There can also be interaction between the inherent characteristics of the building and the performance of the systems. For example, a building with low ceilings, all lightweight finishes, and high internal gains, will place much greater demands on system design and control than a heavyweight building with larger room height to depth ratio. The latter may only need systems to intervene under extreme conditions, giving much less opportunity for the system to indulge in gross inefficiency such as simultaneous heating and cooling.

Thus it may be that strategic decisions for design have much wider implications for the actual performance of buildings over their lifetime. This has largely been missed due to the limitations of building performance models.

16.3.1 Building energy models

The first priority in the development of building models has been in the interaction between the building and climate. This is now achieved to a high level of precision with both simulation and correlation methods. Subsequently, rather fewer models have taken on the modelling of the systems and, to some extent, the interaction between system and building. It is the area of occupant effects that is almost totally absent. This includes a whole set of topics such as operation of heating, cooling and lighting controls, window opening, sunshade operation, perception of comfort in relation to activity, dress, spatial location and room layout, and psychological factors.

It is only in the area of light-switching that some progress has been made. In evaluating the energy saving due to daylight, it has been essential to propose some kind of 'behavioural' algorithm to describe manual switching performance (Hunt, 1981).

16.3.2 The random element

Most of our models are causal or mechanistic, and in the case of building and system models, based upon the principles of physics. Mechanisms and processes which are not built into the model will cause differences between real performance and modelled performance which, because unaccounted for, will be interpreted as 'random error'. Each of the subsystems, building, services systems and occupants will be modelled with this limitation, that is, a given model will only have a limited accuracy due to unforeseen processes which are not included in the model.

Whilst we may not be able to say much about the cause of the error, we may be able to build up a knowledge of the likely random deviation from the predicted performance, not from the conventional error analysis, but from a knowledge of the real subsystem – building, services or occupant – and how they perform and interact in reality. For example, we might be able to say that the performance of a certain kind of office building with a certain air-conditioning system is particularly vulnerable to certain uses, that is, it is highly probable that it will perform non-optimally. This random element will almost certainly be strongly influenced by interaction between the subsystems.

We are now in a position to speculate upon an 'ideal' integrated model. It models the building, the services systems and the occupants with causal models which interact with one another. Overlaid upon this is an empirical 'random' model which predicts the magnitude and direction of probable departures from the causally predicted performance.

16.4 Delivered and primary energy

The way energy use has been presented, both from model predictions and field studies, has been misleading and has led to incorrect design strategies. Figure 16.2(a) and (b) shows the monitored energy use of 14 offices in the UK grouped under naturally ventilated and daylit, and air-conditioned. Note that when presented as delivered energy, heating appears as the largest single demand in non-air-conditioned offices. This is misleading since when presented in terms of primary energy, we see that even in the daylit group, heating is less than lighting by 45%.

Although primary energy presents difficulties of definition in rigorous scientific terms it is far more meaningful than delivered energy. For example it relates well to CO_2 production, where power is generated from fossil fuel, and recent global environmental concerns have focussed attention to this. Primary energy also relates well to money cost. On the other hand summing delivered energy units of the calorific value of fuel to delivered electricity is meaningless. It is rather like trying to define the distance from London to Paris by adding the miles in England to the kilometres in France!

The greatest distortion occurs when we look at the data for air-conditioned, artificially lit buildings. In this case, even the delivered energy for heating is no

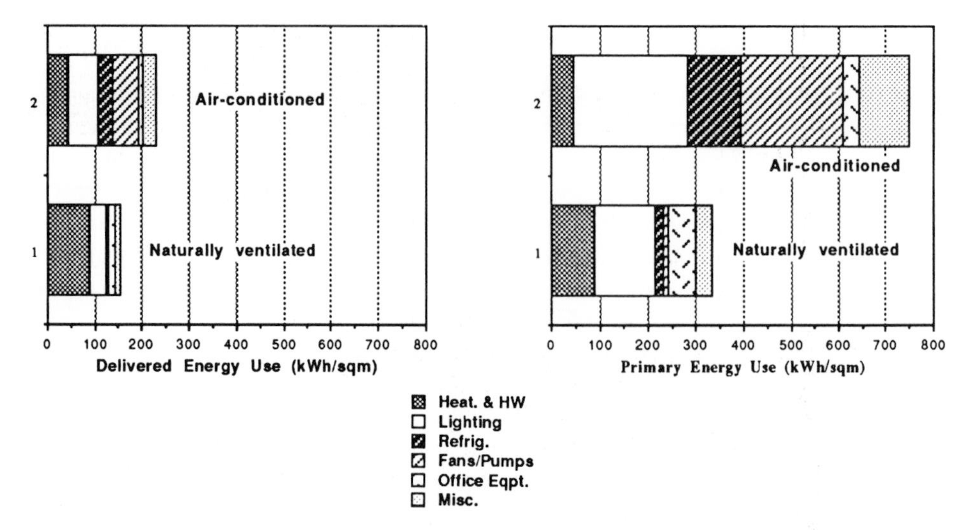

Fig. 16.2 (a) Annual delivered and primary energy use in 14 UK offices. (Source: BRECSU Best Practice Programme, W. Bordass).

Building type		Heat	Cool	Pumps/fans	Light	Total	
BRE low energy office	nv	155	0	9	8	172	Delivered kWh/m^2
		163	0	34	29	226	Primary kWh/m^2
Cornbrook house	nv	85	0	8	50	143	
(condensing boilers)		90	0	29	186	305	
Hereford and Worcester	mm	124	12	19	58	213	
		130	44	70	215	459	
Outer London	ac	81	43	27	70	221	
(heat pumps)		86	158	100	259	603	
Westminster	ac	233	23	124	93	473	
(induct./core)		245	86	459	344	1134	

nv – naturally ventilated, mm – mixed mode, ac – air-conditioned

Fig. 16.2 (b) Breakdown of energy use for five offices nominated to represent 'good practice' showing fivefold variations in primary energy use. (Source: BRECSU Best Practice Programme, W. Bordass.)

longer the largest single component. But when presented as primary energy, heating is twelve times less than the primary energy for lighting, ventilation, pumps and fans.

Historically, this distortion has resulted in a focus of attention towards heating. Technical studies published in the late 1960s and early 1970s promoted the adoption of small windows and deep plans as a means of minimizing surface area to volume ratio, and hence heat losses (Fig. 16.3). We also see a decade of interest in applying passive solar (thermal) techniques to large non-domestic buildings. We now realise that the returns are very small (Dijk & Arkesteijn,

Fig. 16.3 'Office building, Bristol. The Avonbank office block is a thermally efficient building designed with the occupants' needs firmly in mind. A three storey block covering an area of 5264 square metres (56 666 ft^2), it utilizes a heat recovery air-conditioning system to maximize the effects of internal heat gains for heating the building. The design of the building enables a heat balance to be maintained down to an outside temperature of − 4.4°C (24°F). Heat gains from lights, occupants and machinery are recovered through the light fittings, the air being filtered, cooled or heated, and returned to the offices.' (Promotional brochure from the early 1970s.)

1990) and the passive design should have concentrated upon natural ventilation and daylighting as the main measure to save auxiliary energy.

16.5 Strategic design decisions

We now address the question – are there design tools and design guidelines which can assist in making strategic decisions which will increase the probability that a building will perform well as a low-energy user? The difficulty with this kind of tool is that early in the design stage there are few parameters available. If the tool responds to too few parameters, the results may be trivial. If on the other hand the tool requires a large number of parameters both from the building and the services design, the very act of establishing these values will tend to have the effect of freezing the design, even if it is a long way from an optimal solution.

16.5.1 The LT method

An example of a design tool which attempts to satisfy such a brief is the LT method (Lighting and Thermal). This tool responds to parameters available early

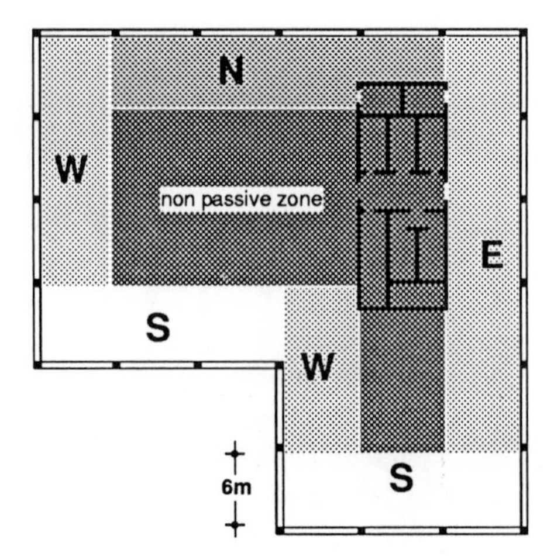

Fig. 16.4 Plan showing passive zones.

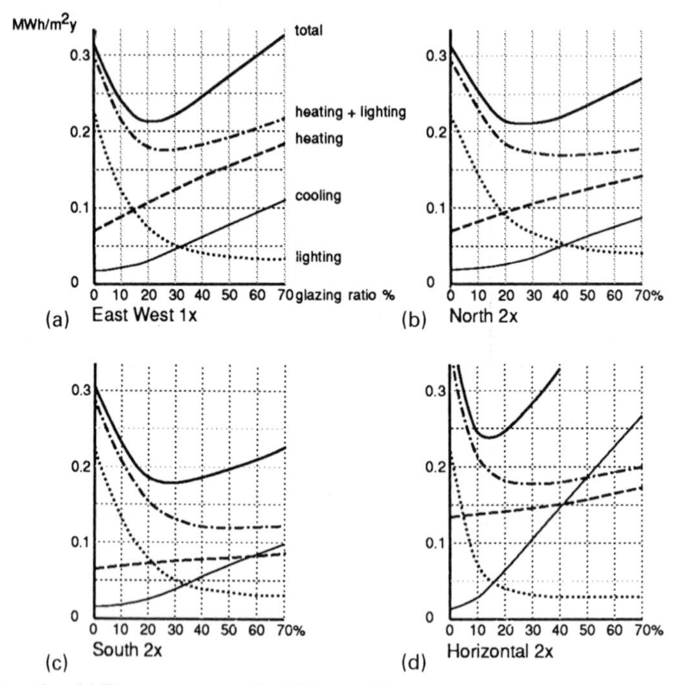

Fig. 16.5 Sample of LT energy curves for UK zone Two.

in design development – plan depth and section, orientation, facade design, the incorporation of an atrium, and the presence of adjacent buildings. It predicts the primary energy consumption for heating, cooling and lighting. It was originally developed for the EC 'Working in the City' Architectural Ideas Competition

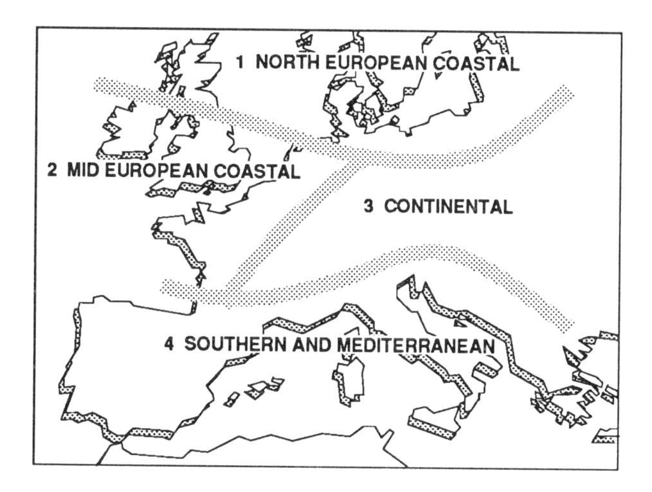

Fig. 16.6 LT climatic zones.

(O'Toole & Lewis, 1990) and it has been described in more detail elsewhere (Baker & Fontoynont, 1990).

Briefly, the method divides the building into passive and non-passive zones (Fig. 16.4); passive zones are those which can benefit from daylighting, natural ventilation, and useful passive solar gain. Non-passive zones have to be artificially lit and ventilated, but do not suffer conductive heat losses or non-useful solar gains. Sets of curves (Fig. 16.5), drawn for different climatic zones (Fig. 16.6), are then used to evaluate an annual primary energy consumption per m^2 for the different zones. These values, after the application of certain correction factors related to overshading, are then used with the appropriate areas to evaluate total primary energy under the catergories of lighting, heating and cooling (Fig. 16.7). The effect of incorporating an atrium, both on heating and lighting energy, can also be evaluated (Fig. 16.8).

16.5.2 Developments of energy design tools

We have earlier established a brief for an ideal strategic design tool – one which not only responds to the initial building design parameters, but would also respond to the interactions between systems and occupant factors as well, and quantify the uncertainty that the particular combination of building, system and occupant will perform as predicted.

The LT method goes only a little way in meeting this brief. With the manual solution it is possible to handle only a very few parameters. However, the model which generates the curves has about 30 input variables, most of which have to be given assumed values. Including the parameters which are input, we have *building* parameters such as room depth, glazing ratio, orientation etc., *system* parameters such as U-values, boiler efficiency and lighting efficacy, and finally

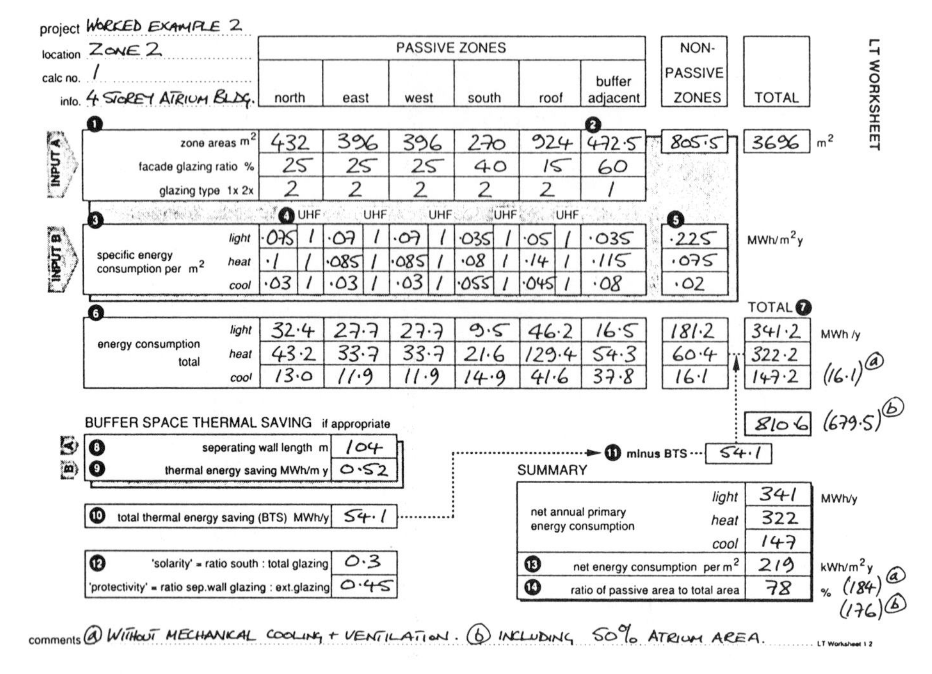

project WORKED EXAMPLE 2
location ZONE 2
calc no. 1
info. 4 STOREY ATRIUM BLDG.

LT WORKSHEET

		north	east	west	south	roof	buffer adjacent	NON-PASSIVE ZONES	TOTAL	
INPUT A	zone areas m^2	432	396	396	270	924	472.5	805.5	3696	m^2
	facade glazing ratio %	25	25	25	40	15	60			
	glazing type 1x 2x	2	2	2	2	2	1			
INPUT B	specific energy consumption per m^2 (UHF) light	·075 / 1	·07 / 1	·07 / 1	·035 / 1	·05 / 1	·035	·225		MWh/m^{2}y
	heat	·1 / 1	·085 / 1	·085 / 1	·08 / 1	·14 / 1	·115	·075		
	cool	·03 / 1	·03 / 1	·03 / 1	·055 / 1	·045 / 1	·08	·02		
	energy consumption total light	32·4	27·7	27·7	9·5	46·2	16·5	181·2	341·2	MWh /y
	heat	43·2	33·7	33·7	21·6	129·4	54·3	60·4	322·2	
	cool	13·0	11·9	11·9	14·9	41·6	37·8	16·1	147·2	(16·1)(a)
									810·6	(679·5)(b)

BUFFER SPACE THERMAL SAVING if appropriate

seperating wall length m	104
thermal energy saving MWh/m y	0·52
total thermal energy saving (BTS) MWh/y	54·1
minus BTS	54·1
'solarity' = ratio south : total glazing	0·3
'protectivity' = ratio sep.wall glazing : ext.glazing	0·45

SUMMARY

net annual primary energy consumption	light	341	MWh/y
	heat	322	
	cool	147	
net energy consumption per m^2		219	kWh/m^{2}y (184)(a)
ratio of passive area to total area		78	% (176)(b)

comments (a) WITHOUT MECHANICAL COOLING + VENTILATION. (b) INCLUDING 50% ATRIUM AREA.

LT Worksheet 1 2

Fig. 16.7 Worked example of the LT method of a 4-storey office block for UK zone Two.

occupant parameters such as occupied hours, and lighting and thermal comfort targets.

The LT model already contains a very rudimentary behavioural element. There is an assumption that as soon as there is no net heating load, shading is deployed so that it eliminates the direct part of the solar radiation. This is in order to keep cooling loads to a realistic value. The model in this version does not, but could, assume that ventilation rate is increased under the same conditions, corresponding to the opening of windows. These are very simple behavioural algorithms but we can speculate about more complex ones.

For example, in a study on simulating the spatial distribution of comfort conditions in a room (Newsham & Baker, 1990) we have found that the number of hours of overheating can be reduced from 514 to 104 per cooling season by simply allowing the occupant to move closer or further from the window. Future versions of the LT model could take account of occupant mobility when evaluating the overheating risk.

16.5.3 Transparency

This issue of behavioural influence and interaction throws up its own particular problems. Any manual method has of course to be severely limited in the number of parameters that it handles. Thus assumptions have to be made. Those concerning building parameters such as U-values etc., are straightforward. But what

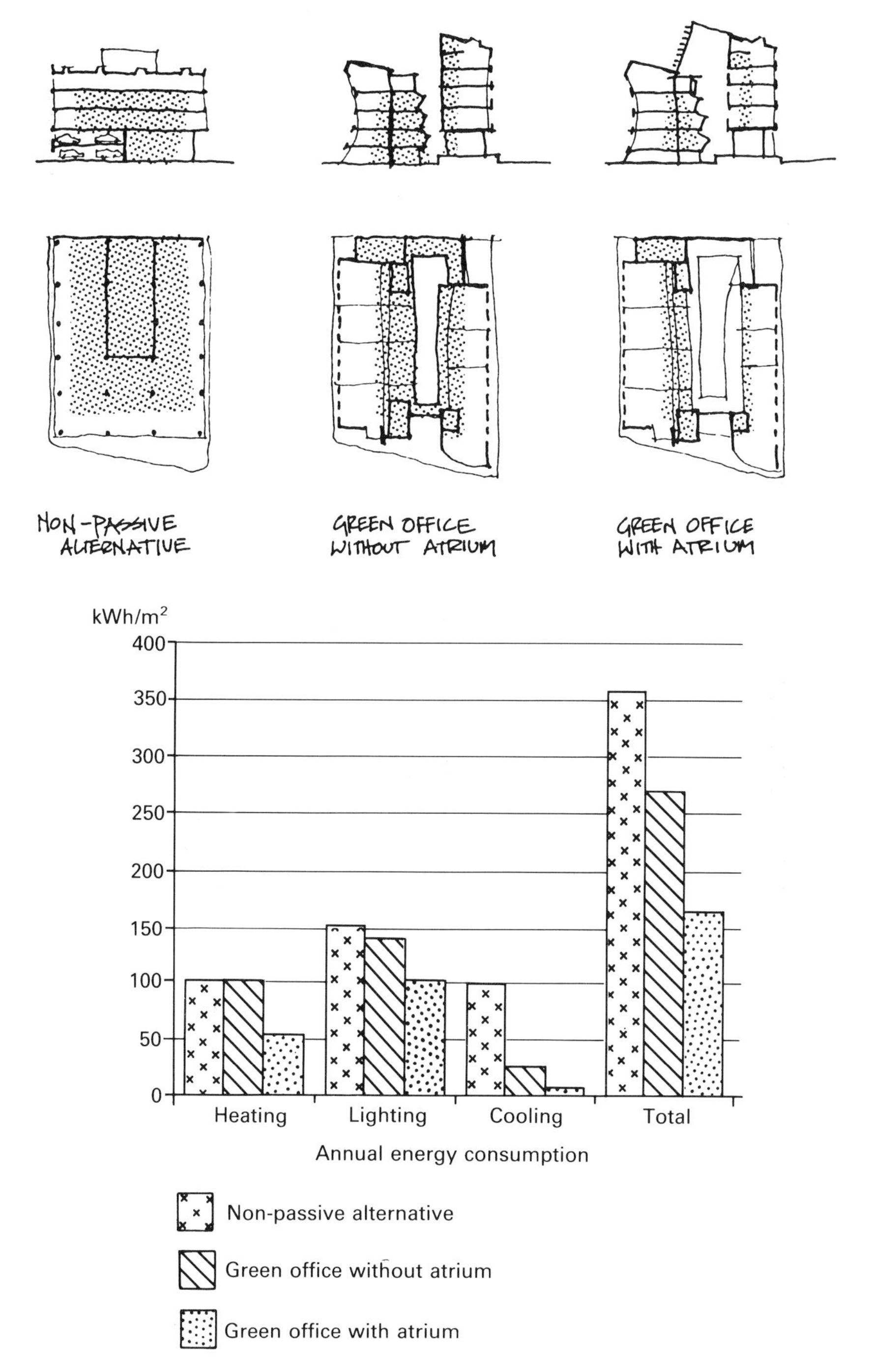

Fig. 16.8 Use of LT method to investigate design proposals. (Source: J. Ouken, Camb MPhil 1991.)

about system operation or behavioural factors? Do we assume that the plant controls are set optimally, and that the occupants do sensible things with the windows and blinds? The danger of making these presumptions is that their importance may be masked from the designer. How can the design tool bring these issues to the designers attention without becoming too specific and thus not truly strategic?

One approach with a manual method is being developed under the BRECSU (Building Research Energy Conservation Support Unit) Best Practice Programme. The project, called Integrated Design Simplified, is to develop the manual LT method further, and to supplement it with detailed design advice. This advice will make reference to LT and thus the two will become mutually supporting. A further feature is that the advice will be building type specific and thus will eliminate much of the redundancy which is characteristic of text book advice.

16.5.4 Multi-parameter design tools

Another approach would be to make the LT model itself (as distinct from the manual method) available as a design tool. But how would the user handle this large number of parameters? The designer would have to provide values for a large number of variables and this would demand that a number of design decisions had already been made. This tends to defeat the object of a strategic model.

A solution to this problem which we are developing is the presentation of results from parametric runs which include the values of the design proposal. For example a three-dimensional graph of energy consumption versus glazing ratio and lighting datum value would be displayed with the design proposal represented as a point on the surface if a lighting value had been input, or a range of points corresponding to default values of lighting datum if no value had been provided (Fig. 16.9).

There would still remain the problem of selecting which variables to display. This could be handled by an expert system which would steer the designer through a number of parametric displays which were appropriate to the particular building use and context. For example on an urban site, it would be useful to see the relationship between lighting energy use, urban horizon angle, and glazing ratio (Fig. 16.9). This would help the designer to zone the use of spaces in the building to relate to lighting requirements.

The three-dimensional graph has a useful analogue. If the vertical axis is redefined to represent a reduction in energy, then we are always trying to move our design to the highest point on the surface. But as in mountain climbing, high places are often dangerously steep – small mistakes mean disaster. Perhaps it is better then to look for a high plateau as a compromise. This is where the random element comes in. Rather than representing our design proposal as a point it should be a fuzzy area. If this area takes us over a precipitous edge, we are in trouble.

We see the two approaches as equivalent, but using different media. The most important common characteristic is that the emphasis is on their *educative* value

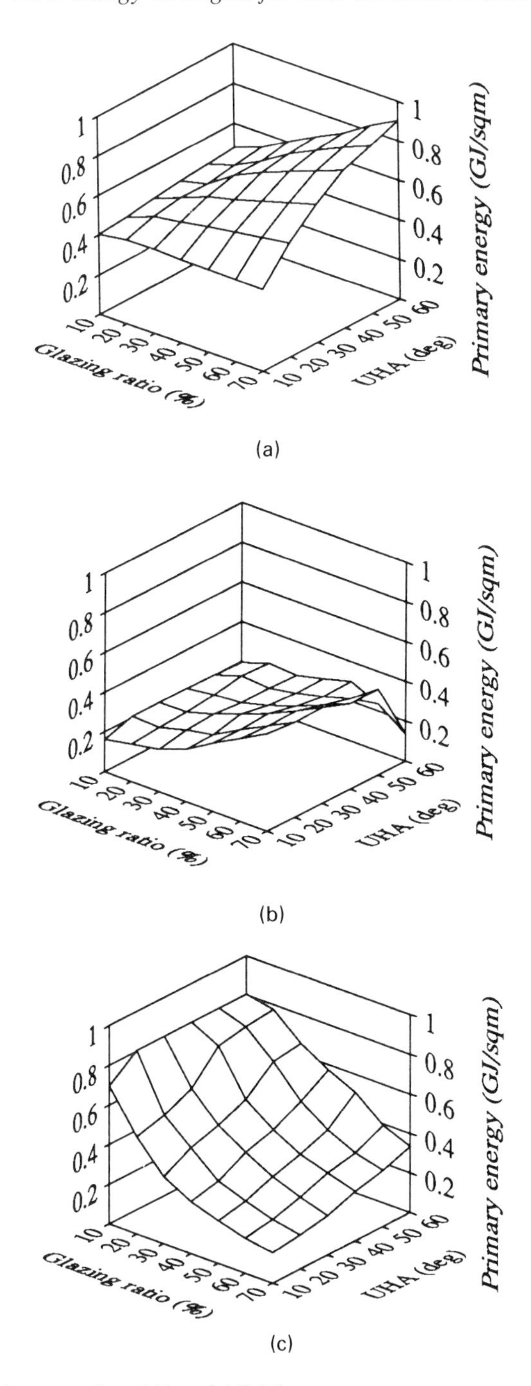

Fig. 16.9 3D graphic output from LT model lighting energy.

rather than their direct *utilitarian* value. It may be stretching the meaning of strategic rather far, but it is probably better that a strategic design tool helps the user to understand broad principles, which he can subsequently apply through intuition and judgement, rather than it being a mechanical exercise which is applied blindly to each project proposal.

16.6 Generic solutions

Both from field studies, and mathematical studies, it is already apparent that certain design strategies are more likely to lead to good energy performance. We can summarize these as follows:

- Avoid the need for air-conditioning.

By –
- Shallow plans.

Allowing –
- Natural ventilation and daylighting (+shading).

Consider –
- Useful solar gains (e.g. vent. preheating).
- Night cooling.

Controls –
- Compatible with occupants requirements.
- Intelligent (e.g caretaker controls).

Plant –
- Low energy lighting and equipment.
- Condensing boilers, heat pumps, heat recovery.
- Combined heat and power.

Operation –
- If mechanical cooling is unavoidable, minimise cooling load by passive means and operate in mixed mode.

It is interesting to speculate just how far the adoption of this simple advice would take the building towards a low energy target. Many researchers and consultants would agree that if these strategies were adopted, the building would almost certainly return energy performance in the top 20% of the range.

What are the problems then in getting these messages across?

Part of the problem seems to be one of conviction on the parts of both the design team and the client. Perhaps simple quantitative energy design tools have some role to play here.

16.7 Conclusion

It seems that for some time we (that is building modellers) have been dreaming of the ultimate model which allows us to optimise every aspect of the design. Due to the complexity of the interactions this is difficult, but the random element, particularly due to occupant influence, makes this impossible. Thus we should stop aiming for *optimization* and instead aim for *safe territory*. Our ideal solution is not a point in solution space but rather a volume.

This is not a new concept – that is the concept of a robust solution. What is perhaps new is the challenge to predict and quantify the robustness from analysis. It would be predicted by modelling the interaction between the building, the systems and the occupants, that is the building as a total system.

The admission of the occupant onto the modelling scene will achieve another important benefit. It will give an opportunity for the health and well-being of the people in response to the building design, to be considered strategically. Fortunately, it is already emerging that the buildings which attain good energy performance by passive means, tend to be those with the least problems of sick building syndrome and other kinds of occupant stress. This will prompt valuable support to passive strategies from clients. Improvements in employee productivity and reduction in absenteeism may well be far more valuable than the reduction in energy cost.

Finally we see energy design tools as having an important educative role and for this reason it is essential that they are transparent and fully supported by explanation. The user must be able to *see* them working. This should increase the body of knowledge in the design of low energy buildings and may ultimately remove the need for such quantitative design tools. After all it is quite possible to design a well insulated house without carrying out a U-value calculation. Instead designers will be able to work from *good advice* generated by specialists and practice experience.

References

Baker, N. & Fontoynont, M. (1990) *Simplified Design Tools for Daylighting and Thermal Design*. Proceedings of the 2nd European Conference on Architecture, Paris 1989, Kluwer.

Dijk, H. & Arkesteijn, C. (1990) *Windows and space heating requirements*. Netherlands Nat. Report from STEP 5 to IEA, Annex XII.

Humphreys, M. & Nicol, J. (1971) *An investigation into the thermal comfort of office workers*, Building Research Station Current Paper CP14/71.

Hunt, D. (1981) Predicting lighting use – A behavioural approach. *Lighting Research Technology*, **13**.

Newsham, G. & Baker, N. (1990) *The role of the occupant in the assessment of building energy performance*. Proceedings of the World Renewable Energy Congress, Pergamon.

O'Toole, S. & Lewis, J. (1990) *Working in the city*. CEC EUR 12919 EN. Eblana Editions.

Steemers, K., Baker, N. & Hawkes, D. (1990) *Parametric Energy Modelling for Urban Situations*, Proceedings of the 2nd European Conference on Architecture, Paris 1989, Kluwer.

Chapter 17

Innovation in the design of the working environment: a case study of the Refuge Assurance building

Michael Corcoran
MSc(Arch), CEng, CIBSE
Director BDP Energy & Environment
AND
John Ellis
CEng, MIMechE, MCIBSE, MInstE
Engineering Partner BDP

With the sharp rise in user dissatisfaction and 'sick building syndrome', in particular, in climate rejecting office buildings, this chapter provides a useful case study of a successful climate accepting building in which the 'passive approach' uses natural ventilation and lighting where possible. The flexible heating, cooling and lighting systems and building energy management system have produced a building whose monitored energy performance compares well with other noteworthy office buildings built in the UK in recent years. The quality of the design and landscaping of the building, as well as the quality of the 'climate accepting' energy strategy, have made this an award winning building.

17.1 Introduction

The last decade has witnessed a number of significant changes in the work environment of offices. These changes reflect a shift in office culture that is largely due to the emergence of the microcomputer.

Continuing concern over energy use, the desire to contain running costs, and opportunities created by developments in energy technologies have also encouraged design innovation. Lately, these influences have been joined by concerns for the health and well-being of office staff, and for the impact that buildings have on the wider environment.

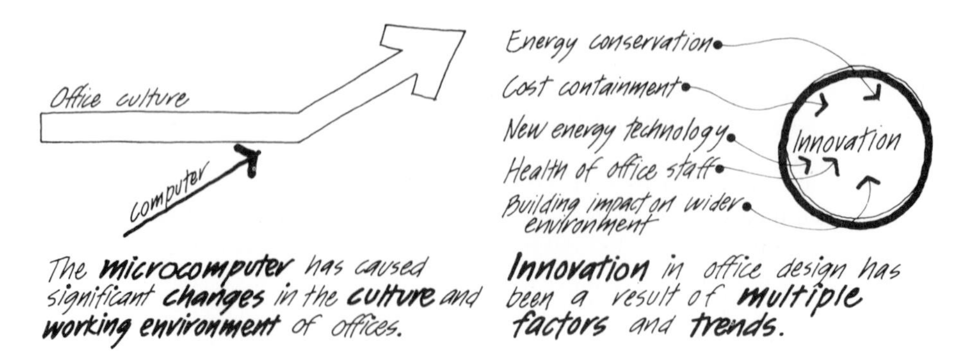

Fig. 17.1 Refuge House in its landscaped setting (Photo: Phil Mastores.)

Designers responding to these influences have been seeking out design solutions that are climate responsive and based on engineering technologies that are no more complex than circumstances demand. This 'passive approach,' as it is sometimes known, is characterized by narrower plan forms that can benefit from daylight and natural ventilation, combined with well-considered orientation and a sheltering landscape. The result is buildings that are more in tune with the environment. This trend is well represented by the new headquarters of Refuge Assurance, shown in Fig. 17.1, which recently won class 1 of the coveted 'Office of the Year' Award of the UK Institute of Administrative Management. Refuge Assurance Plc, with 3800 employees in 160 offices throughout the United Kingdom, is one of Britain's major life and general insurance companies. Its national headquarters have for the past 90 years been on Manchester's Oxford Street.

By the early 1980s, the Oxford Street offices had become inappropriate for Refuge Assurance's mode of operations, and no longer lent themselves to modern staff requirements. Accordingly, Refuge Assurance began an exhaustive appraisal of sites for a new headquarters. The company wished to remain in the Manchester area, and investigated various sites in and around the city. From a short list of five possible available sites, Fulshaw Hall, Wilmslow, was selected and subsequently

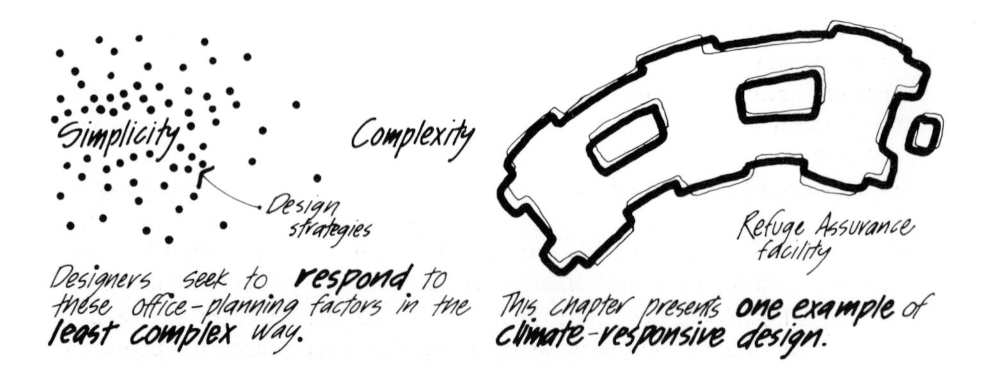

purchased. Building Design Partnership (BDP) was commissioned in 1983 for all building design disciplines, and Refuge Assurance moved into its new headquarters in November of 1987.

17.2 Background

Recent developments in office planning reflect the influence that microcomputers and information technology are having on work patterns. Many organizations are finding that deep-floor office space cannot meet all of their requirements. Meetings are taking an increasing proportion of staff time, and employees who spend long periods staring at computer screens need the opportunity to gaze occasionally at an external scene for visual relief. The *bürolandschaft offices* of the 1970s are proving inflexible in meeting these needs, and the trend in certain office sectors is to return to more narrow-plan forms. These more readily allow a mix of cellular and open-plan office space and provide all staff members with reasonable access to a window.

A further influence is the trend toward multiple tenancies. More than 50% of offices in the United Kingdom are now divided in this way. In anticipation of this demand, office buildings are increasingly being designed with either common or potentially separate access and circulation areas. Even organizations planning to own and fully occupy their own buildings are tending to specify that their buildings should be designed to allow subdivision. This ensures a future place in the market for the building, a factor that, understandably, is of major concern to funding institutions.

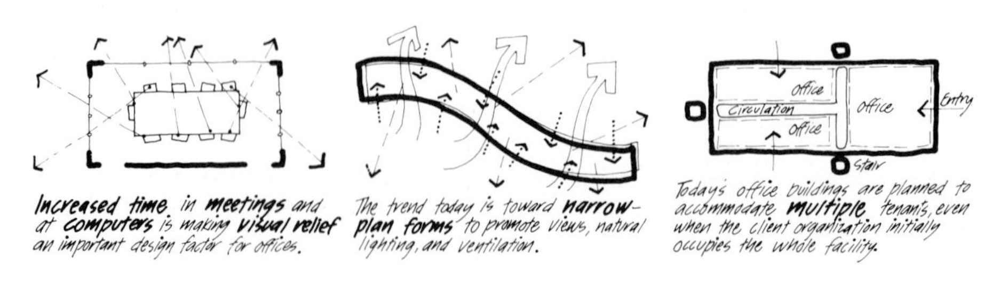

Narrower plan forms have a number of positive attributes in terms of energy efficiency, as they enable more effective use to be made of daylight and, where appropriate, natural ventilation. These positive attributes are to some extent counterbalanced by relatively higher heat losses in winter and higher solar gains in summer. However, the evidence suggests that, in the temperate climate of the United Kingdom, narrow-plan forms are more energy efficient than deep-plan forms.

This was the general conclusion reached in a study of the potential for passive solar energy in non-domestic buildings carried out for the UK Department of Energy, the BDP Energy & Environment and the Martin Centre of Cambridge University. For offices, the potential for displacing energy used for lighting (by effective use of daylight) was found to be almost double the potential for displacing energy used for heating (by useful solar heat gain in winter). Displacement of energy used for lighting is much more valuable than savings in heating energy, due to the relatively high cost of electricity compared to other fuels. In offices, where lighting often represents the most significant and costly use of energy, an overall 20% reduction in lighting energy due to improved daylighting was considered to be achievable.

Improved utilization of daylight, which implies the switching of lighting in relation to daylight availability, is not the only means of reducing lighting energy demand. In recent years, we have seen a reduction in illumination levels, significant improvements in the energy efficiency of lamps, and a departure from general illumination towards more task-related lighting. As a result, lighting loads and their attendant heat gains have reduced from the 50 to 70 W/m^2 of the early 1970s to around 15 to 20 W/m^2 in recent installations.

If this reduction in lighting energy were the only significant influence on internal heat gains, we might have seen a reduction in the demand for air-conditioned offices over the last decade. There are, however, many reasons to specify that a building be air-conditioned, and reducing internal heat gains to levels that can readily be dissipated by natural ventilation is only part of the story.

Counterbalancing the reduction in lighting loads has been an increase in heat gains from office equipment. Information technology has had a significant impact on the office equipment market. A few years ago, electrical loads of about 5 W/m^2 (0.5 W/ft^2) were normally anticipated. Loads of this order corresponded to

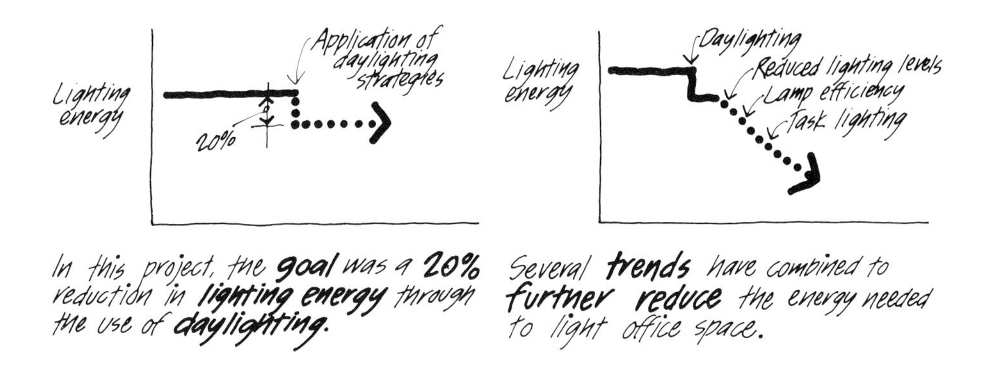

In this project, the **goal** was a **20%** reduction in **lighting energy** through the use of **daylighting**.

Several **trends** have combined to **further reduce** the energy needed to light office space.

equipment such as electric typewriters, photocopiers, and the occasional calculator. These were connected into socket outlets at the perimeter wall, or sometimes into socket outlets on a fixed gird of trunking set into the floor screed. In more recent years, we have seen connected electrical loads rise dramatically, reaching extremes in city dealer rooms with equipment loads of 100 W/m^2 (10 W/ft^2) or more. In more normal office circumstances, we now anticipate the need to cope with equipment loads of up to 50 W/m^2 in certain areas, with a background level of 20 to 25 W/m^2.

We can look forward to the continued growth of information technology. We can also expect to see advances in equipment design, such as liquid crystal displays and CMOS technology, that will reduce electrical consumption and heat release. It is difficult to predict how these two factors will balance out, but for the present the overall effect is that the increase in heat gains from office equipment has substantially eroded the reductions in heat gains from lighting achieved over the last decade.

Another factor that has emerged in recent years is the level of dissatisfaction among building users who have little or no control over their environments. This is now regarded as a contributory factor to that elusive condition known as 'sick-building syndrome.' In deep-plan spaces, it is technically difficult and also expensive to provide individual control over what is essentially a shared environment. Individual control is more easily provided in shallow-planned and more cellular office space. It is also interesting to note that building users are far more tolerant of the wider range of conditions in naturally ventilated buildings than they are of variations in a controlled environment. An illustration of this is that if the temperature in an air-conditioned office were to rise above 25°C in peak summertime conditions, it would probably result in a higher level of dissatisfaction than a temperature of 28°C in a naturally ventilated office.

Many organizations are only just beginning to introduce information technology into their operations, and are unaware of the future implications. These organizations, when contemplating a new office building, are tempted to think that air-conditioning is not warranted in the temperate UK climate. This may indeed be the case in the short term, particularly if the building has been designed with effective solar shading and sufficient thermal mass to suppress peak summertime temperatures. However, it would be a very short-sighted organization these days

that does not at least anticipate that air-conditioning may be needed in the longer term and plan accordingly.

In response to these many influences, a new generation of office buildings is evolving. Pressure on site space in urban centres will continue to result in high-rise buildings, and because of noise, dirt, and wind many of these will need to have sealed, air-conditioned environments. However, away from the pressures of urban centres, a more interesting option is available, as is illustrated by Refuge Assurance's new headquarters.

17.3 Siting

Fulshaw Park, the site selected for the new headquarters, forms a beautiful parkland setting along the A34 road near Wilmslow. Wilmslow is an attractive town with good links to road, rail, and air transportation facilities. It was seen as vital to respond to the exceptional natural quality of the site, and to the architectural qualities of the two main existing buildings there, the 17th century Fulshaw Hall, and the Victorian Harefield House.

BDP carried out surveys on the site conditions, tree planting, and acoustic aspects of the location. The new building was to be sited to have a minimal impact on the local environment and to ensure that its landscaped surroundings would enhance working conditions. As shown in Fig. 17.2, the chosen location, on a curved, open, sloping pasture across a lake from Fulshaw Hall, allowed the greater part of the new building to be hidden from the road and its neighbours. In energy terms, this location also meant a building well sheltered by surrounding mature trees that moderate the wind forces and thereby enhance the prospects for natural ventilation.

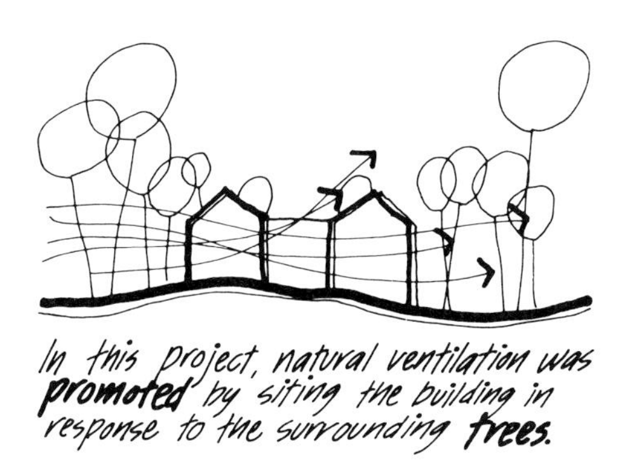

17.4 Planning and built form

The form of the new Refuge Assurance facility, respecting its neighbours, evolved as a low-lying, three-story building tucked into the landscape. Surface parking for 410 cars is placed discreetly to the south. The plan (Fig. 17.3) provides outer and

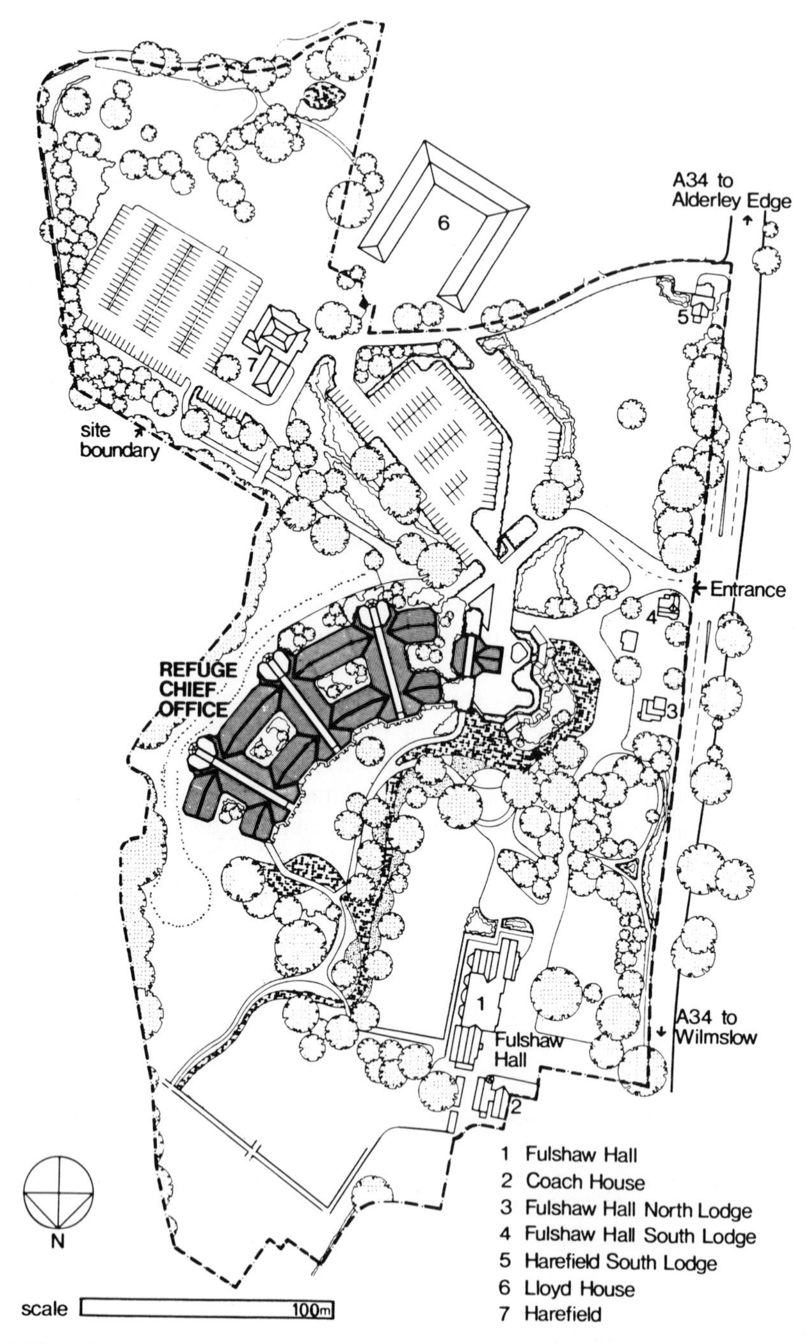

Fig. 17.2 Plan of the Fulshaw Park site. (Illustration: Building Design Partnership.)

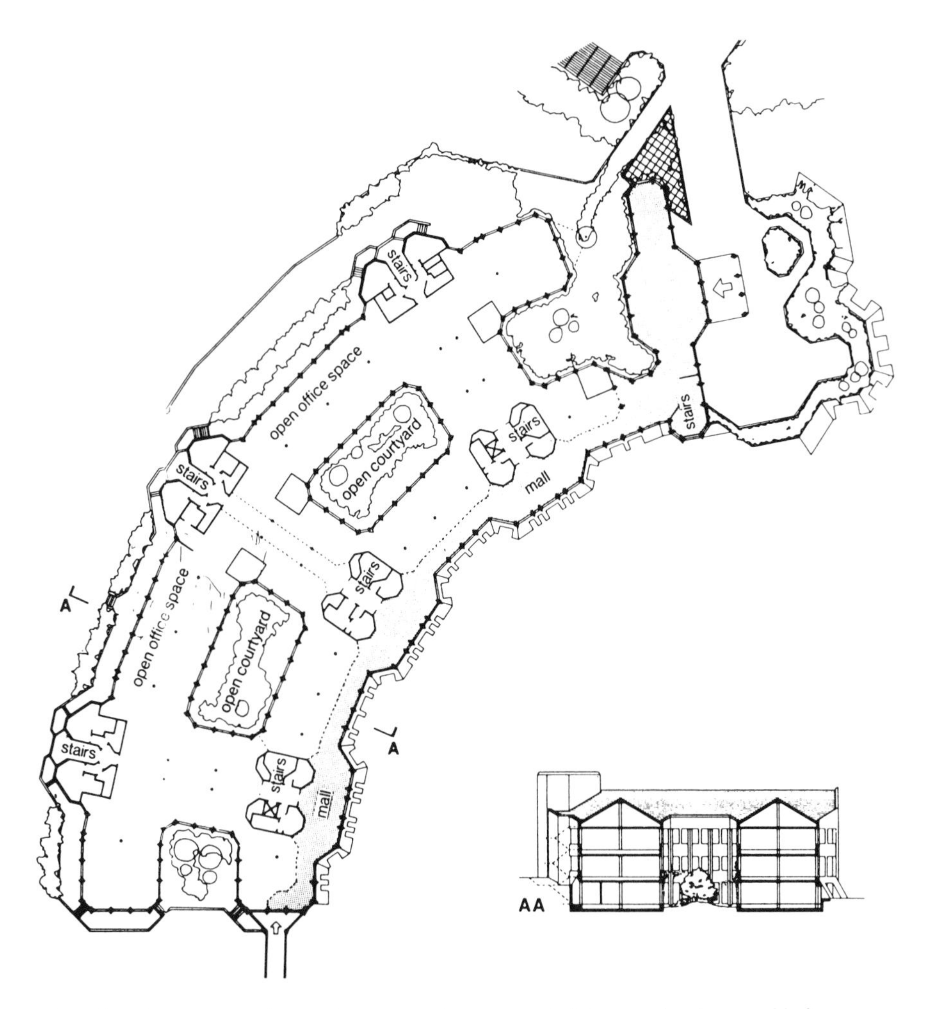

Fig. 17.3 First-floor plan and typical section. (Illustration: Building Design Partnership.)

inner bands of office space, linked at intervals to create enclosed courtyards. Overall, 14 400 m^2 of accommodation is provided. At the lowest level is located a 1000 m^2 computer suite, adjacent to which are a staff dining room, a coffee lounge, and kitchens. The building's external appearance is of a series of linked pavilions, sweeping round in a gentle curve that focuses on Fulshaw Hall. Internal circulation is provided by a mall on the middle level of the inner band of offices, facing the park. Vertical access is via stair towers at back and front. The offices are 12 m wide along the bands, and 15 m wide along the links, providing good daylighting. The planning module is 1500 mm, allowing practically any type of interior layout, including cellular offices. A raised floor with a 600 mm cavity to accommodate cabling and air-handling systems extends throughout all office spaces.

17.5 Fabric

Pitched roofs of natural blue-grey slate, stone copings and sills, and predominantly red brick facades help establish the building in its setting. The structure is *in situ* concrete columns supporting 300 mm-thick concrete floor slabs. There are no suspended ceilings in the office areas. Instead, the undersides of the floor slabs are finished with a textured plaster that acts as an acoustic absorber and diffuser to break up the direct reflections of noise within the open-plan offices. The omission of a suspended ceiling allows a generous 3 m floor-to-ceiling height, which considerably enhances the effectiveness of daylight. Exposing the undersides of the concrete floor slabs allows the structure to act as a thermal flywheel, absorbing internal heat gains and thereby reducing or eliminating the need for cooling.

Insulation standards are by no means exceptional; the walls met the UK Building Regulations requirement, which at the time of design was 0.6 W/m^2/°K. Windows average 40% of the wall area (as measured internally) and are of high quality. The window size was chosen after an analysis of the conflicting issues of daylighting and thermal performance. The units selected incorporate a manually controlled Venetian blind between panes of clear glass set in a rigid, well-sealed frame. They are of a tilt and turn design that allows controllable, draught-free natural ventilation and safe access for cleaning from inside the building.

17.6 Environmental systems

Summertime temperature calculations indicated that orientation was a significant factor. The calculations also demonstrated the degree to which the results were sensitive to assumptions about equipment heat gains. Consultations with the client established that, whereas initial equipment power density might only be in the order of 5 W/m^2, this was expected to rise to 15 to 20 W/m^2 in the medium term. Taking the immediate situation, naturally ventilated offices facing south were expected to peak at temperatures in the region of 28° to 30°C, while more northerly facing offices would peak at 25° to 27°C. The thermal flywheel effect of the exposed undersides of the floor structure played a part in this by suppressing

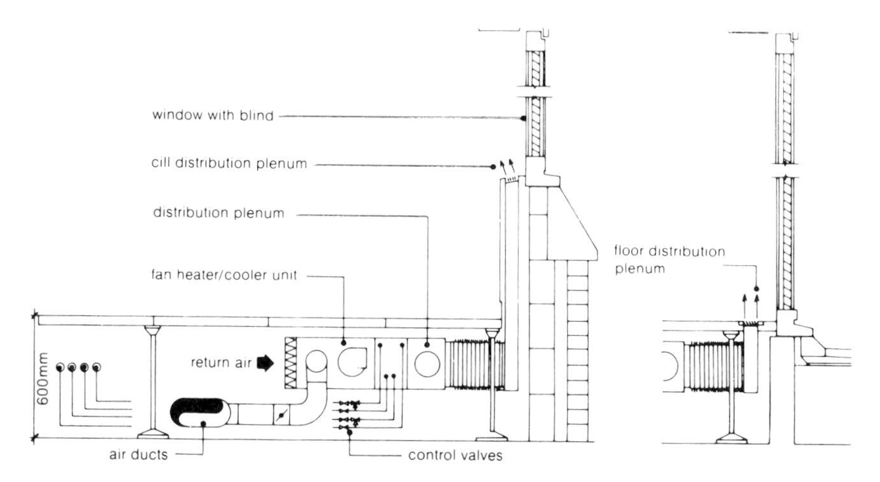

Fig. 17.4 Section through floor showing air distribution. (Illustration: Building Design Partnership.)

peak temperatures by 1 to 2°K less than would have occurred with a conventional suspended ceiling. Perhaps a more telling factor was that temperatures in the southerly facing offices would only exceed 27°C for a relatively short part of the working year – between 100 and 300 hours per year.

Not all office spaces need cooling, at least not initially. Depending on orientation, the breakpoint beyond which cooling is required seems to be when equipment power density exceeds 20 W/m^2. Below this level, natural ventilation will generally be sufficient in a narrow-floor building.

The system chosen was a four-pipe fan-coil unit installation with primary air plants supplying ducted fresh air to the fan-coil units. Throughout the offices, the air-distribution ducts, water circuits, and some 400 fan-coil units are located within the floor cavity (Fig. 17.4).

The floor void also forms a return-air plenum for the fan-coil units. Air is discharged from the units at a rate of six to eight air changes per hour via diffusers at the window sill and circular diffusers set into the floor. At current levels of heat gain, the fan-coil units operate at the lowest of their speed settings, which means that there is substantial spare capacity available to cope with future increases. The primary air plants have been set initially to provide approximately one air change per hour of fresh air (10 l/s/person) but can be adjusted to increase this to a maximum of two air changes per hour if desired.

All fan-coil units are fitted with individual control valves on the heating and chilled water connections, which operate in response to local thermostats. The units can also be controlled from the computer-based building management system. From his or her own office, the building manager is thereby able to adjust temperature-control settings and prevent chilled-water valves from opening in areas that can be adequately cooled by natural ventilation.

The system allows a flexible, three-mode arrangement for environmental control. Winter requires controlled mechanical ventilation with heating. The windows

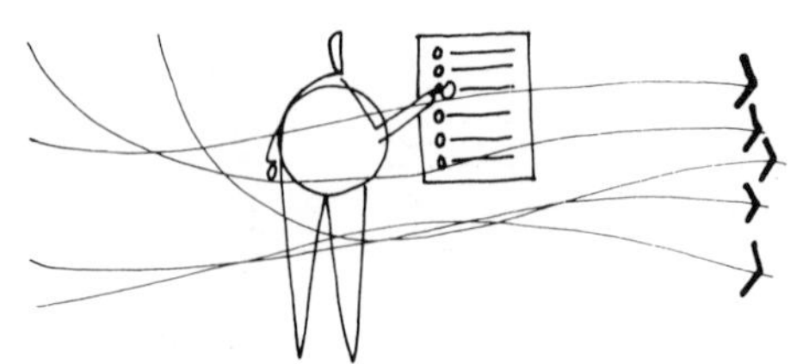

The building manager can control temperature-control settings in the building to promote the use of ***natural*** *ventilation by the building occupants.*

would normally be closed. In mid-season, the system is in a free-running mode with natural ventilation. Summer requires controlled mechanical ventilation with a provision for partial cooling.

Two independent cooling systems serve the office spaces and the computer suite. The office system comprises two 360 kW screw-compressor refrigeration machines, each capable of meeting 50% of the anticipated maximum cooling load. Heat is rejected through a single open-circuit evaporative cooling tower. The computer suite system also uses two screw-compressor machines, but each of these is capable of meeting the full anticipated cooling load of 140 kW. Heat is rejected via two open-circuit evaporative cooling towers, each of which again is capable of full duty.

Both cooling systems are equipped to operate a 'free-cooling' cycle. In the office system, this consists of a plate heat exchanger installed in parallel with the chiller. Due to there being a single cooling tower, the arrangement is 'all or nothing,' in that cooling at a given time is either provided by the chillers or by 'free cooling.' The chilled-water circuit operates at a relatively high flow temperature of 12°C to avoid condensation at the fan-coil units and to maximize the period of free-cooling operation. However, the internal load patterns and the effectiveness of natural ventilation principally determine the duration of this period. An assessment based on typical office conditions suggests that, over a year, 30% of what would otherwise be cooling load is dissipated by natural ventilation; approximately 40% is matched by free-cooling capacity, and only about 30% is dealt with by the chillers.

The free-cooling arrangement for the computer suite differs from that of the offices, in that a plate heat exchanger is connected to one of the two cooling-tower circuits, upstream of the chillers in the chilled-water circuit. Advantage is taken of a standby cooling tower for a free-cooling contribution to be made whenever the outside temperature is low enough. The computer suite chilled-water circuit, at full load, operates at 10°C flow and 15°C return. The standby cooling tower can therefore start to make a contribution to the cooling duty whenever it can provide water cooled down to 14°C, leaving a 1°C margin for system heat gains. This can be achieved when the outside temperature is less than 12°C. The proportion of the load that can be matched by free cooling increases at lower temperatures so that the full cooling load can be met at around 6°C. Over

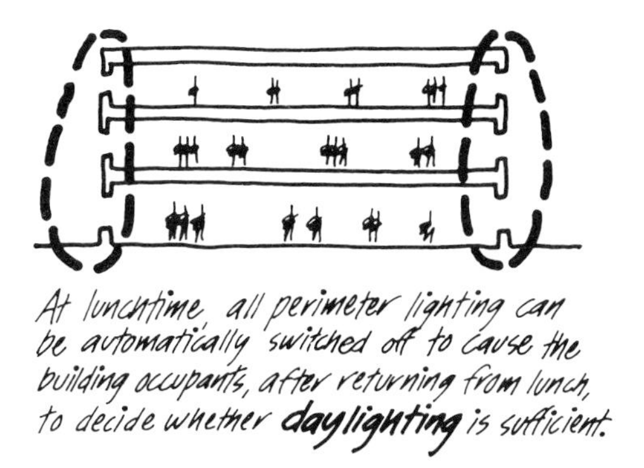

the year, approximately 65% of the cooling requirements of the computer suite are met by free cooling, leaving only 35% to be dealt with by the chillers.

Lighting to the office spaces, which received the accolade of being 'highly commended' in the 1987 EMILAS Lighting Awards, is by purpose-made 250 W metal halide uplighters. These are column mounted in most instances, and are designed to provide an average 400 lux on the working plane. This is achieved at a power density of about 18 W/m^2.

Uplighting is ideally suited to offices such as this in which many people work at computer screens for lengthy periods. Troublesome reflections in the screens are avoided, while the ceiling luminance is high enough to offset the gloomy appearance that so often results when ceiling-mounted low-glare light fittings are used. The 3 m floor-to-ceiling height makes the most of the uplighting. Ceiling luminance is by no means uniform, and is none the worse for that. The textured plaster finish diffuses the fringing effect that sometimes occurs with uplighters due to imperfect reflector optics.

A design illumination level of 400 lux, together with windows on 40% of the facade, should enable good use to be made of daylight. Unfortunately, current technology does not provide an effective means of dimming metal halide lamps in relation to daylight availability, and the run-up time of these lamps does not encourage casual switching. A lighting control system has been incorporated by which coded signals can be transmitted to each individual uplighter. The system can be programmed to switch off all perimeter uplighters at lunchtime so that, on return from lunch, staff members can judge whether they wish to switch their local uplighter back on or whether daylight is sufficient.

17.7 Energy performance

Before the building was occupied, a detailed assessment was made of the building's likely energy performance and annual fuel costs. The building has now been in use for over a year, and feedback has been obtained on how well the predicted performance compares to reality. As might be expected, there are differences. Some of these differences can be ascribed to assumptions made in the assessment

(such as occupancy periods) that do not correspond to actuality. Other differences are perhaps due to a 'settling-in' factor as the client, and particularly the building-management team, learn how to get the best out of the new building.

Unfortunately, the building energy-management system does not incorporate the degree of energy monitoring that would allow a full audit of energy use patterns. Investigation of the differences between predicted and actual energy performance is therefore not straightforward. However, we are helped in this task by the interest being shown by Dr Bill Bordass who, together with Quantity Surveyors Davis, Langdon & Everest, is carrying out a program of case studies on energy-efficient offices for the UK Department of Energy. The Refuge Assurance building has been included in this program. This enables the comparison of predicted and actual energy performance to be made (Fig. 17.5).

Before commenting further, it must be emphasized that, although the total gas and electricity consumption figures are metered values, the distribution is an

	Predicted	Actual or assessed
Gas		
Heating	1719	1223
Hot water	107	150
Catering	229	132
Total gas	2055	1505 (metered)
Electricity		
Lighting	402	631
Cooling	101	500
Fans and pumps	365	200
Catering	91	99
Office equipment	95	99
Central computer	438	780
Other	–	91
Total electricity	1492	2400 (metered)
Total delivered energy	3547	3905

Fig. 17.5 Annual energy use (10^3 kWh).

Building	With computer load	Without
BRE low energy office	135.6	135.6
Hereford & Worcester HQ	170.2	153.2
NFU Mutual & Avon HQ	263.5	166.3
Refuge Assurance HQ	241.8	171.8
South Staffs Water Co	180.5	180.5

Fig. 17.6 Annual energy use.

assessment. Efforts continue to verify the electrical aspect of this assessment by the use of portable recording ammeters. Some initial conclusions can, however, be drawn.

It is interesting to note that, whereas the balance between gas and electricity usage seems to be very different, the actual total delivered energy is only 10% higher than predicted. This may in large part be due to higher lighting heat gains reducing the heating demand. It does appear that lighting is in use for longer periods than anticipated, due perhaps to a combination of flexible working hours, a high incidence of overtime working, and ineffective switching in relation to daylight availability.

The other major difference between predicted and assessed performance is in relation to the computer load. The assessment implies that computer heat gains had been underestimated or that the diversity factor for computer equipment in simultaneous use is higher than was expected. No problems have arisen in maintaining conditions in the computer suite, which implies that the error is principally in the diversity assumption. Irrespective of this, it does not seem that fully effective use has been made of the free-cooling arrangement. This can probably be ascribed to the settling-in factor for the building-management team.

If these general conclusions are correct, Refuge Assurance can look forward to improvement in energy efficiency and reduced energy costs as it fine tunes the building. It is, however, worth setting the performance of this building into the context of other recent, noteworthy, energy-efficient office buildings in the United Kingdom. To do this fairly, the energy used in relation to central computer operations needs to be separately accounted for, as this varies considerably depending on the significance of central computing to the organizations occupying these buildings.

The data in Fig. 17.6 show annual total delivered energy per m^2 of gross floor area, with and without computer power and the associated cooling loads.

Conclusions

Refuge Assurance now has a building that respects and gains considerably from its rural surroundings. The building has been designed to be highly flexible to the demands the future will bring, and yet (or perhaps as a result) it is able to offer an energy performance that compares favourably to any other noteworthy office building completed in the UK in recent years. While the solution adopted here may not be applicable to buildings in a more urban setting, it provides a useful pointer to what can be achieved.

Acknowledgements

The authors wish to acknowledge the contribution that their colleagues in BDP have made to this chapter. Particular thanks are due to the job team led by James Chapman (architect partner), Peter Shuttleworth (job architect), Geoff Spiller (services engineering partner), Joe Sweeney (mechanical services engineer), Dave Belton (electrical services engineer), and Peter Smethurst (quantity surveyor). Acknowledgements and thanks are also owed to Dr Bill Bordass for permission to use information gathered in the course of his work on energy efficiency in offices, being carried out for the UK Department of Energy.

Chapter 18

Low energy student residences at the University of East Anglia

Paul Ruyssevelt
BA, BArch, RIBA, PhD
Manager, Energy and Environment Division, Halcrow Gilbert Associates Ltd

Energy simulation models now provide a useful method of assessing the relative performances of different design strategies in buildings and predictions of the relative cost and energy savings for alternative designs. This is demonstrated by a new student residential block at the University of East Anglia where simulations provided evidence that by superinsulating the building, significant savings in the annual energy running costs could be made with an acceptable extra capital costs being incurred in the construction of the building. On the basis of the results of these simulations the building was superinsulated.

18.1 Design background

In January 1990, energy consultants Halcrow Gilbert Associates (HGA) were called in to work alongside Rick Mather Architects and Fulcrum Engineering on the design of two groups of student residences for the University of East Anglia (UEA). At the time HGA were asked to assess the energy and environmental performance of the proposed student residences; the design comprised a series of terraces containing a number of 10 person units each having three floors of accommodation. Since this time the design has been extended to include an additional floor of single person accommodation as illustrated in the section shown in Fig. 18.1.

18.2 Assessment method

From the outset the members of the design team had been debating with the client the relative merits of differing levels of insulation and optional strategies for ventilation. HGA proposed a series of detailed computer simulation runs to provide an assessment of the performance of the various options from which energy cost and cost effectiveness could be calculated.

In order to provide an efficient and rapid assessment at an appropriate level of detail the ten person unit was divided into four zones as illustrated in Fig. 18.2.

Little was known regarding the typical occupancy pattern of a student residence, and hence sensitivity studies were undertaken for both normal and high occupancy levels. In order to assess occupant effects a model was constructed for

Halcrow Gilbert Associates

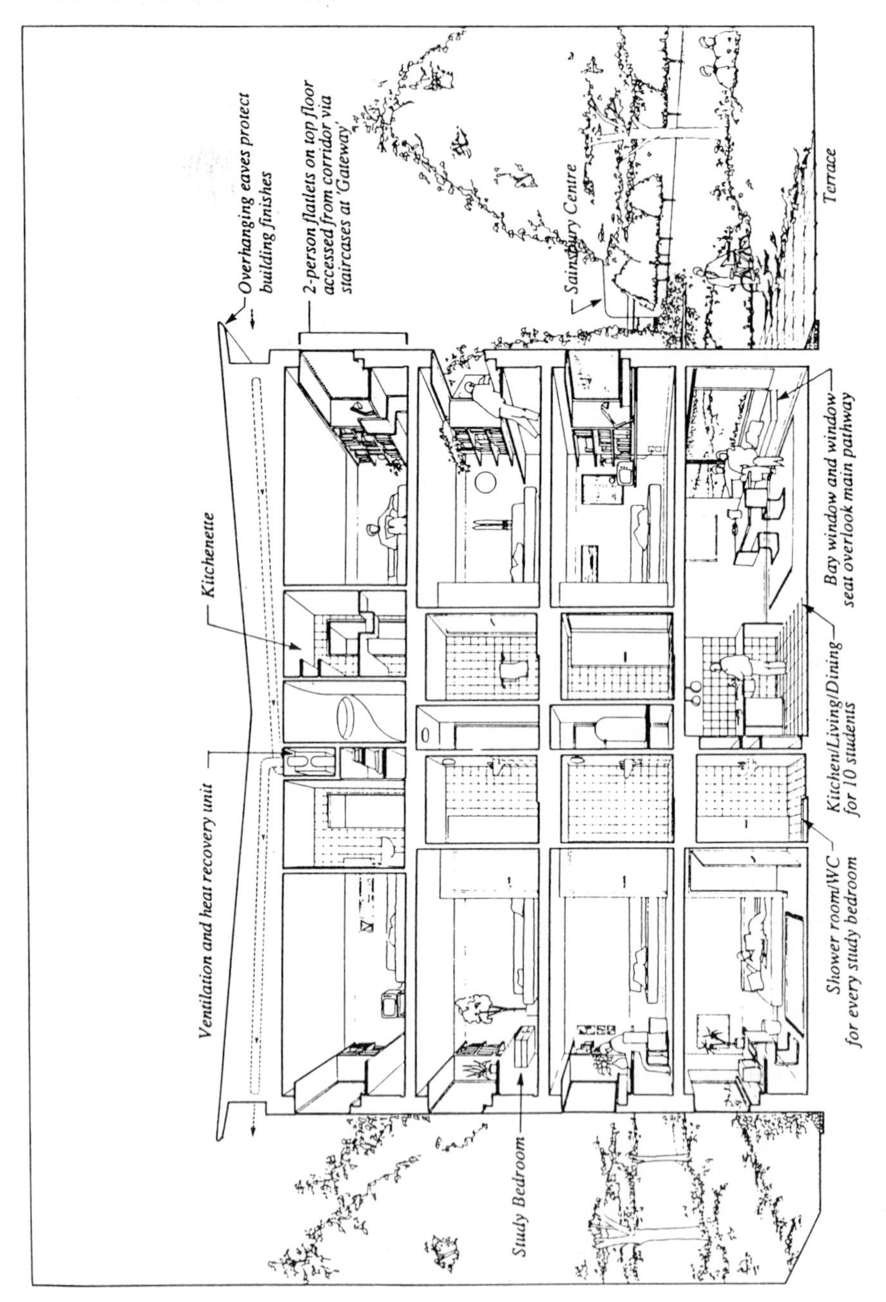

Fig. 18.1 Plans and sections of current four storey design.

	second:	Four beds	
	first:	Four beds	
south	groundfront: living	groundback: Two beds	north

Fig. 18.2 Building zones.

	1 Building Regulations 1990	2 Superinsulated with passive ventilation	3 Superinsulated with mechanical ventilation
U-values (W/m^{2}K)			
Walls	0.45	0.22	0.22
Roof	0.25	0.15	0.15
Floor	0.3	0.18	0.18
Windows	2.8	2.8	2.8
Ventilation (ac/h)			
Winter	1.0	1.0	0.5*
Summer**	6.0	6.0	6.0

* Heat recovery in the ventilation system reduces the effective heat loss due to ventilation by over 50%.
** Ventilation rates are achieved by opening windows to provide cross-flow ventilation.

Fig. 18.3 Design options assessed.

each zone which accounted for the internal heat gains arising at different times of day from people, lights, cooking, hot water and other sources.

Initially many different thermal insulation strategies were considered. However, HGA suggested that a rapid approach could be taken by assessing only the extreme scenarios. To this end, three basic options were examined and these are presented together with a number of the principle input variables in Fig. 18.3.

In addition to the three basic simulations, two variants of the superinsulated unit employing low energy lights and Kappafloat double glazing were examined.

18.3 Design energy performance

An analysis of the heating season heat loss of the design options as illustrated in Fig. 18.4 shows a 5000 kWh reduction when comparing the superinsulated option to the Building Regulations standard. The reduction is limited by the minimal area of exposed wall and roof, but this also means that the cost of superinsulation would be low. Figure 18.4 shows that a further substantial reduction in heat loss is achieved through the introduction of a mechanical ventilation system which recovers over 50% of the heat from the exhaust air.

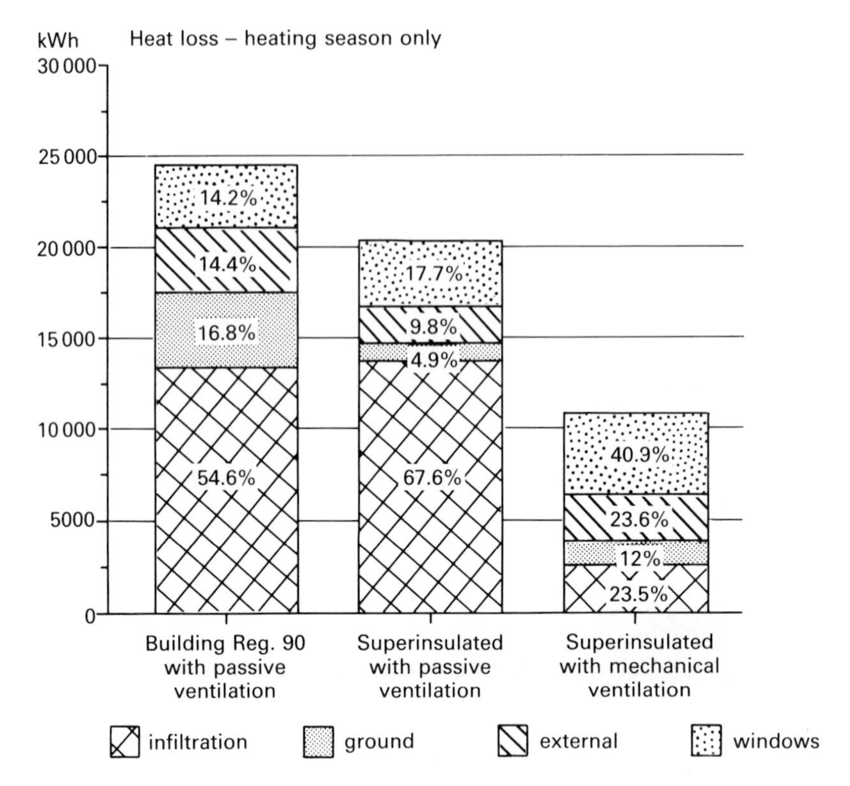

Fig. 18.4 Heat loss from ten person unit – heating season only.

	1 Building Regulations 1990	2 Superinsulated with passive ventilation	3 Superinsulated with mechanical ventilation
Annual Auxiliary Heating			
Energy (kWh)	12 265	8196	333
Gas cost	£151	£101	£4
Electricity cost	£538	£360	£15
Addition for low energy lights (kWh)		500	
Reduction for kappafloat (kWh)		547	

Fig. 18.5 Auxiliary heat requirements.

The auxiliary heating requirements of the three options are presented in Fig. 18.5. The results are also presented in graphical form in Fig. 18.6 which shows a breakdown of auxiliary energy by zone.

The results in Fig. 18.5 show a significant saving in energy as a result of using

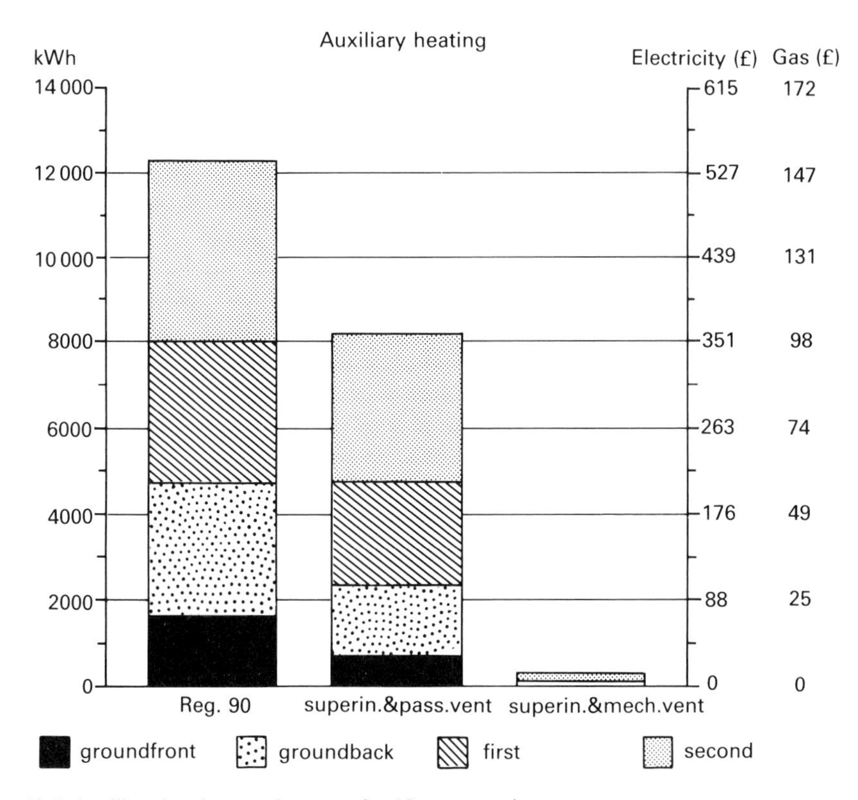

Fig. 18.6 Auxiliary heating requirements for 10 person unit.

superinsulation and mechanical ventilation with heat recovery. It is important to note that the majority of the saving is actually attributable to heat recovery in the mechanical ventilation system.

18.4 Summertime conditions

A natural concern with superinsulated buildings is the likelihood of summertime overheating. To investigate potential problems in this area output was generated from the computer simulations for a summer design day with high levels of solar radiation and high temperatures. Figure 18.7 shows the mean temperatures for both superinsulated and Building Regulations design options with normal and high occupancy patterns. The simulation runs are coded as follows:

- UEASUBNO: Building Regulations, normal occupancy.
- UEASUBHO: Building Regulations, high occupancy.
- UEASUSNO: Superinsulated, normal occupancy.
- UEASUSHO: Superinsulated, high occupancy.

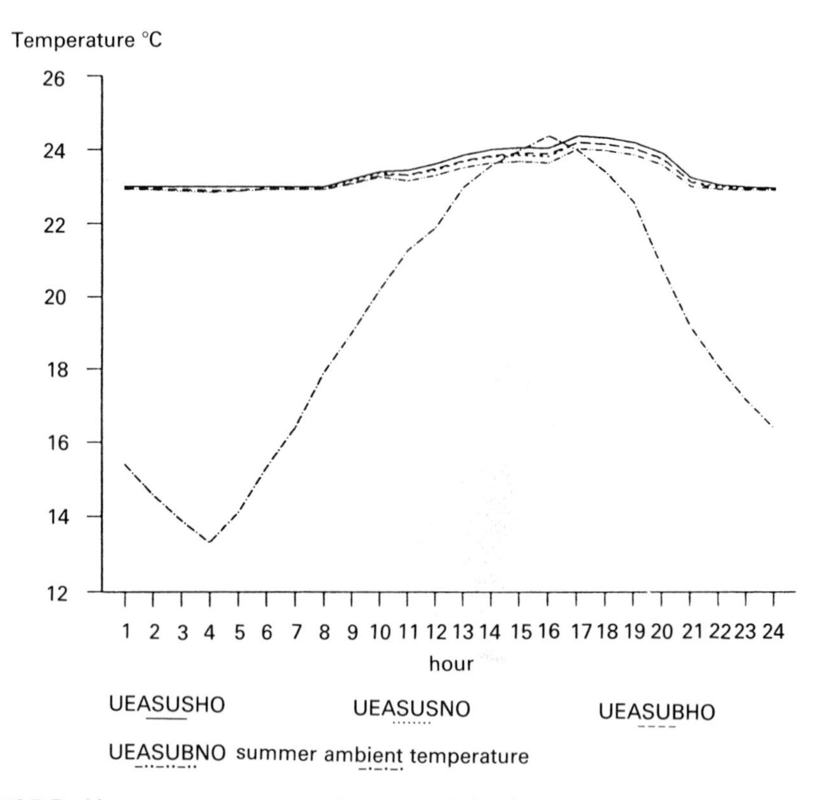

Fig. 18.7 Residence mean temperatures for summer design day.

It can readily be seen from Fig. 18.7 that there is very little difference between the temperatures experienced in the two design options. In fact, due to the substantial thermal capacity of the construction, the maximum temperature does not exceed the external temperature in either case.

18.5 Conclusions

Savings approaching £150 per year for heating by gas are predicted for a 10 person residence by employing superinsulated construction standards and mechanical ventilation with heat recovery. Furthermore, internal environmental conditions in the wintertime will be substantially enhanced by draught-free ventilation, and summertime conditions will be similar to those experienced in other well ventilated residences.

Since it will be possible to meet the minimal heating requirements of the superinsulated design with only 200 W heaters in each of the rooms, a substantial saving may be made on the cost of the heating system. Balancing the various additions and reductions in capital costs attributable to the superinsulated option resulted in a small overcost compared to the Building Regulations option. This overcost proved acceptable to the University and the scheme is now proceeding as a superinsulated design.

Conclusions

Chapter 19

Buildings and energy: an overview of the issues

Loren Butt
BSc(Eng), CEng, MIMechE
Director of Loren Butt Consultancies

In the face of enormous energy challenges such as pollution, waste, diminishing fossil fuel resources, global warming and the depletion of the ozone layer, designers have a key role to play in the meeting of these challanges. A fundamental change in our fuel consumption patterns is required away from a dependence on finite and polluting fossil fuels towards clean and limitless renewable energy sources such as the sun, the winds and the tides. While there is an urgent need for more research into renewable energy, in many cases the technology already exists to put into place such changes, and what is now required is the foresight and resolve to take action while there is still time. The survival of our species may depend upon it.

Discussion

Having looked at some of the particular environmental aspects of buildings and energy we should not forget that the design decisions we make at the drawing board have a direct bearing on the wider context of energy in our society.

Energy can be seen to lie at the root of many of the serious global issues we are facing today – pollution, waste, diminishing resources of fossil fuels, and climate change. Such considerations have made some politicians and businessmen adopt a firm stance on energy issues, or at least appear to do so. Individual issues such as the 'gas-guzzling' car or the highly energy consuming building become the focus of moral concern, and yet energy has, throughout history, also played a very positive role in the development of our civilisation.

During the last 200 years the rapid increase in the availability and use of fossil fuels is related to two marked changes in world population trends; first, there has been a dramatic improvement in life expectancy and the quality of life in the industrialized nations, and secondly there has been a population explosion in the less developed nations.

For the industrialized nations energy has been a catalyst for the remarkably diverse expansion of activities that have characterized these centuries. Energy consumption has tended to grow at an ever increasing rate in these priviledged areas of the world. Energy has both fuelled and fed off the inflationary spiral of consumption. In the West our trading structure enables most people to have sufficient food, clothing and shelter as well as, in many cases 'luxuries' such as land and sea travel and the enjoyment of the arts and humanities. Energy has made all this possible for us.

Compare this with the plight of the Third World: disease, famine, disasters and deprivation are widespread and worsen continually with the rapidly growing population.

If the experience of the industrialized nations is any guide the only hope for the Third World is through greater availability and use of energy leading to higher standards of living, standards of living on a par with those in the West which have resulted from 200 years of development. There is no doubt that the transfer of Western technology to the less technologically developed societies of the Third World could speed up the rate of their development, and energy there too will act as a catalyst. This will inevitably lead to a rapid increase in the energy consumption levels in the Third World.

In order to meet the growing needs of the Third World the industrialized nations will have to make a greater share of global non-renewable fossil fuel resources available to them, so creating more pressure on the West to conserve and use efficiently what fuels we have.

This highlights the urgency with which we should tackle research into renewable energy resources, such as those from the sun, the wind and the tides.

Any consideration today of energy must take account of pollution, both local and global. The burning of fossil fuels with today's technology generally involves emissions of numerous atmospheric pollutants, but this is an avoidable problem. We have the technology to produce clean emissions from power stations, but they are not applied yet – the time will come, sooner or later, when they have to be.

Many industrial processes seriously pollute the water, land and air around them. This too is entirely avoidable and should be legislated against. It is chastening to compare Western Europe with its apparently severe problems of environmental pollution and acid rain, to the recently opened up Eastern block which has graphically revealed the horrendous scale of industrial pollution which we in the West also considered acceptable as little as 50 years ago.

In the last decade CO_2 and CFCs have become the main target of environmental concern because of the enormity of the changes that may affect the planet through the 'greenhouse effect' and the depletion of the ozone layer. International agreements have placed considerable controls on the emissions of CFCs but CO_2 emissions are not yet touched by 'global limitation' treaties despite the fact that CO_2 emissions resulting from fossil fuel burning may present a serious threat to our future. For instance global temperature increases threaten to melt the polar ice-caps, and flood the world's low lying land. It is only to be hoped that we have sufficient time to adapt the defences necessary to protect the most vulnerable populations of the world.

If and when climate changes occur they may also be beneficial such as the parts of the world where higher rainfall or higher temperatures are predicted suggesting a potential for higher food production levels in these areas. The capacity of the oceans, covering more than half the world, to absorb CO_2 may be greater than we think and could yield an increase in plankton bloom populations on which fish feed providing increased food supplies from fishing. In such ways climate changes may benefit some regions of the world.

Due to the uncertainty of the climatic prediction techniques used we need at least another ten years of research before we can determine more precisely what impact the 'greenhouse effect' may have on global climate, but this is no justification for complacency. There is no doubt that we should be cautious and legislate to control such emissions while we conduct research into the nature and impact of global climate change.

So energy conservation really does matter. There is much we should do about it at global, national and local levels, there is much we can do about it, and indeed there is much we are doing about it. The UK has achieved a great deal over the past twenty years. Annual primary energy consumption has remained stable over this period while the Gross Domestic Product (GDP), a measure of our activities, has increased by 50% in real terms. This has come about through steady improvements in fuel efficiency in all areas of consumption, including building construction and usage.

Again we cannot be complacent. Investment in research and development, and innovative applications of energy efficiency in design will have to be intensified to continue stable energy consumption levels alongside economic growth, as there are diminishing returns in energy saving measures as standards of energy efficiency increase.

What might the long term future of energy be? If, as seems reasonable to assume, the world's fossil fuels are a finite resource, we must expect them to run out. That day appears to be a long way off. However, we must use this resource sparingly and with a care for the legacy we will leave for future generations. Research and development into renewable energy – limitless and non-polluting sources of energy – must expand.

We have the energy technologies of the future at hand already, albeit in their infancy. Photovoltaic and solar wind turbine generators are on line in the USA already. In France the Odeillo 'four solaire' and the Rance river tidal power scheme have shown this can be done. Other available technologies include fuel cells and the future will probably reveal further renewable energy sources as yet untapped or even imagined. Meanwhile in the last throes of the fossil fuel energy age we are now desperately seeking to conserve energy in every conceivable way. How can this best be done?

First, we should look carefully at the choice of the fuels we use and try to use our fossil reserves in the most efficient way. One could, for example, suggest that a world policy should be introduced to use oil only for road, air and sea transport, gas solely to produce high temperatures for industrial processes, and coal only to generate electricity for stationary machinery, tracked transport and the environmental control of buildings. The consequences of such a policy could be far reaching but the logic of it needs to be examined.

Oil is a concentrated fuel and the only practical option present for road and air travel. It is far too precious a commodity to be burned in a heating boiler. The boiler itself is an anachronism in a sense. The burning of oil produces flame temperatures of 3000°C, which is used to heat water to temperatures of only around 100°C. Acidic and toxic pollutants are discharged into the immediate vicinity of the building. These emissions can be cleaned, but with difficulty on a small scale – and all building boilers are small scale.

The same thermodynamic considerations apply to gas fired boilers, though emissions are cleaner. The ability of gas to be delivered by fixed infrastructure piping, and to produce flame temperatures of 3000°C is surely best suited to industrial processes, in the manufacture of products to enhance the lives of the people that industry serves.

And electricity? At present most electricity is produced by fossil fuel burning boilers producing steam to drive the turbines powering electricity generators. The process at present is extremely inefficient and polluting to the atmosphere. But power stations are large and capable of significant technological improvement on the current best standards of fossil fuel powered stations which today only deliver 38% of the energy content of the imput sources.

Electricity can be generated from either of the three main fossil fuels, oil, gas or coal, but coal is perhaps the best option, and particularly if the power station is situated near the coal source. This is because the other two fuels should strategically be reserved for the functions outlined above. Unfortunately fuel price structures, political pressures and pollution issues often conspire to force the hand of the electricity generating companies to use oil and gas today, a choice we may have cause to regret in the future.

Almost every building uses electricity – for lighting and the powering of machinery and equipment. Electricity is used to drive the pumps, fans, lifts and escalators that service buildings. It is used for air-conditioning systems and refrigeration equipment, and it can also meet other needs such as provision of domestic hot water, cooking and space heating.

Space heating by electricity can take two main forms: direct conversion of the electrical imput to heat, or by the more energy efficient utilisation of heat pumps. Heat pumps extract heat from one or several sources and deliver it to the user. The sources can be outdoor air, rivers, lakes, or even heat produced by the activities of the occupants of the building. By so doing an efficient heat pump can deliver up to five times more energy to the user than it consumes in electricity, so cancelling out the inefficiency of the power stations. Heat pumps can also produce mechanical summer cooling of buildings – which some buildings will always require.

For many years the environmental engineers have been attracted to another energy efficient method of converting fossil fuel into energy – the combined heat and power (CHP) systems. These use fossil fuel, usually gas, to power engines which drive electrical generators. The engines may be of different kinds, including gas turbines, and the waste heat from the process is then used for other purposes such as district heating. One problem in this system is that it relies on co-incident and parallel demands for electricity and heat but in some instances CHP has proved extremely successful.

There is tremendous potential for the application of thermal storage technology. There are already techniques for smoothing the peaks of energy demands created by variations in local seasonal and diurnal climate and building usage patterns over 24 hours. Thermal storage opens up the possibility of storing energy between seasons, capturing the summer sun to warm buildings in winter. More research is needed into the possibility of greater utilization of this free, and non-polluting energy.

Natural daylight is free. However it is very variable in intensity and it carries large amounts of invisible infra-red energy. With good design and control mechanisms natural daylight can save significant amounts of energy in buildings but with bad design and poor controls daylight can lead to serious problems of excessive heat within buildings making excessive demands on energy consuming air-conditioning systems.

Artificial lighting is probably the area in which there is probably the most research and development in the building industry today. New light sources are becoming available that promise significant energy savings, but again control is important because an energy efficient light is not that when it is left on when not required. But the energy consumption of a building is a complex equation balancing interrelating factors.

The daylight potential of a building for example is affected by the depth of the building plan, and deep-plan forms require the substantial use of artificial lighting, but the deep-plan can also save energy in space heating requirements due to the low external surface area that results from the deep-plan. Here is where the new dynamic computer simulation techniques can help us to evaluate the comparative merits of different design decisions, and to eliminate the energy waste inherent in the oversizing of services plant that is necessary only when simple analysis is used. The use of computer modelling may perhaps offer enormous potential for energy savings in the future.

The foregoing concentrates on energy consumption in building use, and that is important, but it is also necessary to consider the energy consumed in the making of buildings. Some building products and materials are inherently high energy consumers and this has been pointed out in the chapter by John Connaughton. Industry will have to look at ways in which the energy requirements of the production processes can be reduced, but we too as designers must look at ways in which the energy costs of the materials in our buildings can be reduced.

Many years ago the then President of the RIBA, Alex Gordon prophetically invented the design philosophy of 'long-life, loose-fit, low-energy'. It is a philosophy we should encourage today.

Long-life – buildings should be built, and products chosen to last, to optimize on the energy content of the materials of which the building is made.

Loose-fit – buildings should be adaptable and flexible. They should be capable of being used for a variety of functions during their life span without incurring energy wastage through rebuilding for successive reorganizations of space.

Low-energy – we have passed the point when a building can be called 'low-energy' simply by looking at its annual energy bills. A low-energy building should have low recurring energy needs, but a full audit of its life-cycle consumption of energy will present a truer picture.

Energy savings are currently only calculated on 'financial pay-back period' criteria. A quite different and preferable approach is to compare the energy content of the capital investment of the energy systems of the building with the annual energy savings they produce. A financial pay-back period of ten years would probably be deemed to be not viable, but the scheme may have an energy pay-back period of only one to two years. Clearly there is a need for enforceable measures and incentives to enhance the importance of energy in this equation.

Appropriately formulated Building Regulations can have a beneficial effect, but perhaps a more effective measure would be the introduction of a 'carbon tax' which would result in a change in the way building costs are assessed. The equation would then widen to include not only material and direct energy costs, but also a measure of the load on the environment.

Conclusion

Energy is the fuel for all life on this planet. It has been a key factor in the evolution of our species into the advanced industrialized civilization of the 1990s. To misuse it may lead to destruction. Man is an infinitely adaptable species now facing enormous challenges in the energy sphere, of how to limit global emissions of climate-threatening gases, to limit pollution, to redress the balance between energy consumption and development in the First and Third Worlds, and to find alternatives to our finite fossil fuel reserves.

There is no doubt that the solution to these enormous challenges lies in our ability to adapt from dependence on a fossil fuel economy to one powered by clean, renewable energy, to tap into sources such as the limitless flow of radiation from the sun. Only through the realization of the enormity of these challenges and the sensible rethinking and adaptation of our current technology, will we be able to keep open our option to continue our tenancy of the planet.

Useful addresses for energy designers

Australia	Coal Research and Energy Technology Branch Department of Primary Industries and Energy GPO Box 858, Canberra ACT 2601 Australia (Tel: 61-62-724781)
Canada	Energy Mines and Resources (CANMET) 2082 Marie Victorin, Suite 210 Varennes, Quebec J3X 1R3 Canada (Tel: 1-514-652-4621)
Denmark	Danish Energy Agency Landemaerket 11, DK-1119 Copenhagen Denmark (Tel: 45-33-926700)
	NOVA PRO 25 Sophienholmvej, DK-4340 Toelloese Denmark (Tel: 45-59-186999)
	SBI, Dr. Neergaards Vej 15 Postboks 119, 2970 Hørsholm Denmark (Tel: 45-42 86 55 33)
Finland	CADDET-Suomi, Energy Information Centre PO Box 289, SF-00181 Helsinki Finland (Tel: 358-0-6848423)
	Ministry of Trade and Industry Pohj. Makasiinikatu, SF-00130 Helsinki Finland (Tel:358-016-05236)
France	CSTB, 4 Avenue du Recteur Poincaré 75782 Paris Cedex 16 France (Tel: 33-1-40 50 28 28)
Germany	1fBt Reichpietschufer 74-76 1000 Berlin 30 Germany (Tel: 49-30-25 03 211)
Ireland	EOLAS, Construction Industry Glasnevin, Dublin 9 Ireland (Tel: 353-1-37 01 01)

Italy	Dipartemento di Energetica Politecnico di Torina Corso Duca degli Abruzzi 24 Torino 10129, Italy (Tel: 39-11-5567456)
	ICITE Via Lombardia 49 Frazione Sesto Ulteriano 20098 Wqn Tiuliqno Milanese (MI) Milano, Italy (Tel: 39-2-98061)
Japan	Energy Conservation Center Svax Nishi-Shinbashi Bldg. 39-3 2-Chome Nishishinbashi Minolo-ku Tokyo, 105 Japan (Tel: 81-3-433-0311)
	NEDO Sunshine 60, 30F 1-1 3-Chome, Higoshi-Ikebukuro Toshima-ku Tokyo, 170 Japan (Tel: 81-3-987-9412)
New Zealand	Ministry of Energy PO Box 2337, Wellington New Zealand (Tel: 64-4-727044)
Norway	Institute for Energy Technology PO Box 40, NO-2007 Kjeller Norway (Tel: 47-6-806000)
	Royal Norwegian Council for Technology for Scientific and Industrial Research PO Box 70, Tasen, 0801 Oslo Norway (Tel: 47-2-237685)
Portugal	LNEC, Avenida do Brasil 101 1799 Lisboa Codex Portugal (Tel: 351-1-84 82 131)
Spain	ICCET CSIC, Serrano Galvache s/n Apartado de Correos 19002 28033 Madrid Spain (Tel: 34-1-202 04 40)
Sweden	Chalmers Industrit-knik Chalmers Teknikpark, S-41288 Gothenburg Sweden (Tel: 46-31-724226)

	Swedish Council for Building Research Senkt Göransgatan 66, Stockholm Sweden (Tel: 46-8-6177300)
Switzerland	info-Energie, PO Box 310, CH 5200 Brugg Switzerland (Tel: 41-56-417771)
	Office Féderal de L'Energie Belpstrasse 36, CH-3003 Bern Switzerland (Tel: 41-31-615662)
The Netherlands	Novem, PO Box 17, 6130 AA Sittard The Netherlands (Tel: 31-4490-95234)
	TNO IBBC, Lange Kleiweg 5 2288 GH Rijswijk, PO Box 49, 2600 AA Delft The Netherlands (Tel: 31-15-842 000)
United Kingdom	Building Research Energy Conservation Support Unit (BRECSU), Garston Watford, Herts WD2 7JR United Kingdom (Tel: 44-923-664258)
	Energy Efficiency Office, Eland House Stag Place, London SE1E 5DH United Kingdom (Tel: 44-71-273-3000)
	Energy Technology Support Unit (ETSU) Building 149, Harwell Laboratory Oxfordshire OX1 ORA
	Friends of the Earth (FoE) 26-28 Underwood Street, London N1 7IQ (Tel: 44-71-4901555)
	The National House-Building Council (NHBC) Chiltern Avenue, Amersham, Bucks HP6 5AP (Tel: 44-494-434477)
	Timber Research and Development Association Stocking Lane, Hughenden Valley High Wycombe, Buckinghamshire HP14 4ND (Tel: 44-240-243091)
USA	Department of Energy 1000 Independence Avenue SW Washington DC 20585 USA (Tel: 1-202-586-1660)

Oak Ridge National Laboratory
Energy Division, PO Box 2008
Oak Ridge, TN 37831-6206
USA
(Tel: 1-615-576-8152)

Index

Page numbers in *italics* indicate illustrations